Lecture Notes in Electrical Engineering 1134

The book series *Lecture Notes in Electrical Engineering* (LNEE) publishes the latest developments in Electrical Engineering—quickly, informally and in high quality. While original research reported in proceedings and monographs has traditionally formed the core of LNEE, we also encourage authors to submit books devoted to supporting student education and professional training in the various fields and applications areas of electrical engineering. The series cover classical and emerging topics concerning:

- Communication Engineering, Information Theory and Networks
- Electronics Engineering and Microelectronics
- Signal, Image and Speech Processing
- Wireless and Mobile Communication
- Circuits and Systems
- Energy Systems, Power Electronics and Electrical Machines
- Electro-optical Engineering
- Instrumentation Engineering
- Avionics Engineering
- Control Systems
- Internet-of-Things and Cybersecurity
- Biomedical Devices, MEMS and NEMS

For general information about this book series, comments or suggestions, please contact leontina.dicecco@springer.com.

To submit a proposal or request further information, please contact the Publishing Editor in your country:

China

Jasmine Dou, Editor (jasmine.dou@springer.com)

India, Japan, Rest of Asia

Swati Meherishi, Editorial Director (Swati.Meherishi@springer.com)

Southeast Asia, Australia, New Zealand

Ramesh Nath Premnath, Editor (ramesh.premnath@springernature.com)

USA, Canada

Michael Luby, Senior Editor (michael.luby@springer.com)

All other Countries

Leontina Di Cecco, Senior Editor (leontina.dicecco@springer.com)

**** This series is indexed by EI Compendex and Scopus databases. ****

Jason C. Hung · Neil Yen · Jia-Wei Chang
Editors

Frontier Computing on Industrial Applications Volume 4

Proceedings of Theory, Technologies and Applications (FC 2023)

Springer

Editors
Jason C. Hung
Department of Computer Science
and Information Engineering
National Taichung University of Science
and Technology
Taichung City, Taiwan

Neil Yen
School of Computer Science and Engineering
University of Aizu
Aizuwakamatsu, Japan

Jia-Wei Chang
Department of Computer Science
and Information Engineering
National Taichung University of Science
and Technology
Taichung City, Taiwan

ISSN 1876-1100 ISSN 1876-1119 (electronic)
Lecture Notes in Electrical Engineering
ISBN 978-981-99-9341-3 ISBN 978-981-99-9342-0 (eBook)
https://doi.org/10.1007/978-981-99-9342-0

This Springer imprint is published by the registered company Springer Nature Singapore Pte Ltd.
The registered company address is: 152 Beach Road, #21-01/04 Gateway East, Singapore 189721, Singapore

Paper in this product is recyclable.

Contents

Case Study in Developing Extensible Virtual Assistant Using Genie Framework

Yi-Ting Wu[1], Albert Chang[2], Yu Hung Tsai[1], Po-Chuan Wang[1], Tinghao Chen[1], and Jeng-Wei Lin[1](✉)

[1] Tunghai University, Taichung 407224, Taiwan
{g10490001,s08490007,s08490037,g11490031,jwlin}@thu.edu.tw
[2] BSI Pacific, Taiwan Branch, Taipei 11492, Taiwan
albert.chang@bsigroup.com

Abstract. Deep learning has made significant improvement in natural language processing. Nowadays virtual assistants or chatbots attract attention of many researchers and are expected to be applied in more and more areas. We had designed and implemented an extensible financial virtual assistant using Genie framework. A new device (or skill) is developed to offer financial services in backend server cloud. The device and supported APIs (Application Programming Interface) are registered in an open repository Thingpedia. When Genie receives user utterances, it translates them into ThingTalk programs using a large deep-learning neural networks. Then, Genie executes the ThingTalk programs, which may invoke the financial services through the registered APIs. ThingTalk is a declarative programming language. Domain experts can easily describe financial services in high-level viewpoint with minimal knowledge and experiences of computer programming and system development, while complex services are implemented in backend servers and access through APIs. As a result, domain experts and computer engineers together can fast and easily build a virtual assistant that support natural language interface.

Keywords: Extensibility · Virtual Assistant · Chatbot · NLP · Large Language Model · Genie · ThingTalk

1 Introduction

Financial technology, abbreviated as FinTech, is emerging in recent years. Combined together with multiple information and communication technologies (ICT), such as mobile communications, social media, cloud services, and big data analysis, FinTech is expected to significantly change how financial services are provided and consumed. For example, in a user behavior survey of a very popular social app in Taiwan, more than half of its users had accessed official accounts operated by financial services [1, 2]. Chatbots, or virtual assistants, for different financial services become popular in our daily life. However, most of them can understand only common and simple questions and provide answers according to some predefined rules. Using traditional ICT technologies, it is not easy to create a virtual assistants that can interact with its users in natural languages [3].

J. C. Hung et al. (Eds.): FC 2023, LNEE 1134, pp. 1–10, 2024.
https://doi.org/10.1007/978-981-99-9342-0_1

Nowadays, various applications have been emerging due to the development of artificial intelligence (AI). Deep learning [4] promotes natural language and speech signal processing significantly. Chatbots and virtual assistants have gained a lot of attention and are applied in many scenarios, such as education, health care, entertainment, and so on [5]. In new scenarios, the dialogue systems of these virtual assistants are improved to make their replies more like human replies [6], and furthermore the way users interact with them is more similar to the way with humans [7].

Virtual assistants, such as Amazon Alexa [8, 9], Google OpenWeave [10], Apple Homekit [11], Samsung SmartThings [12], were designed and developed for these big giant companies. On the other hand, Genie (previously Almond) is an open-source virtual assistant that takes several important issues into consideration, such as privacy, extensibility, and programmability [13–16].

In this paper, we presents an extensible virtual assistant for financial services using Genie framework. Figure 1 shows the scenario of the proposed virtual assistant. It acts as a chatbot that can provide historical and real-time information of Taiwan Stock market in some chatrooms in a social media. We must note that in practice, there will be user questions that the chatbot cannot understand. It has to handover these questions to the customer support team and technical support team. The latter has to extend the capacity of chatbot while the services cannot stop.

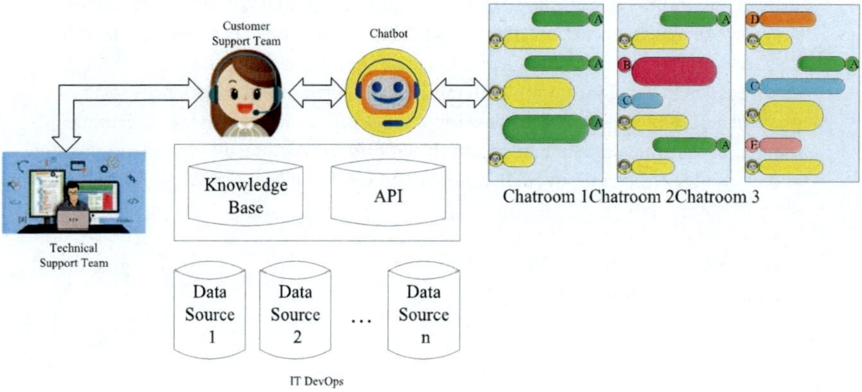

Fig. 1. Scenario of the proposed virtual assistant

We have created the chatbot based on Genie framework [3, 13–16]. A new device (or skill) for Taiwan Stock market is developed to offer stock information in backend server cloud. The device and supported APIs (Application Programming Interface) are registered in an open repository Thingpedia. When Genie receives user utterances, it translates them into ThingTalk programs using a large deep-learning neural networks. Then, Genie executes the ThingTalk programs, which may invoke the financial services through the registered APIs. Figure 2 shows the system architecture of the proposed virtual assistant.

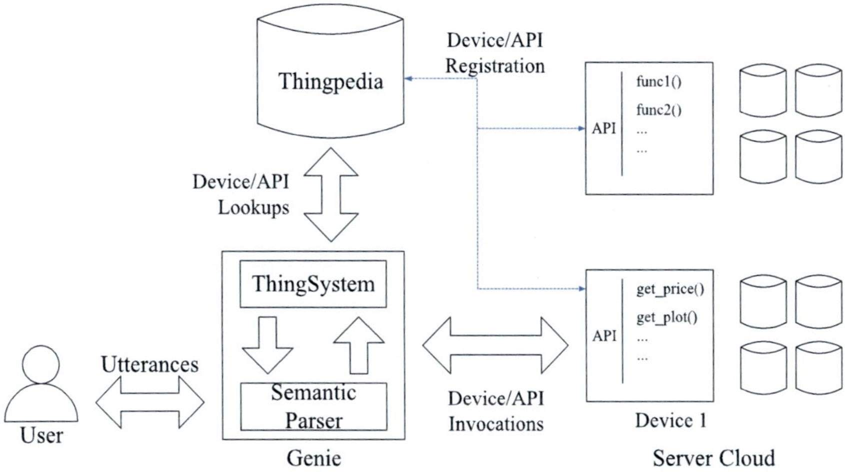

Fig. 2. Proposed virtual assistant architecture based on Genie framework

ThingTalk is a high-level declarative programming language. With a basic level of knowledge and experiences of computer programming and system development, domain experts can conceptually describe financial services in ThingTalk in a high-level viewpoint. Function blocks that together support the financial services physically are implemented by the technical team in backend server cloud. The APIs for these function blocks are registered in Thingpedia. When Genie executes ThingTalk programs, it can look up the requested APIs in Thingpedia and then invoke the corresponding function blocks to realize the financial services. As a result, domain experts and technical team can collaboratively develop and extend chatbot fast and easily.

In the remaining of this paper, we will describe the proposed virtual assistant in Section 2, present the preliminary experiment results in Section 3, and give the conclusions in Section 4.

2 Proposed Virtual Assistant

In this session, we first investigate the possible function set of the proposed virtual assistant. As well, we will simply describe ThingTalk programming language. Then, we present the design and implement of the device for Taiwan Stock.

2.1 Function Set of the Virtual Assistant

First, we collected news, reports, press releases, and articles from various platforms for Taiwan Stock. Keywords were identified, such as price, stock names, stock codes, weighted index, and so on. User utterances to query information of these keywords were manually generated, such as the following query utterances.

- Check the trading volume/opening price/highest price/… of today's market?
- Give me the company information of XYZ/ABC/…*?

- Query XYZ/ABC/... stock code?
- XYZ/ABC/... percentage change today
- XYZ/ABC/... daily/weekly/monthly line
- stock (highest, average, and/or lowest) price of XYZ/ABC/... today/yesterday/last week?

  *XYZ/ABC/... refer to an abbreviation, full name, or nickname of a stock.

We most note that this collection of user intentions and corresponding utterances is typically incomplete. As we have stated above, there will be always a need to extend the capacity of the proposed chatbot for Taiwan Stock market. Thus, in the beginning, we developed a small set of functions for stock information queries.

2.2 ThingTalk

ThingTalk [13–16] is a high-level declarative language designed to access Internet services and IoT devices. ThingTalk is domain-specific and data focused. It has a very simple construct of three types of clauses: ***stream*** (s), ***query*** (q), and ***action*** (a). The construct follows.

$$s[\Rightarrow q]? \Rightarrow a; \tag{1}$$

In a ThingTalk program, s is a ***stream*** clause that determines when the rest of the program runs. It can be a periodic timer, or it can monitor the result of a ***monitorable query*** function defined in Thingpedia for changes. The optional ***query*** clause (q) specifies what data should be retrieved. Results of ***queries*** can be filtered. They can also be used as an input parameter in a subsequent function invocation. The ***action*** clause (a) specifies what the program should do.

For example, a user command "notify me when I receive a text" can be done by the following ThingTalk program.

```
monitor (@org.thingpedia.builtin.thingengine.phone.sms())
                          => notify
```
(2)

To be short, we omit the grammar details of ThingTalk in this paper.

2.3 Device for Taiwan Stock Market

In this study, we registered a new device named as TaiwanStock in Thingpedia, as well as the APIs for the function blocks required to support intended stock services. As shown in Fig. 3, a ***query*** function `get_price()` is declared as an API of this device. Again, we omit the detail of the syntax in this paper.

The three types of clauses in ThingTalk can usually be mapped to three type of phrases in natural languages, and vise versa. For example, a ***stream*** clause can be mapped to ***when*** phrase (***WP***), a query clause to ***noun*** phrase (***NP***), and an ***action*** clause to ***verb*** phrase (***VP***), respectively.

```
Service:TaiwanStock
class @com.TaiwanStock {
    query get_price(
        in opt company : String
          #_[canonical={
            default="base",
            base=["stocknumber", "number","stock code",
                   "stock symbol","ticker symbol"],
            property=["# stock","# stock price"]
          }],
        out stockNo : String
          #_[canonical="stock symbol"],
        out stockName : String
          #_[canonical="stock name"],
    ...)
    ...}
```

Fig. 3. A snapshot of the TaiwanStock device

These phrases are used as ***primitive templates*** in genie-toolkit [13–16]. Table 1 shows some mapping examples used in this study, where genie-toolkit considers ${ } as a holder of a parameter. Combined with ***constructive templates*** for the nature language, English in this study, genie-toolkit can generate a large number of user utterances and their corresponding ThingTalk programs. Thus, we can train the semantic parser in Genie to translate user utterances to ThingTalk programs.

Table 1. Some phrases in nature language and ThingTalk clauses.

Phrases	Type of phrases	ThingTalk clauses
price of ${p_code}	***NP***	`@taiwanstock.get_price (company=p_code)`
${p_company}'s K line	***NP***	`@taiwanstock.get_Kplot (company=p_company)`
When ${p_company} rose by more than ${p_change}	***WP***	`edge (monitor (@taiwanstock.get_price (company=p_company))) on change >= p_change`
Call ${p_number}	***VP***	`@org.thingpedia.builtin. thingengine.phone.call (number=p_number)`

3 Preliminary Experiment

3.1 Data Source

In this study, we downloaded the historical data of 2022 Taiwan Stock market from Taiwan Economic Journal (TEJ) [17]. The data was preprocessed and then stored in a SQL database in the backend server cloud.

3.2 Backend Servers

There are two types of backend servers: database servers and application servers. All servers are generic personal computers running Ubuntu 18.04.6 LTS. MariaDB [18] is adopted in the database server. Functions declared in the TaiwanStock device are implemented using Python and Flask [19] according to Restful API design in the application servers.

3.3 Genie Server

Genie server accepts user utterances, translate them into ThingTalk programs by the semantic parsers, and execute the program to fulfill user requests. Genie accesses the TaiwanStock device via the APIs registered in Thingpedia and implemented in the backend application servers.

Genie server is also a generic personal computers running Ubuntu 18.04.6 LTS. It is equipped with AMD® Ryzen 5 2600 six-core processor$\times$12, 64 GB DRAM, and NVIDIA GeForce RTX 2070.

Currently, Genie server adopts BART-base [20] as its semantic parser.

3.4 Experiment

In this preliminary study, we carefully collected 80 user utterances, and designed their corresponding ThingTalk programs. Some user utterances are simple sentences, while others are dialogs of several sentences. Table 2 shows some of the user utterances and their corresponding ThingTalk programs.

In order to train the semantic parser, BART-base. Phrases were identified as ***WP***, ***NP***, and ***VP***, and mapped to ThingTalk clauses. We used genie-toolkit to generate the training data to train BART-base.

Finally, for the 80 user utterances, we randomly pick up some predicted ThingTalk programs for manual evaluation.

3.5 Experiment Results

It took one hour or so for genie-toolkit to generate the training data and then train semantic parser, currently BART-base.

Table 2. Some user utterances and their corresponding ThingTalk program.

User utterances	ThingTalk program
What is the stock price of 2318?	`$dialogue @org.thingpedia.dialogue.transaction .execute; @com.TWstock.get_price(query="2318") ;`
Check 2318 stock price on April 15	`$dialogue @org.thingpedia.dialogue.transaction .execute; @com.TWstock.get_price(query="2318", date = new Date("2022-04-15"))`
1101 Trends	`$dialogue @org.thingpedia.dialogue.transaction .execute; @com.TWstock.get_plot(stockname ="1101");`

Figure 4 shows a snapshot when chatting with Genie, where the user requests a K line graph of stock No. 2317.

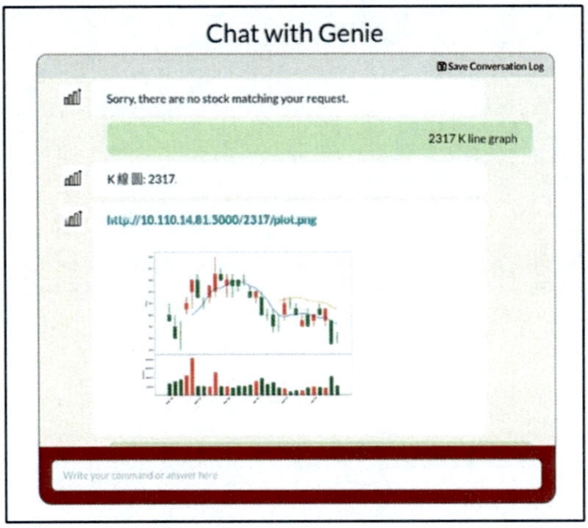

Fig. 4. Snapshot of chatting with Genie

Table 3. Some results of the evaluation

No	User utterances	Target ThingTalk program	Predicted ThingTalk program	Evaluation
1	hello	`$dialogue @org.Thingpedia.dialogue. transaction.greet;`	`$dialogue @org.Thingpedia.dialogue. transaction.greet;`	ok
42	tsmc stock price on DATE_0	`$dialogue @org.Thingpedia.dialogue. transaction.execute; @com.TaiwanStock.get_price( date = new Date (2022,3,10, new Time( 8 , 0)), query = " tsmc ") ;`	`$dialogue @org.Thingpedia.dialogue. transaction.execute; [stockName]of @com.TaiwanStock.get_price( date = DATE_0 );`	ok_ function
61	I want to query the stock price of 0050 stock code	`$dialogue @org.Thingpedia.dialogue. transaction.execute; @com.TaiwanStock.get_price( query = " 50 stock code " );`	`$dialogue @org.Thingpedia.dialogue. transaction.execute; @com.TaiwanStock.get_price( query = " 50 " );`	ok_ without_ param
68	query the stock price of 1101	`$dialogue @org.Thingpedia.dialogue. transaction.execute; @com.TaiwanStock.get_price( query=" 1101 " );`	`$dialogue @org.Thingpedia.dialogue. transaction.execute; @com.TaiwanStock.get_price( query = " 1101 " );`	ok

3.6 Evaluation

For the 80 user utterances, we randomly pick up some generated ThingTalk programs for manual evaluation.

We carefully compare the predicted and target ThingTalk programs [16]. The result of the each comparison could be ok, ok_without_param, ok_function, ok_device, ok_num_function, ok_syntax, or wrong_syntax.

Table 3 show some results of the evaluation. The evaluation result shows 76 of 80 predicted ThingTalk programs are effective.

4 Conclusions and Future Works

In this study, we had designed and implemented an extensible chatbot for Taiwan Stock market based on Genie framework. Genie server accepts user utterances, translate them into ThingTalk programs by the semantic parsers, and execute the program to fulfill user requests. Genie accesses the TaiwanStock device via the APIs registered in Thingpedia and implemented in the backend application servers.

ThingTalk is a declarative programming language. Domain experts can conceptually describe various services in ThingTalk in a high-level viewpoint. They need just a basic level of knowledge and experiences of computer programming and system development. On the other hand, functions actually carrying out the complex computing logics are declared as APIs, and implemented by the technical team in backend server cloud. With ThingTalk programming language as a bridge, domain exports can focus on the service logics in high level, while technical teams can focus on implementing the APIs in the backend servers. As a result, it is much easier to extend the capacity of the chatbot than earlier approaches.

4.1 Future Works

Currently, several new functions of the chatbot have been identified. New APIs are under construction to extend the capacity of the chatbot. As well, new large language models (LLM), such as ChatGPT [21], are under investigation for translation from user utterances into ThingTalk programs.

Acknowledgment. This work was supported in part by National Science and Technology Council, Taiwan, under Grant MOST 111-2221-E-029-018-.

References

1. King, B.: Bank 4.0: Banking Everywhere, Never at a Bank. Wiley (2018)
2. 2021 LINE User Usage Survey. Nielsen (2021). (in Chinese). https://linecorp.com/zh-hant/pr/news/zh-hant/2021/4000. Accessed 20 May 2023
3. Wu, Y.-T.: Design and implementation of extensible financial chatbot. Master Thesis. Department Information Management, Tunghai University (2023)
4. LeCun, Y., Bengio, Y., Hinton, G.: Deep learning. Nature **521**(7553), 436–444 (2015)

5. Liao, S.-W., Hsu, C.-H., Lin, J.-W., Wu, Y.-T., Leu, F.-Y.: A deep learning-based Chinese semantic parser for the Almond virtual assistant. Sensors **22**(5), 1891 (2022)
6. Shah, H., Warwick, K., Vallverdú, J., Wu, D.: Can machines talk? Comparison of Eliza with modern dialogue systems. Comput. Hum. Behav. **58**, 278–295 (2016)
7. Adamopoulou, E., Moussiades, L.: An overview of chatbot technology. In: Maglogiannis, I., Iliadis, L., Pimenidis, E. (eds.) AIAI 2020. IFIP Advances in Information and Communication Technology, vol. 584, pp. 373–383. Springer, Cham (2020). https://doi.org/10.1007/978-3-030-49186-4_31
8. Amazon Alexa Voice AI. https://developer.amazon.com/alexa. Accessed 20 May 2023
9. Goyal, A., Metallinou, A., Matsoukas, S.: Fast and scalable expansion of natural language understanding functionality for intelligent agents. In: Proceedings of the 2018 Conference of the North American Chapter of the Association for Computational Linguistics: Human Language Technologies, New Orleans, LA, USA, 1–6 June 2018 (2018)
10. OpenWeave. https://openweave.io/. Accessed 20 May 2023
11. HomeKit Overview. https://developer.apple.com/apple-home. Accessed 20 May 2023
12. SmartThings Developers. https://developer.smartthings.com/. Accessed 20 May 2023
13. Campagna, G., Ramesh, R., Xu, S., Fischer, M., Lam, M.S.: Almond: the architecture of an open, crowdsourced, privacy-preserving, programmable virtual assistant. In: Proceedings of the 26th International Conference on World Wide Web, pp. 341–350 (2017)
14. Campagna, G., Xu, S., Moradshahi, M., Socher, R., Lam, M.S.: Genie: a generator of natural language semantic parsers for virtual assistant commands. In: Proceedings of the 40th ACM SIGPLAN Conference on Programming Language Design and Implementation, pp. 394–410 (2019)
15. Campagna, G., Semnani, S., Kearns, R., Sato, L.J.K., Xu, S., Lam, M,: A few-shot semantic parser for wizard-of-oz dialogues with the precise ThingTalk representation. In: Findings of the Association for Computational Linguistics, ACL 2022, Dublin, Ireland, pp. 4021–4034 (2022)
16. Genie Wiki. https://wiki.genie.stanford.edu/. Accessed 20 May 2023
17. Taiwan Economic Journal. https://www.finasia.biz/. Accessed 20 May 2023
18. MariaDB. https://mariadb.org/. Accessed 20 May 2023
19. Flask-RESTful. https://flask-restful.readthedocs.io/en/latest/. Accessed 20 May 2023
20. Lewis, M., et al.: BART: denoising sequence-to-sequence pre-training for natural language generation, translation, and comprehension. In: ACL (2020)
21. Introducing ChatGPT. https://openai.com/blog/chatgpt. Accessed 20 May 2023

LNMER-Net: A Metabolically Enhanced Lymph Node Metastasis Recognition Model Based on Lung Lymph Nodes and Microenvironment

Lingyun Wang[1], Huiyan Jiang[1(✉)], Yang Zhou[1], Qiu Luan[2], Bulin Du[2], Yaming Li[2], Xuena Li[2], and Yan Pei[3]

[1] Software College, Northeastern University, No. 195, Chuangxin Road, Hunnan District, Shenyang 110169, China
hyjiang@mail.neu.edu.cn

[2] Department of Nuclear Medicine, The First Affiliated Hospital of China Medical University, Shenyang 110001, China

[3] Computer Science Division, The University of Aizu, Aizu-Wakamatsu, Fukushima 965-8580, Japan

Abstract. PET/CT is the preferred device for lung cancer and lymph node metastasis diagnosis, and mining effective features from PET/CT images to identify lung lymph node metastasis has important research significance and application value. Multi-phase PET/CT has temporal properties that can better represent changes in lesions' structural and metabolic properties. Early-phase PET images can show a wide range of lesion areas. Delayed-phase PET images can show the high uptake properties of ^{18}F-FDG in malignant tumor cells. Thus, multi-phase PET represents the variability of benign/malignant lesions better in the temporal dimension. This paper first proposes a metabolic enhancement method for lung lymph nodes and their microenvironment, a lymph node metastasis recognition network (LNMER-Net). The network has three branches: multi-modal early-phase feature fusion channel, multi-modal delayed-phase feature fusion channel, and single-modal metabolic decay channel. To enhance the feature of the lymph node region, a multi-receptive field-based feature extraction and feature space optimization (MRFO) method is proposed to extract lymph node features by multi-scale convolution operations and embed them in the multi-modal fusion channel. To exploit the information on the metabolic changes of the lesion in the early-phase and delayed-phase, differential results of the multi-phase PET images are fed into the single-modal metabolic decay channel to enhance the microenvironmental features. To verify its effectiveness, a multi-phase PET/CT dataset from China Medical University is used. The proposed method achieves 84.5%/82.9% in Accuracy/Recall, which is better than SOTA methods such as Res2Net, Comformer, and NextViT.

Keywords: Lung lymph node metastasis recognition · Metabolic enhancement · Multi-phase PET/CT · Feature optimization

J. C. Hung et al. (Eds.): FC 2023, LNEE 1134, pp. 11–22, 2024.
https://doi.org/10.1007/978-981-99-9342-0_2

1 Introduction

Lung cancer is one of the most common malignancies worldwide and has a high mortality rate among cancers [1]. Cancer can metastasize to nearby lymph nodes, tissues, organs, and other parts of the body, and metastasis is a common cause of death from cancer [2, 3]. Early detection of lung lymph nodes and rapid identification of their benign and malignant nature is crucial for patient survival [4]. Currently, in clinical practice, Computed Tomography (CT) and Positron Emission Tomography (PET) are important and advanced imaging tools for the diagnosis of cancer. CT imaging extracts detailed anatomical high-resolution information, and PET imaging extracts metabolic and functional information about organs [5]. Due to the variable signs and small diameter of lung lymph nodes [6], they are more disturbed by other normal tissues. Moreover, the amount of slice data obtained from PET and CT is huge, and it is time-consuming and difficult to ensure that small nodes are not missed if physicians directly analyze and identify the lesion areas in each slice [7]. Therefore, it is worthwhile to diagnose the lung lymph nodes accurately identified by computer based on PET/CT images.

Currently, the existing classification algorithms related to lung lymph nodes and pulmonary nodules can be divided into two main categories. One is based on traditional feature descriptors to extract shallow manual features to identify malignant and benign nodules; the other is to design various deep learning methods to extract deeper abstract semantic features of images for classification. The traditional methods extract features (including texture, shape, intensity, and morphological features) from nodules manually, reflecting the heterogeneity of nodules. Then feed the features into a classifier to predict the class of nodules. Many experimental results demonstrate that traditional methods can obtain good classification results [8–14].

Despite the popularity of handcrafted features for classification, many limitations remain. For example, handcrafted features cannot fully characterize heterogeneous lung lymph nodes, and it can be difficult to filter key features among the numerous feature information. Researchers have recently started utilizing convolutional neural networks (CNNs) for medical image tasks. Compared with traditional methods, CNNs are automatic and adaptive models that learn highly discriminant features from various image data for classification, and omit feature design and other processes. Initially, Hua et al. [15] showed that the classification of nodules in CT images using CNNs and deep confidence networks (DBNs) both outperformed traditional methods. He et al. [16] proposed a deep residual network (ResNet), which introduced shortcut connections in deep learning models to alleviate the problem of excessive depth of neural networks. Huang et al. [17] created dense convolutional networks (DenseNet) that can enhance feature propagation, generalize better, and prevent overfitting. Chen et al. [18] proposed a dual path network (DPN) that incorporates ResNet, which focuses on feature reuse, and DenseNet, which focuses on feature generation. Gao et al. [19] designed Res2net based on ResNet, which extracts global and local features of images more comprehensively and effectively through multi-receptive fields. In addition to CNNs, other neural network structures have been used in image classification studies, and Vaswani et al. [20] proposed the Transformer method that relies on an attention mechanism to accomplish the classification task. Peng et al. [21] fused CNN and Transformer models and proposed the first parallel hybrid network of the two, which combines the advantages of both and improves the classification effect without adding more computation. Li et al. [22] designed a new hybrid model of CNN and Transformer, which model can achieve significant advantages in the classification task.

At present, the above studies are mainly based on single-phase CT images, and the existing works lack multi-modal multi-phase studies. To make lung lymph node identification more accurate and efficient, this paper combines the advantages of multi-modality and multi-phase, digs deeper into the temporal information, and introduces metabolic decay information into the network to better assist the network in diagnosis. The contributions of this paper are mainly in the following three aspects:

1. A novel model is proposed for multi-modal multi-phase data, involving three input channel branches, namely multi-modal early-phase feature fusion channel, multi-modal delayed-phase feature fusion channel, and single-modal metabolic decay channel.
2. A single-modal metabolic attenuation channel is proposed for PET images imaged using ^{18}F-FDG as a tracer, which is the first current branch of application for tumor characterization of multi-phase PET images.
3. A multi-receptive field-based feature extraction and feature space optimization method is designed, using convolutional blocks of multi-scales for feature extraction, and then filtering out the feature information that is more decisive for classification after corresponding operations.

2 Recognition of Lung Lymph Node Metastasis

CT images contain textural information about the lesion area, and PET images with ^{18}F-FDG as a tracer have metabolic information. Early-phase imaging and delayed-phase imaging have a temporal relationship, and the PET images showing metabolic information are significantly different.

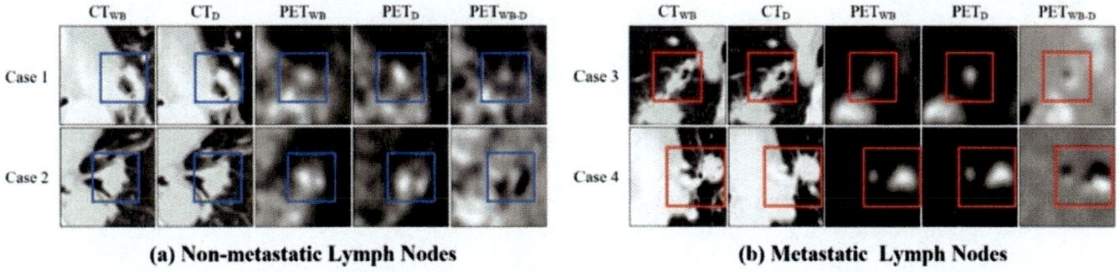

Fig. 1. Schematic diagram of lung lymph nodes in different cases. (a) The diagram shows an example of non-metastatic lymph nodes in lung lymph nodes (benign). (b) The diagram shows an example of metastatic lymph nodes in lung lymph nodes (malignant). (a)(b) Each row of both figures shows the lymph nodes area of different patients (Case1–4), and each column from left to right shows the early-phase CT image, delayed-phase CT image, early-phase PET image, delayed-phase PET image, and early-phase PET - delayed-phase PET difference image, respectively, and the rectangular box indicates the lymph nodes area, blue is benign and red is malignant.

The difference between the early-phase image and the delayed-phase image is obtained as a differential image. Some of the lymph nodes are shown in Fig. 1. The different images visualize the metabolic differences in time series and show the microenvironmental information of metabolic decay. To better extract the lymph nodes and their surrounding features, this paper uses multi-modal multi-phase PET/CT images and multi-phase PET differential images.

2.1 LNMER-Net

To enhance the metabolic characteristics of lung lymph nodes and their microenvironment, this paper proposes the LNMER-Net, a lymph node metastasis recognition network model involving three channel branches. The network architecture is shown in Fig. 2 and is described as follows:

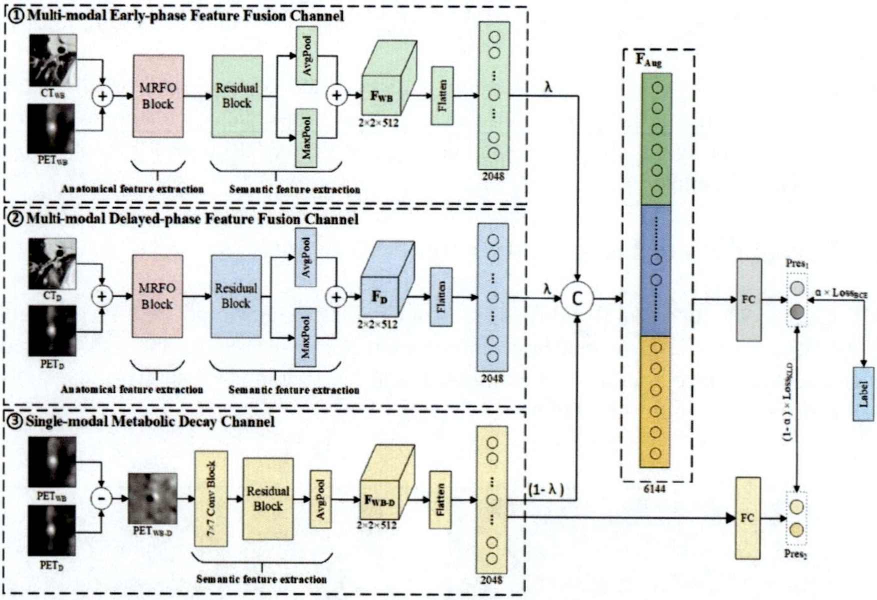

Fig. 2. Architecture of metabolically enhanced lymph node metastasis recognition network based on lung lymph nodes and their microenvironment (LNMER-Net)

① The first channel is called the multi-modal early-phase feature fusion channel, and the input of this channel is the fused image (I_{WB}) of early-phase PET (PET_{WB}) and early-phase CT (CT_{WB}). At first, this branch extracts the underlying anatomical features by the multi-receptive field-based feature extraction and feature spatial optimization method (MRFO) proposed in this paper. The shape features and texture features are extracted using multi-scale convolution blocks, and the features are spatially optimized by the corresponding operations. Then the deep semantic features are extracted after residual blocks, and flattened after compressing features in pooling layers to obtain early-phase lymph node features, denoted as F_{WB}.

② The second channel is the multi-modal delayed-phase feature fusion channel, and the input of this channel is the fused image (I_D) of delayed-phase PET (PET_D) and delayed-phase CT (CT_D). Similar to the multi-modal early-phase feature fusion channel, the input is flattened after the MRFO block, residual block, and pooling layer to obtain the delayed-phase lymph node feature, which is noted as F_D.

③ The third channel is a single-modal metabolic decay channel, where the input is the difference image ($I_{WB\text{-}D}$) between the early-phase PET (PET_{WB}) and the delayed-phase PET (PET_D), and the microenvironmental features are highlighted by metabolic decay. This metabolic decay branch focuses more on minute detail information, so MRFO containing multiple large convolutional blocks is not used. Structural features are extracted using a 7×7 convolution kernel, and deep semantic features are extracted using residual blocks, which are subsequently pooled and then flatten to obtain lymph node microenvironment features, denoted as $F_{WB\text{-}D}$. The features F_{WB}, F_D, and $F_{WB\text{-}D}$ of the three channels are concatenated to obtain the region of interest enhancement features, denoted as F_{Aug}. The cascade is augmented by setting learnable adaptive weights in the network for the three features. The weights of F_{WB} and F_D are λ and the weight of $F_{WB\text{-}D}$ is $(1 - \lambda)$. The F_{Aug} goes through the fully connected layer, resulting in the prediction $Pred_1$. The $F_{WB\text{-}D}$ alone goes through the fully connected layer, resulting in the prediction $Pred_2$.

In this paper, the base-stem networks of all three paths use ResNet50 [16], which is the mainstream model in current classification networks with a wide range of applications and strong applicability. However, its receptive field is small and the long-distance dependence of spatial pixels is lost. Therefore, this method improves the lung lymph node classification task to make the network achieve better results.

2.2 Multi-receptive Field-Based Feature Extraction and Feature Space Optimization Method (MRFO)

Compared with small convolution kernels, large convolution kernels have larger receptive fields, and using large receptive fields has an irreplaceable effect compared to superimposing multiple small receptive fields. There is a higher shape bias and a stronger dependence on the target shape using large receptive fields, and a stronger texture bias and a higher dependence on the image texture for small receptive fields [23]. CT images can show clear texture information and PET images can show metabolic range. For the complex features of PET/CT fusion image information, this paper designs a feature extraction and feature space optimization module based on multi-receptive fields, denoted as MRFO. The architecture diagram is shown in Fig. 3.

The PET images (I_{PET}) and CT images (I_{CT}) are fused and inputted, and the image shape features and texture features are first extracted by a multi-receptive field-based feature extraction module, i.e., five convolutional blocks of different scales ($C_{1\text{–}5}$) and a dilation convolution (DC). The large convolution kernel has a large effective field of perception, can examine the feature map of a wider area, and the obtained features have global characteristics and pay more attention to the shape features of the fused image; the small convolution kernel is less computationally intensive, less likely to ignore local features, and focuses on the image texture information. Using multi-scale receptive fields, the original image is probed with multiple filters with complementary effective fields of view to obtain useful image contextual information at multiple scales, allowing feature extraction of regions of interest to ensure both shape features and texture features. Then the feature information extracted by multi-scale convolution is concatenated to obtain the feature F_C, as in *Eqs.* (1)–(3).

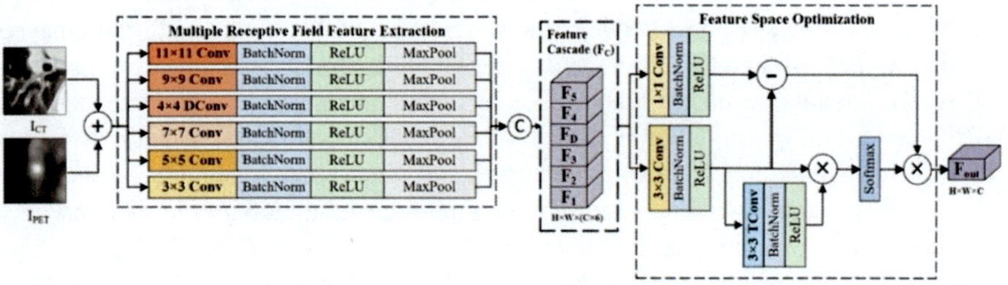

Fig. 3. Multi-receptive field-based feature extraction and feature space optimization module (MRFO) architecture diagram

The F_C enters the feature space optimization module and performs the corresponding operations in the spatial dimension to extract the spatial relationship features. The F_C is first convolved by 1 × 1 convolution ($C_{1\times 1}$) to obtain feature F_{C1} and by 3 × 3 convolution ($C_{3\times 3}$) to obtain feature F_{C2}, respectively. Feature F_2 is again convolved by 3 × 3 transposition ($TC_{3\times 3}$) to obtain feature F_{C3}, as in *Eq.* (4). The F_{C2} is multiplied with the F_{C3} and then passed through Softmax and then multiplied with the difference between the F_{C1} and the F_{C2} to output feature F_{out}, as in *Eq.* (5). Among them, $C_{1\times 1}$ serves to reduce the dimensionality and perform the convolution operation only in the channel direction to achieve the reduction of the number of channels and reduce the number of parameters without changing the other dimensional information. After $C_{3\times 3}$, $TC_{3\times 3}$ and their multiplication operations, the features are further extracted and the planar correlation of the features is found at the same time. The difference operation is performed between the F_{C1} and the F_{C2} to find the different features and filter out the features that are more decisive for subsequent image recognition. The MRFO is calculated as follows:

$$F_D = P_{max}(\sigma(BN(DC(I_{\text{PET}} + I_{\text{CT}})))) \tag{1}$$

$$F_n = P_{max}(\sigma(BN(C_n(I_{\text{PET}} + I_{\text{CT}})))) \tag{2}$$

$$F_C = Cat(F_1, F_2, F_3, F_4, F_5, F_D) \tag{3}$$

$$F_{C1} = C_{1\times 1}(F_C), F_{C2} = C_{3\times 3}(F_C), F_{C3} = TC_{3\times 3}(F_{C2}) \tag{4}$$

$$F_{out} = (F_{C1} - F_{C2}) \odot Softmax(F_{C2} \odot F_{C3}) \tag{5}$$

where $C_n(\cdot)$, $C_{1\times 1}(\cdot)$, $C_{3\times 3}(\cdot)$ denote convolutional operations, $DC(\cdot)$ denotes dilated convolutional operation, BN denotes batch normalization operation, $\sigma(\cdot)$ denotes ReLU function, $P_{max}(\cdot)$ denotes maximum pooling, $Cat(\cdot)$ denotes concatenate operation, $TC(\cdot)$ denotes transposed convolutional operation, $+$ denotes addition of the corresponding elements of an array, and $\odot$ denotes multiplication of the corresponding elements of an array, n $\in$ [1,5].

3 Results and Discussion

3.1 Dataset and Preprocessing

The private dataset contains 99 lung lymph nodes from 51 patients from the Department of Nuclear Medicine, The First Hospital of China Medical University. All lung lymph node regions are manually outlined by experienced radiologists as the ground truth, and all lymph node metastases are determined by pathological puncture examination. The age of the patients ranged from 38 to 75 years, and the number of men and women was 28 and 23 cases, respectively. The PET image planar voxel size is 2.03642 mm $\times$ 2.03642 mm $\times$ 5.00 mm, the planar resolution is 400 $\times$ 400; the CT image planar voxel size is 0.976563 mm $\times$ 0.976563 mm $\times$ 2.00 mm, the planar resolution is 512 $\times$ 512. Each case had early-phase PET images and CT images, and delayed-phase PET images and CT images. The early-phase images were taken normally after the patient was injected with ^{18}F-FDG, and the delayed-phase images were taken about two hours after the early-phase medical images were taken.

The data is preprocessed, and the CT images are converted to Hu value truncation and then normalized. For PET images, resampling is first performed to rigidly align with CT images, which are truncated to SUV values and then standardized. PET and CT slices containing the lymph nodes are cropped to 64 $\times$ 64 centered on the lymph nodes according to the ground truth. A total of 959 sets of images (each set includes 4 images) are finally used for the experiments. 720 sets are used for the training set and 239 sets for the test set in the experiments. And data enhancement is performed on the data in the training set, with horizontal and vertical flips and random rotation of the images at certain angles.

3.2 Experimental Details and Evaluation Metrics

This work is conducted on an NVIDIA GeForce RTX 3060 12G server under Windows 11 operating system, based on the PyTorch framework. The Adam optimizer is set with a learning rate of 0.0001. The batch size is set to 8. The image data are preprocessed as described in Sect. 3.1. The network input size is 64 $\times$ 64, the network task is to determine whether the region is a lung lymph node metastasis, and the network output results in probability values for both categories. The experiments use loss functions including BCELoss, KLDivLoss, and total Loss, see *Eqs.* (6)–(8).

$$L_{BCE} = y_n \times \ln(x_n) + (1 - y_n) \times \ln(1 - x_n) \tag{6}$$

$$L_{KLD} = y_n(logy_n - x_n) \tag{7}$$

$$L_{\text{Total}} = \alpha * L_{BCE} + (1 - \alpha) * L_{KLD} \tag{8}$$

where x_n denotes the predicted value, and y_n denotes the actual label. In the network, the enhancement feature F_{Aug} enters the fully connected layer to get the prediction result $Pred_1$ which is used with the true label to calculate L_{BCE} using *Eq.* (6). The microenvironmental feature $F_{WB\text{-}D}$ enters the fully connected layer to get the prediction result $Pred_2$ which is used with $Pred_1$ which is treated as a weak label, to calculate L_{KLD} using *Eq.* (7). L_{BCE} and L_{KLD} set weights and sum to get L_{Total}, see *Eq.* (8). The network works best when $\alpha = 0.7$ by experiment.

3.3 Comparison with Other Methods

The lymph node metastasis recognition model based on the metabolic enhancement of lymph nodes and their microenvironment proposed in this paper is compared with other advanced networks. In ResNet50 [16], Densenet121 [17], DPN92 [18], Res2Net50 [19], Conformer [21] and Next-ViT [22] methods, multi-modal early-phase fusion images are concatenated with multimodal delayed-phase fusion images using a single branch channel input network for experiments. The results are shown in Table 1.

Table 1. Comparison with other advanced methods.

Method	Accuracy	F1-score	Precision	Recall
ResNet50 [16]	0.741	0.740	0.750	0.757
Densenet121 [17]	0.745	0.729	0.738	0.725
DPN92 [18]	0.778	0.768	0.771	0.766
Res2Net50 [19]	0.741	0.725	0.733	0.721
Conformer [21]	0.766	0.752	0.760	0.747
Next-ViT [22]	0.766	0.764	0.767	0.777
LNMER-Net (Ours)	**0.845**	**0.836**	**0.848**	**0.829**

Note: The best results are shown in bold black font

As can be seen in Table 1, our proposed network can achieve the best results in terms of Accuracy, F1-score, Precision, and Recall, whether compared with ResNet50 [16], Densenet121 [17], DPN92 [18] base network, or Res2Net50 [19], Conformer [21], Next-ViT [22] the latest methods are improved compared with each other. The proposed method deeply explores the data features of the multi-modal single-phase and single-modal multi-phase and designs multi-modal feature extraction modules and metabolic attenuation channels that facilitate lesion identification and enhance lymph node features and microenvironmental features closely around lesion features. The current SOTA methods are poorly adapted to the data characteristics and lesion features and thus perform generally in the problem of lung lymph node metastasis identification.

3.4 Ablation Experiments

In this paper, a multi-channel branching network is designed to input multi-modal early-phase fusion images, multi-modal delayed-phase fusion images, and single-modal metabolic attenuation information into the network separately, and the extracted early-phase lymph node features, delayed-phase lymph node features, and microenvironmental features are enhanced in cascade. The early-phase and delayed-phase images extract information to enhance the lymph node features, and the multi-phase PET image difference highlights the lymph node metabolic attenuation information to enhance the microenvironmental features. For multi-modal images, PET images contain metabolic information but have low resolution, while CT images can more clearly represent the lymph nodes and the surrounding texture information, and the two are fused and input to the network to enhance the lymph node features. So, the MRFO is proposed to extract the shape and texture information from the corresponding fused images by using multi-scale receptive fields and to obtain the underlying anatomical structure features and enhance the lymph node features by filtering the feature information that is more decisive for classification through spatial optimization of the corresponding operations. In this paper, ablation experiments are performed on the network model, and the results are shown in Table 2 and Table 3.

Multi-channel Ablation Experiment. In this section, this work performs experimental comparisons by varying the number of branches of the input network, with the network branches cascaded before the fully connected layer. The experimental results for the single-branch network are shown in the first three rows of data in Table 2, the data for the two-branch network results are shown in rows 4–6 of Table 2, and the last row shows the experimental results using three channels.

Table 2. Ablation experiments of LNMER-Net under different channel inputs

Method	Channel			Accuracy	F1-score	Precision	Recall
	WB	D	WB-D				
1 Channel	√	-	-	0.766	0.759	0.757	0.760
	-	√	-	0.736	0.713	0.735	0.708
	-	-	√	0.669	0.592	0.697	0.609
2 Channels	√	√	-	0.828	0.814	0.839	0.805
	√	-	√	0.770	0.744	0.785	0.736
	-	√	√	0.791	0.772	0.800	0.764
3 Channels (Ours)	√	√	√	**0.845**	**0.836**	**0.848**	**0.829**

Note: The best results are indicated in bold black font, √ indicates that the branch is added to the network, and - indicates that the branch is not used

From the data in the table, we can see that the network simultaneously sets up a multi-modal early-phase feature fusion channel, multi-modal delayed-phase feature fusion channel and single-modal metabolic decay channel to extract and enhance lymph node features and microenvironmental features to make the network fit better, and the best results can be obtained with Accuracy reaching 0.845 and F1-score reaching 0.836. Using the single-modal metabolic decay channel alone is the least effective, which indicates that the lymph node features extracted from multi-modal fused images are essential for image classification and that the network is not sufficient to accurately classify images by considering only learning the metabolic decay information in single-modal. The experimental results using dual channels show that enhancing lymph nodes or microenvironmental features improves the network effect.

MRFO Ablation Experiment. In this section, ResNet50 with three branch inputs is chosen as the baseline network (BL) for this work, and the results are shown in row 1 of Table 3. Each channel is selected to add or not an MRFO block before entering the residual block. Adding one MRFO block to the network results in rows 2–4 of Table 3, adding two MRFO blocks results in rows 5–7 of Table 3, and adding three MRFO blocks results in row 8 of Table 3.

Table 3. MRFO ablation experiment

Method	Channel			Accuracy	F1-score	Precision	Recall
	WB	D	WB-D				
ResNet50 (BL)	-	-	-	0.749	0.745	0.745	0.753
BL+MRFO (1)	√	-	-	0.770	0.744	0.785	0.736
	-	√	-	0.799	0.777	0.819	0.767
	-	-	√	0.766	0.752	0.760	0.747
BL+MRFO (2)	√	-	√	0.762	0.761	0.801	0.794
	-	√	√	0.824	0.820	0.818	0.826
	√	√	-	**0.845**	**0.836**	**0.848**	**0.829**
BL+MRFO (3)	√	√	√	0.736	0.724	0.727	0.723

Note: The best results are indicated in bold black font, √ indicates that the module is added to the network, and - indicates that no action is taken

As can be seen from the data in the table, using MRFO in the two channels of multi-modal early-phase feature fusion and multi-modal delayed-phase feature fusion in the network gives the best results, with a 10.4% improvement in Accuracy and a 9.6% improvement in F1-score compared to BL. The priority of using MRFO in different branches also varies, with the most significant improvement in the multi-modal delay-phase channel. It is worth noting that using MRFO in all three channels becomes less effective instead. The reason is that for the single-modal metabolic decay channel, small and fine structural information needs to be learned from the input data, while the multi-modal early-phase channel and multi-modal delayed-phase channel have complex input fusion image information and need to extract lymph nodes and their background information. MRFO combines large receptive fields with small receptive field convolution kernels, which is more conducive to the extraction of structural information of complete large regions, and therefore the use of MRFO in multi-modal early-phase passages and multi-modal delayed-phase passages is more effective in enhancing the network.

In summary, the method proposed in this paper has good performance in all metrics, but there is still some space for improvement before it is put into clinical application. Therefore, to improve the accuracy and stability of the model performance, we need to further explore the definition of a priori features, effective feature extraction, and feature optimization, to build a model with excellent performance and realize intelligent diagnosis in the real sense.

4 Conclusion

In this paper, based on a multi-phase PET/CT dataset, the proposed network extracts multi-modal early-phase lymph node features, multi-modal delayed-phase lymph node features, and microenvironmental features. Applying the multi-modal multi-phase data, this method enhances the lymph node features while significantly enhancing the tiny fine microenvironmental features by metabolic attenuation information to assist the network in better classification. The multi-modal data combine the advantages of PET images containing metabolic information and CT images containing texture information to provide the network with more comprehensive and complementary features of lymph nodes. The proposed MRFO extracts complementary features from multi-modal fused images uses multi-receptive fields to extract more image contextual features, captures both global and local information of images, and optimizes shallow features to extract more effective deep semantic features. The experimental results show that the method in this paper improves by 6.7%, 6.8%, 7.7%, and 6.3% over the optimal method in Accuracy, F1-score, Precision, and Recall. In comparison with ResNet50 [16], with the addition of the proposed metabolic decay channel and MRFO module alone, Accuracy is improved by 7.9% and 9.6%, respectively. Therefore, the proposed network is superior and promising for research in lung lymph node classification.

Acknowledgement. This work was supported by the Natural Science Foundation of Liaoning Province (No. 2021-YGJC-07).

References

1. Gridelli, C., et al.: Non-small-cell lung cancer. Nat. Rev. Dis. Primers. **1**(1), 1–16 (2015)
2. Pham, T.D.: Classification of Benign and Metastatic Lymph Nodes in lung cancer with deep learning. In: 2020 IEEE 20th International Conference on Bioinformatics and Bioengineering (BIBE), pp. 728–733. IEEE (2020)
3. Mehlen, P., Puisieux, A.: Metastasis: a question of life or death. Nat. Rev. Cancer **6**(6), 449–458 (2006)
4. Klik, M.A.J., v Rikxoort, E.M., Peters, J.F., Gietema, H.A., Prokop, M., v Ginneken, B.: Improved classification of pulmonary nodules by automated detection of benign subpleural lymph nodes. In: 3rd IEEE International Symposium on Biomedical Imaging: Nano to Macro, pp. 494–497. IEEE (2006)
5. Xia, X., Zhang, R.: A novel lung nodule accurate segmentation of PET-CT images based on Convolutional neural network and Graph Model. IEEE Access **11**, 34015–34031 (2023)

6. Rami-Porta, R., Asamura, H., Travis, W.D., Rusch, V.W.: Lung cancer—major changes in the American Joint Committee on Cancer eighth edition cancer staging manual. CA: Cancer J. Clinicians **67**(2), 138–155 (2017)
7. Moltz, J.H., et al.: Advanced segmentation techniques for lung nodules, liver metastases, and enlarged lymph nodes in CT scans. IEEE J. Sel. Topics Signal Process. **3**(1), 122–134 (2009)
8. Madero Orozco, H., Vergara Villegas, O.O., Cruz Sánchez, V.G., Ochoa Domínguez, H.D.J., Nandayapa Alfaro, M.D.J.: Automated system for lung nodules classification based on wavelet feature descriptor and support vector machine. Biomed. Eng. Online **14**(1), 1–20 (2015)
9. Akram, S., Javed, M.Y., Hussain, A., Riaz, F., Usman Akram, M.: Intensity-based statistical features for classification of lungs CT scan nodules using artificial intelligence techniques. J. Exp. Theor. Artif. Intell. **27**(6), 737–751 (2015)
10. Kaya, A., Can, A.B.: A weighted rule based method for predicting malignancy of pulmonary nodules by nodule characteristics. J. Biomed. Inform. **56**, 69–79 (2015)
11. De Carvalho Filho, A.O., Silva, A.C., Cardoso de Paiva, A., Nunes, R.A., Gattass, M.: Computer-aided diagnosis of lung nodules in computed tomography by using phylogenetic diversity, genetic algorithm, and SVM. J. Digital Imaging **30**, 812–822 (2017)
12. Li, X.X., Li, B., Tian, L.F., Zhang, L.: Automatic benign and malignant classification of pulmonary nodules in thoracic computed tomography based on RF algorithm. IET Image Proc. **12**(7), 1253–1264 (2018)
13. Gong, J., Liu, J.Y., Sun, X.W., Zheng, B., Nie, S.D.: Computer-aided diagnosis of lung cancer: the effect of training data sets on classification accuracy of lung nodules. Phys. Med. Biol. **63**(3), 035036 (2018)
14. Wu, W., Hu, H., Gong, J., Li, X., Huang, G., Nie, S.: Malignant-benign classification of pulmonary nodules based on random forest aided by clustering analysis. Phys. Med. Biol. **64**(3), 035017 (2019)
15. Hua, K.L., Hsu, C.H., Hidayati, S.C., Cheng, W.H., Chen, Y.J.: Computer-aided classification of lung nodules on computed tomography images via deep learning technique. OncoTargets Therapy 2015–2022 (2015)
16. He, K., Zhang, X., Ren, S., Sun, J.: Deep residual learning for image recognition. In: Proceedings of the IEEE Conference on Computer Vision and Pattern Recognition (CVPR), pp. 770–778. IEEE (2016)
17. Huang, G., Liu, Z., Van Der Maaten, L., Weinberger, K.Q.: Densely connected convolutional networks. In: Proceedings of the IEEE Conference on Computer Vision and Pattern Recognition (CVPR), pp. 4700–4708. IEEE (2017)
18. Chen, Y., Li, J., Xiao, H., Jin, X., Yan, S., Feng, J.: Dual path networks. In: Advances in Neural Information Processing Systems, vol. 30 (2017)
19. Gao, S.H., Cheng, M.M., Zhao, K., Zhang, X.Y., Yang, M.H., Torr, P.: Res2Net: a new multi-scale backbone architecture. IEEE Trans. Pattern Anal. Mach. Intell. **43**(2), 652–662 (2019)
20. Vaswani, A., et al.: Attention is all you need. In: Advances in Neural Information Processing Systems, vol. 30 (2017)
21. Peng, Z., et al.: Conformer: local features coupling global representations for visual recognition. In: Proceedings of the IEEE/CVF International Conference on Computer Vision, pp. 367–376 (2021)
22. Li, J., et al.: Next-ViT: next generation vision transformer for efficient deployment in realistic industrial scenarios. arXiv preprint arXiv:2207.05501 (2022)
23. Ding, X., Zhang, X., Han, J., Ding, G.: Scaling up your kernels to 31 $\times$ 31: Revisiting large kernel design in CNNs. In: Proceedings of the IEEE/CVF Conference on Computer Vision and Pattern Recognition (CVPR), pp. 11963–11975. IEEE (2022)

Key Factors for Unsubscribing from YouTube Channels: A Study of YouTubers in Taiwan

Hsuan-Che Yang[1(✉)] and Wen-Chih Chang[2]

[1] Department of Information Management, Chihlee University of Technology, New Taipei City 220, Taiwan
hc_yang@mail.chihlee.edu.tw

[2] International Master Program of Information Technology and Applications, National Pingtung University, Pingtung City, Taiwan
yilan.earnest@mail.nptu.edu.tw

Abstract. YouTube has become a major advertisement media for industries. People watch TV shows, movie trailers, news and episodes, on-line shows streaming as well. When people want to purchase something, they can find some unboxing reviews, experiences shares, using instructions and services introductions on YouTube before making orders. Traveling guides and educational tutorials are also can be found on YouTube. Thousands of YouTube channels need channel's subscribers directly or indirectly affect the revenue of channel owners. In addition to YouTube's own profit from watching, the subscribers are also an important reference when manufacturers want to choose key opinion leaders (KOL) or internet celebrities to promote their products or services. Since, maintaining a stable growth of channel subscribers and reducing the occurrence of unsubscribing are the key issue that every YouTuber needs to understand urgently. In this study, we used a survey to understand subscribers and YouTubers on unsubscribing, and analyzed some key factors for unsubscribing. The data analyses include descriptive analysis, factor analysis, reliability analysis, t-test, one-way ANOVA, and Pearson's correlation analysis. Through this study, people can understand key factors for unsubscribing reasons on YouTube platform and the priority of these factors. Whether there is special unsubscribe factors for different channel types and attributes. The key contribution of this study is: a questionnaire on key factors for unsubscribing on the YouTube platform was designed, and it could be utilized for further use in related studies in the future. We also summarize the responses and suggestions of we-media channel owners on the factors for unsubscribing.

Keywords: We-media · YouTuber · Unsubscribe

1 Introduction

With the impact of Web 2.0 and the socialization of the internet, internet has gradually replaced traditional media such as television, radio, and newspapers, which were the dominant media in the past. The current mainstream media has gradually evolved from a top-down business model of content-centered, top-down information dissemination to

J. C. Hung et al. (Eds.): FC 2023, LNEE 1134, pp. 23–38, 2024.
https://doi.org/10.1007/978-981-99-9342-0_3

today's decentralized, bottom-up content sources, with emphasis on sharing, collaboration and community relations among people [1–3]. YouTube is now a day the leading video-sharing platform in the world, with over two billion monthly active users in internet era [4, 5]. YouTube was founded in 2005, and started out as a platform for broadcasting people's life among friends. After its acquisition by Google in 2006 for US$1.65 billion, YouTube has become a multipurposed videos platform for users [6]. According to the annual report from We Are Social (https://wearesocial.com), which is an advertising agency in the United Kingdom that focuses on the use of online communities around the world. The report shows that the total population in Taiwan was 23.87 million in January 2022, and 21.72 million people used internet. The internet penetration rate stood at 91% in Taiwan [7, 8]. The social media users in Taiwan at the start of 2022 was equivalent to 89.4% of total population, about 21.35 million population. People spent around 8 h and 7 min using the internet each day. Social media was accessed for 2 h and 4 min in one day [8]. The ranking of the most visited websites according to Alexa Internet, (which was acquired by Amazon and discontinued on May 1, 2022). YouTube.com is the second most visited site in Taiwan behind Google.com, with 13 h and 31 min a day and 9.64 daily pageviews per visitor [9].

YouTube Economics is a subfield of the economics that focuses on the study of the economic aspects of the YouTube platform. YouTube provides a unique environment for content creators, advertisers, and consumers to interact with one another. Some of the specific areas of study in YouTube Economics include the effects of ad revenue on content creation, the impact of algorithm changes on viewership and engagement, and the role of YouTube in shaping online culture. We call people who owned YouTube channels and created video contents YouTubers. Nowadays, those videos creators gradually evolved into key opinion leaders (KOLs) in cyber world. YouTube provide the revenue sharing business model, which encourage YouTubers to generate and upload videos. When people watch the advertisements provided by YouTube during watching videos, the channel owners earned the advertising revenue, which was shared from YouTube. Besides advertising revenue, YouTube also provide several ways for creators gaining revenue, including channel members, which can watch special contents prepared by channel owners. When YouTubers live streaming in channel, the audiences are also able to donate in chat room.

Channel subscribers are crucial to YouTubers because subscribers represent a loyal audience that regularly watches their videos and engages with their content. When viewers subscribe to channels, they receive notifications when new videos are uploaded, and these notifications can prompt them to watch and engage with the content. Having a large number of subscribers also help YouTubers attract new viewers and build their brand. When potential viewers see that a channel has a substantial subscriber base, they may be more likely to check out the content and subscribe themselves. Additionally, subscribers play a significant role in determining the success of a YouTube channel. YouTube's algorithm takes into account the number of subscribers a channel has, as well as how frequently they engage with the content, when recommending videos to users. This means that channels with more subscribers and higher engagement rates are more likely to appear in search results and be recommended to viewers, leading to increase views and revenue for YouTubers.

On the other hand, when viewers unsubscribe from channels can have several negative effects on YouTubers. Losing subscribers can be discouraging for creators, as it indicates that some viewers are no longer interested in their content or have found other channels new to watch. From a practical standpoint, a loss of subscribers can also lead to a decrease in views and revenue. When subscribers unsubscribe, they are less likely to receive notifications about new videos, which means they may not watch and engage with as much content. This can result in a decrease in views and ad revenue, as well as a decrease in engagement metrics like likes and comments. In addition, YouTube's algorithm takes into account the number of subscribers and their engagement when recommending videos to viewers. If a channel loses a significant number of subscribers, this can signal to the algorithm that the content is not as engaging or relevant, and may lead to a decrease in visibility and recommended videos. Furthermore, a loss of subscribers can also affect the perception of the channel among advertisers and potential sponsors. Brands often look for channels with a large and engaged audience when considering sponsorship opportunities, so a drop in subscribers can make a channel less attractive to potential partners.

As we mentioned above, it is not difficult to find that subscribers are not only a necessary requirement to become a YouTube partner, but the number of subscribers is also a key factor in the latter sources of revenue [1]. In today's we-media inception, producing new content and topics, improving filming and editing techniques is certainly a concern for creators, but maintaining and expanding the number of subscribers is also a growing concern for YouTubers. In this study, the term unsubscribing refers to subscribers of a YouTube channel who have unsubscribed from channels for one reason or others and no longer follow any videos or messages posted by subscribed channels. This study intends to conduct questionnaire to understand subscribers and YouTubers on unsubscribing, and then summarize key factors for unsubscribing.

2 Relevant Research

We-media marketing is a marketing strategy that leverages social media platforms and user-generated content to promote products, services, or brands. In this approach, the emphasis is on creating content that is shared among users and encouraging the creation of user-generated content to promote a brand or product. We-media refers to a group of individuals or entities that produce and distribute content through social media platforms like Facebook, Instagram, Twitter, and YouTube. In 2018, 92% of small and medium-sized enterprises increased their investment in social media. Multi-platform operation can expand the brand of enterprises to multiple social platforms [10, 11] The Swedish creator PewDiePie, whose channel [12] is divided into gaming and entertainment, initially focused on producing unboxing videos for games. Later, he gradually shifted his focus to videos about proving internet rumors, internet memes, and even producing a single [13]. On August 25, 2019, he became the world's first YouTuber with over 100 million subscribers and the first to receive the Red Diamond Creator Award. Although higher subscription numbers do not necessarily equal higher view counts and incomes, the number of YouTube channel subscribers directly affects the channel's rating. vidIQ (https://vidiq.com/) is an online tutorial website that mainly provides video tutorials on how to grow a YouTube channel. vidIQ also offers a Chrome browser extension that allows users to analyze related data of YouTube channels. According to vidIQ's report, it takes time and effort to establish a stable loyalty between the YouTuber and the audience, so YouTubers often ask their audience to give likes, comments, and subscribe and hope viewers will not miss new videos uploaded on the channel. The website also points out six factors that can cause fans to unsubscribe from YouTube channels [14]:

1. YouTube's review team actively detects whether there are fake or bot accounts, and will stop the subscription function or even delete the account if such accounts are found.
2. Posting content that is always the same can make subscribers feel tired and unsubscribe.
3. If the audience's interests and preferences grow beyond what the channel offers, they may unsubscribe and find another channel that meets their needs.
4. Posting outdated content that is no longer popular can cause subscribers to leave.
5. YouTube is a highly competitive platform with thousands of related channels with similar content and themes. Subscribers may choose to unsubscribe to watch channels that are more suitable for their interests.
6. When a YouTuber does not focus on a particular niche market and the content is all over the place without a central theme or unique features, viewers may choose to leave the channel.

NoxInfluencer (NoxInfluencer.com) is a marketing platform based in Beijing, they collect KOLs worldwide. It was created in 2017. The NoxInfluencer tool supports ten different languages globally and can be used on both PCs and smartphones. Its primary coverage includes countries in Asia, like Singapore, Malaysia, Thailand, Myanmar, Vietnam, as well as Japan, Korea, Hong Kong, Macau, Taiwan, Russia, the Middle East, and Europe. They target the YouTube, Instagram, TikTok, and Twitch social platforms. Its purpose is to address the challenges encountered in influencer marketing processes, such as influencer screening, fake follower identification, influencer promotion tracking, and influencer pricing, etc. They use their marketing experience, big data and AI technologies combining them with smartphone advertising and application alliance programs. Users can log in via a Google account to use the platform.

In the YouTube "Influencer Rankings" disclosed by NoxInfluencer, six rankings were included. Such as Top 100 Influencers with the most followers, Top 100 Influencers with the fastest growth of followers, Top 100 Influencers with the highest average views, Top 100 Influencers with the highest Nox rating, Top 100 Influencers with the fastest drop in followers, and Top 100 Influencers with the highest views in 30 days. Among them, the top 100 influencers in Taiwan with the fastest drop in followers were the main research target in our study.

According to the data obtained on February 2, 2021, the top 100 YouTubers in Taiwan with the fastest losing fans, in order of the number of channel types, are: Vlogs, with 25 channels; Gamming, with 23 channels; Entertainment, with 17 channels; Movies and Animations, with 7 channels; Tips & Knowledge and Comedy, with 6 channels each; Music, with 4 channels; News and Politics, with 3 channels; Sciences & Skills and Educations, with 2 channels each; and finally, there are 1 channel each for Welfare

Issues, Cars & Transportation, Traveling, and Pets. We compiled Fig. 1 to show the above data for the top 100 YouTube channels in Taiwan who have lost the most fans.

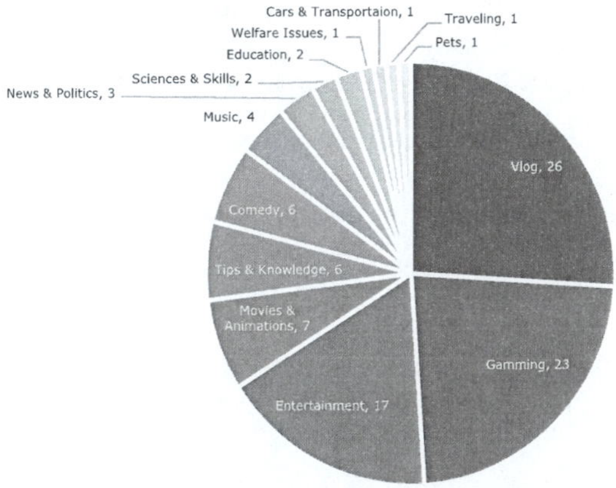

Fig. 1. YouTube channel unsubscribe top 100 category distribution

In the study on the influence of YouTubers on followers' use intention, the authors used the Stimulus-Organism-Response (S-O-R) model to investigate how buzz affects business performance [15]. The results of the study showed that an increase in the frequency of posting videos on YouTube channels would have two adverse effects on YouTube channel buzz:

1. Dilution: The more videos a YouTube channel publishes each month, the more each video's buzz will be diluted by many videos, resulting in lower viewership, comments and likes.
2. Decrease in quality: With the increase in the number of videos posted each month, the increase in the number of videos posted will decrease the quality of a single video when the YouTuber's body, mind, strength and energy are fixed.

According to an article published by Han [16], the number of messages and comments watched affects YouTuber revenue, but the length of time the channel has been around does not have a positive effect on revenue. Viewers' desire for new and topical content may be a factor in this result. In Jin-Chia Chang's master's thesis, he used the Analytic Hierarchy Process (AHP) to explore the needs of the top social media fans using YouTube as an example. The results of the study showed that the importance of the components and attributes were stress relief, emotional, cognitive, social integration, and personal integration in order [17].

3 Experiment Design

In this study, we designed a questionnaire based on the literature and determined the content of the questionnaire through pre-test and exploratory factor analysis for reliability and validity analysis to achieve the purpose of the study. The structure of the study is shown in Fig. 2. There are two hypotheses in the study, namely:

(1) Individuals' characteristic variables have significant differences on "Channel types and Behaviors".
(2) There is a significant difference or correlation between individual characteristics and "Unsubscribe Factors".

A total of 450 questionnaires were randomly distributed and 445 copies were returned, of which 420 were valid, with a valid return rate of 93.3%. The first part of the questionnaire is basic information, second part is channel types and behaviors and the third part is the analysis of the factors for unsubscribing.

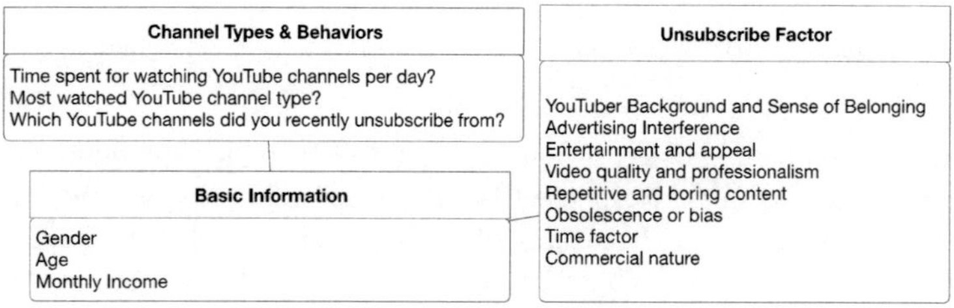

Fig. 2. Research Structure

4 Experiment and Analysis

After experiment, the data analyses include descriptive analysis, factor analysis, reliability analysis, t-test, one-way ANOVA, and Pearson's correlation analysis. We collect some interesting result as follow.

4.1 Basic Information

The number distribution and percentage of gender, age, and monthly incomes are shown in Table 1.

Table 1. Times and percentage of gender, age and monthly incomes

Items	Options	Times	%	Items	Options	Times	%
Gender	Male	152	36.2	Monthly Incomes ($NT)	Below 5000	44.0	10.5
	Female	268	63.8		5,001–10,000	40.0	9.5
Total		**420**	**100.0**		10,001–15,000	52.0	12.4
Age	15–24	260	10.5		15,001–20,000	20.0	4.8
	20–24	216	51.4		20,001–25,000	60.0	14.3
	25–29	56	13.3		25,001–30,000	68.0	16.2
	30–34	56	13.3		30,001–35,000	56.0	13.3
	35–39	4	1.0		35,001–40,000	32.0	7.6
	40–44	8	1.9		40,001–45,000	20.0	4.8
	45–49	12	2.9		45,001–50,000	4.0	1
	50–54	16	3.8		50,001–55,000	8.0	1.9
	55–59	4	1.0		65,001–70,000	8.0	1.9
	60–64	4	1.0		Above 80,000	8.0	1.9
Total		**420**	**100.0**	**Total**		**420**	**100.0**

4.2 Channel Types and Behaviors

Time Spent for Watching YouTube Channels per Day

Among respondents, 94 (44.8%) watched clips for less than 1 h per day, 78 (37.1%) spent 2–3 h per day, and the rest are shown in Table 2.

Table 2. Time spent for watching YouTube Channels per Day

Options	Times	%	Options	Times	%
Less than 1 h	188	44.8	5–6 h	12	2.9
2–3 h	156	37.1	Over 6 h	4	0.9
3–4 h	48	11.4	**Total**	**420**	**100.0**
4–5 h	12	2.9			

Most Watched YouTube Channel Type

Among respondents, 112 (26.7%) watched YouTube most channel often for "Entertainment", and 56 (13.3%) each for "Music" and "Movies & Animation", as shown in Table 3.

Table 3. Most Watched YouTube Channels

Channel Types	times	%	Channel Types	times	%
Entertainment	112	26.7	News & Politics	16	3.8
Movies & Animations	56	13.3	Cars & Transportations	16	3.8
Music	56	13.3	Educations	12	2.9
Vlog	36	8.6	Pets	12	2.9
Comedy	36	8.6	Traveling	8	1.9
Gamming	32	7.6	Sciences & Skills	4	1.0
Tips & Knowledge	24	5.7	**Total**	**420**	**100**

Latest YouTube Channel Type of Unsubscribing from
The top 5 channel types of unsubscribing from are: Vlog; Entertainment; Gamming; News & Politics; and Tips & Knowledge. And the others are listed in Table 4.

Table 4. Which YouTube channels did you recently unsubscribe from?

Channel Types	times	%	Channel Types	times	%
Vlog	120	28.6	Educations	12	2.9
Entertainment	80	19.0	Movies & Animations	8	1.9
Gamming	64	15.2	Cars & Transportations	8	1.9
News & Politics	60	14.3	Traveling	8	1.9
Tips & Knowledge	20	4.8	Pets	8	1.9
Comedy	16	3.8	Music	4	1.0
Sciences & Skills	12	2.9	**Total**	**420**	**100**

4.3 Unsubscribe Factors

We used factor analysis to obtain the factor load values, which were summarized into eight dimensions, and these eight dimensions are shown in Table 5. The original 32 questions were eliminated due to the low factor load of questions: 2, 13, 17, and 22, and others were organized into eight dimensions: Background and Sense of Belonging; Advertising Interference; Entertainment and Appeal; Videos Quality and professionalism; Repetitive and Boring Content; Obsolescence or Bias; Time Factor, and Commercial Nature. The overall Cronbach's alpha (α) was .923, and the eight dimensions were all above .7. Based on Ong Choon Hee's [18] suggestion, the acceptable range of Cronbach's alpha was .7 or higher. Our study would show the reliability.

Table 5. Cronbach's alpha of eight dimensions in questionnaire

Factors		#	Items	Load	Variance %
Factor 1 (α = .836)	YouTubers' Background & Sense of Belonging (F1)	14	YouTuber's social role change. For example, marriage or childbirth	.749	12.388
		23	No like-minded people in the comments section	.694	
		26	The original recommended friends have unsubscribed	.674	
		32	No longer share situation with outsiders	.574	
		06	The video content no longer has the function of cultivating the body and soul	.553	
		24	Many differences in cultural backgrounds	.545	
Factor 2 (α = .861)	Advertising Interference (F2)	18	Too much advertising leads to dislike	.861	12.288
		20	Advertising time is too long	.855	
		16	Comments from other audiences full of vulgar and indecent content	.648	
		19	Advertising cannot be ignored to cause displeasure	.614	
Factor 3 (α = .774)	Entertainment and appeal (F3)	15	The content of the channel is no longer entertaining	.685	8.776
		11	Other channels could to be more interesting with similar videos	.666	
		09	Scandalous rumors about channel owners are offensive	.579	

(*continued*)

Table 5. (*continued*)

Factors		#	Items	Load	Variance %
		10	The image of the channel owner has changed, causing it to not meet its own expectations	.515	
		08	Channel owners' realistic social performance is no longer attractive	.483	
Factor 4 (α = .728)	Video quality and professionalism (F4)	07	Decline in film quality	.768	8.074
		12	Gradually lacking of professional	.655	
		31	Changes in personal preferences and interests	.606	
		21	Other channels provide more interesting content	.466	
Factor 5 (α = .749)	Repetitive and boring content (F5)	01	Repetitive and boring contents	.749	7.910
		04	Just tried it originally, but sure not interested in now	.665	
		30	Minority channels	.563	
Factor 6 (α = .713)	Obsolescence or bias (F6)	03	Cannot learn new knowledge any more	.752	7.577
		05	Increasingly biased content and position	.660	
Factor 7 (α = .705)	Time factor (F7)	28	No time to stay tune	.779	6.114
		27	Changes in lifestyle or work	.655	

(*continued*)

Table 5. (*continued*)

Factors		#	Items	Load	Variance %
Factor 8 (α = .700)	Commercial nature (F8)	25	Film is full of commercial sales and not interested	.587	6.078
		29	No desire to subscribe, only to be influenced by the recommendations of others	.532	

4.4 Variance and Correlation Analysis

The results of collating the variables with significant variances and correlations are listed as follows (Table 6):

Table 6. Analysis of gender in individual variables

	Male (n = 152)	Female (n = 152)	F-test	p-value
Most Watched Channel Type	u = 5.500	u = 4.850	25.344	.000***
Which YouTube channels did you recently unsubscribe from?	u = 4.631	u = 3.950	31.864	.000***
Entertainment and appeal (F3)	u = 3.173	u = 3.268	11.907	.01**
Commercial nature (F8)	u = 3.013	u = 3.238	10.726	.01**

*p < 0.05, **p < 0.01, ***p < 0.001

A significant difference between gender and channel types and behaviors (F = 25.344; p = .000). There were differences in the channel types and behaviors among respondents of different genders. The most watched channel type among females were entertainment (29.9%), followed by movies and animations (16.4%). The entertainment (21.1%) and gamming (13.2%) for males, as shown in Table 7.

A significant difference between gender and latest channel unsubscribing from (F = 31.864; p = .000). The top two types of latest channel unsubscribing from for females were vlog at 29.9% and entertainment at 19.4%. For males, 26.3% were also vlog and 21.1% were gamming. And the rest are also shown in Table 7.

Table 7. Cross Table between genders and Most watched Channels and Latest channel unsubscribing from

			Vlog	Gamming	Entertainment	Movies & Animations	Tips & Knowledge	Comedy	Music	News & Politics	Sciences & Skills	Educations	Cars & Transportation	Traveling	Pets	Total
Most Watched Channel Type	M	n	16	20	32	12	4	12	16	8	4	8	12	4	4	152
		%	10.5	13.2	21.1	7.9	2.6	7.9	10.5	5.3	2.6	5.3	7.9	2.6	2.6	100.0
	F	n	20	12	80	44	20	24	40	8	0	4	4	4	8	268
		%	7.5	4.5	29.9	16.4	7.5	9.0	14.9	3.0	0.0	1.5	1.5	1.5	3.0	100.0
Which YouTube channels did you recently unsubscribe from?	M	n	40	32	28	0	8	0	0	8	4	12	8	8	4	152
		%	26.3	21.1	18.4	0.0	5.3	0.0	0.0	5.3	2.6	7.9	5.3	5.3	2.6	100.0
	F	n	80	32	52	8	12	16	4	52	8	0	0	0	4	268
		%	29.9	11.9	19.4	3.0	4.5	6.0	1.5	19.4	3.0	0.0	0.0	0.0	1.5	100.0

A significant difference (p = .001) between gender and entertainment and appeal in factor 3 of the unsubscribe factors, with females (u = 3.268) being higher than males (u = 3.173), indicating that females place more importance on this issue. There is also a significant difference between females (u = 3.238) and males (u = 3.013) on the commercial nature factor.

Elder audiences had a positive relation with unsubscribe factor (F = 5.482; p = .000). The four factors that were significantly correlated were YouTubers' Back-ground & Sense of Be-longing (F1); Advertising Interference (F2); Repetitive and boring content (F5), and Obsolescence or bias (F6). Detailed information is shown in Table 8.

Table 8. The correlation analysis between unsubscribe factors and age

	R^2	F	p-value	Beta	Correlation
F(Unsubscribe factors)	0.079	5.482	0.000***	3.080	+
YouTubers' Background & Sense of Belonging (F1)	0.036	16.619	0.000***	2.841	+
Advertising Interference (F2)	0.02	9.600	0.002**	3.042	+
Entertainment and appeal (F3)	0.002	0.929	0.336	3.613	–
Video quality and professionalism (F4)	0.001	0.001	0.973	3.904	–
Repetitive and boring content (F5)	0.017	7.336	0.007**	3.096	+
Obsolescence or bias (F6)	0.025	3.113	0.000**	10.936	+
Time factor (F7)	0.001	3.988	0.079	0.779	–
Commercial nature (F8)	0.009	1.794	0.182	0.092	–

*p < 0.05, **p < 0.01, ***p < 0.001

More monthly incomes had a positive correlation with unsubscribe factor (F = 5.507; p = .000). The five factors that were significantly correlated were YouTubers' background and sense of belonging (F1), advertising interference (F2), entertainment and appeal (F3), repetitive and boring content (F5), and time factor (F7). Detailed information will be shown in Table 9.

Table 9. Regression Correlation Analysis between unsubscribe factors and monthly incomes

	R^2	F	p-value	Beta	Correlation
F (Unsubscribe factors)	0.072	5.074	0.000***	2.764	+
YouTubers' Background & Sense of Belonging (F1)	0.057	25.173	0.000***	3.125	+
Advertising Interference (F2)	0.029	12.449	0.000**	3.712	+
Entertainment and appeal (F3)	0.011	5.617	0.018*	4.145	–
Video quality and professionalism (F4)	0.004	3.167	0.076	4.463	–
Repetitive and boring content (F5)	0.034	15.525	0.000***	3.387	+
Obsolescence or bias (F6)	0.001	.524	0.469	5.105	+
Time factor (F7)	0.015	6.375	0.012*	4.285	–
Commercial nature (F8)	0.017	7.101	0.008**	4.219	–

*p < 0.05, **p < 0.01, ***p < 0.001

5 Conclusions and Discussion

YouTube has become an important part of people's daily lives and has become a huge influence in we-media. Many people watch clips in YouTube every day, and many YouTubers become KOLs in our daily live. This study found that the most watched channel types in Taiwan were: entertainment; music; and movies and animations. While the channels of unsubscribing from in order were: vlogs; entertainment and gamming. In this study, we used factor analysis to classify the factors for unsubscribing into eight major dimensions: YouTubers' background and sense of belonging; advertising interference; entertainment and appeal; video quality and professionalism; repetitive and boring content; Obsolescence or bias; time factor and commercial nature. In the following, we discuss and suggest some important findings that are significantly related to the study.

According to our study, there were differences in channel types and behaviors across gender respondents. Women accounted for 29.9% of the most watched channels in entertainment compared to 21.1% for men, indicating that women have a greater preference for entertainment channels, while men have a greater interest in gamming channels, with a higher percentage of 13.2% compared to 4.5% for women. This shows that there is a significant difference between men and women in terms of channel preferences. This can be used as a reference for YouTubers to promote or give it a try in their channels.

There was also a significant difference between gender and "Which YouTube channels did you recently unsubscribe from?" (F = 31.864; p = .000). It was found that the percentage of vlogs that were unsubscribed was slightly higher among females than males at 29.9% and 26.3% respectively. The results of the factor analysis showed that the change of Youtuber's social role was one of important factors. As for the second place, the type of female unsubscribe from is entertainment, which accounts for 19.4%. From the results of the study, it is inferred that the content of the channel is no longer entertaining or other channels could to be more interesting with similar videos. As for the second type of male unsubscribe, gamming accounted for 21.1%. We found that advertising interference and commercial nature had a certain influence on unsubscribing. It would be a matter of choice for all channel owners. There was also a significant difference (p = .001) between genders and entertainment and appeal (F3), with females (u = 3.268) caring more about this issue than males (u = 3.178). Women (u = 3.238) were also more likely than men (u = 3.013) to have a greater interest in the commercial nature (F8) factor. It was found that there was a positive correlation between age and unsubscribe factor (F = 5.482; p = .000). In other words, the elder audiences have higher demand for every key factor, especially for the four factors including: YouTubers' background & sense of belonging (F1); advertising interference (F2); repetitive and boring content (F5) and obsolescence or bias (F6). Elder audiences have higher expectations of the social role of YouTuber and do not want to be distracted by ads that are not related to the video content. Elder audiences also do not like repetitive and boring contents. They care about learning something new from watching clips as well. More monthly incomes had a positive correlation with unsubscribe factor (F = 5.507; p = .000). The findings revealed that audiences with higher income were similar to those of older in three factors including: YouTubers' background & sense of belonging (F1), advertising interference (F2), and repetitive and boring content (F5), it demanded more from YouTubers. They are also more demanding in terms of entertainment and appeal (F3) and time factor (F7), and are quick to unsubscribe from channels that are no more entertaining in their limited time after their working.

References

1. Cheng, M., Qiu, X.: Research on We-media marketing in Web3.0 environment. Manage. Eng. **29**, 8 (2017). https://doi.org/10.5503/J.ME.2017.29.003
2. Murugesan, S.: Understanding Web 2.0. IT Prof. **9**, 34–41 (2007). https://doi.org/10.1109/MITP.2007.78
3. O'Reilly, T.: What is Web 2.0. O'Reilly Media, Inc., Sebastopol (2009)
4. McAnany, E.G.: Saving the World: A Brief History of Communication for Development and Social Change. University of Illinois Press (2012)
5. Crowley, D., Heyer, P.: Communication in History: Technology, Culture, Society. Routledge, London (2015)
6. Richards, J.: Google: we don't know how to make money from YouTube (2008). https://web.archive.org/web/20080821152232/http://technology.timesonline.co.uk/tol/news/tech_and_web/article4120860.ece
7. McMullan, J.: A new understanding of 'New Media': online platforms as digital mediums. Convergence **26**, 287–301 (2020). https://doi.org/10.1177/1354856517738159
8. Kemp, S.: Digital 2022: Taiwan. https://datareportal.com/reports/digital-2022-taiwan. Accessed 18 Apr 2022
9. Alexa: Top Sites in Taiwan. https://www.alexa.com/topsites/countries/TW. Accessed 18 Apr 2022
10. Heinig, I.: 5 Steps to Create a Multi-Platform Social Media Strategy|The Manifest. https://themanifest.com/digital-marketing/5-steps-create-multi-platform-social-media-strategy. Accessed 2 Feb 2022

11. Brown, N.: Why Multi-Platform Content Is Key for Marketing to Younger Generations. https://www.skyword.com/contentstandard/multi-platform-content-key-marketing-younger-generations/. Accessed 2 Feb 2021
12. PewDiePie: PewDiePie – YouTube. https://www.youtube.com/user/PewDiePie. Accessed 28 Jan 2021
13. PewDiePie: Unboxing 100 MIL YouTube AWARD!! https://www.youtube.com/watch?v=DYlesHOaPkY. Accessed 7 May 2023
14. Sweatt, L.: 6 Honest Reasons You Might Be Losing YouTube Subscribers. https://vidiq.com/blog/post/losing-youtube-subscribers/. Accessed 25 Jan 2021
15. Corrêa, S.C.H., Soares, J.L., Christino, J.M.M., de Sevilha Gosling, M., Gonçalves, C.A.: The influence of YouTubers on followers' use intention. JRIM **14**, 173–194 (2020). https://doi.org/10.1108/JRIM-09-2019-0154
16. Han, B.: How do YouTubers make money a lesson learned from the most subscribed YouTuber channels. IJBIS. **33**, 132 (2020). https://doi.org/10.1504/IJBIS.2020.104807
17. Chang, C.-C.: Demands of Top Fans in Social Media - A Study of Youtube (2020). https://www.airitilibrary.com/Publication/alDetailedMesh?docid=U0002-0107202013210400
18. Hee, O.C.: Validity and reliability of the customer-oriented behaviour scale in the health tourism hospitals in Malaysia. Int. J. Caring Sci. **7**, 771–775 (2014)

Practical Research on AI Visual Focus Analysis in Online Teaching

Ming-Feng Lee[1], Guey-Shya Chen[2], Ming-Zhi Cheng[2(✉)], Hui-Chien Chen[2], and Jian-Zhi Chen[3]

[1] Department of Information Management, National Taichung University of Science and Technology, Taichung, Taiwan
[2] Institute of Educational Information and Statistics, National Taichung University of Education, Taichung, Taiwan
grace@mail.ntcu.edu.tw, cms109101@gm.ntcu.edu.tw
[3] Corporate Synergy Development Center (CSD), Taipei, Taiwan

Abstract. This project aims to study learners' visual focus in online learning and their behavior patterns when engaging with different learning media and cognitive processes. It uses AI models and data analysis methods to evaluate learners' online preferences and information processing.

The project consists of three stages:

(1) Collecting and Integration: Use visual movement devices to collect learners' visual movement data and other data to analyze visitors' visual behavior and identify key eye movement factors.

(2) AI Applications: Use AI data mining technologies to identify the visual focus of the target learners through machine learning and data analysis.

(3) Integrated Analyses of Online Learning: Optimize online learning design by combining visual focus analysis and information collected in visual mode. Collaborate with companies to establish an "adoption model for visual focus analysis of online education," which will be verified by the online teaching platform.

Keywords: Visual focus · Online learning · Artificial intelligence · Data analysis · behavior patterns

1 Introduction

This project aims to achieve two outcomes:

(1) thorough research on the visual focus process of learners in online learning
(2) creation of an "application model of online teaching visual focus analysis" in collaboration with industry, to enhance the online learning experience for learners.
   1. This project analyzes learners' eye movement patterns to categorize their learning situations and develop practical applications. It also aims to improve deep learning and data mining models through a visual focus analysis system. The project builds upon previous media detection applications and expands its model.

J. C. Hung et al. (Eds.): FC 2023, LNEE 1134, pp. 39–44, 2024.
https://doi.org/10.1007/978-981-99-9342-0_4

2. This project aims to study the differences between online and physical learning by conducting eye movement observation research on online learning. The research results will be shared through academic speeches, workshops, and industry-government-academia lectures to promote interdisciplinary concepts and the transformation of research innovations into interactive online education displays and technology. The project will also integrate various resources to work together.
3. Developmental psychology studies show that the human brain analyzes gaze to predict mental behavior and manipulate colors to attract consumers. Studies are conducted using questionnaires, stimuli, and research methods. In addition, psychologists have found that visual behaviors may also show a degree of preference, for example: they will look twice as long at favorite items (Li Suxin and Li Jimian, 2001), or their pupils will dilate significantly when they see favorite items (Janisse & Peavler, 1974).

   This study aims to use a webcam eye tracker and questionnaire survey to monitor real-time cognitive processes and attention, providing a reference for teachers designing teaching materials in hybrid teaching. The study has two purposes: exploring the influence of different design styles of pictures and texts on subjects' reading and exploring the relationship between graphics and cognition using the eye movement data of subjects.

2 Literature Discussion

2.1 A Subsection Sample

Eye Movement Research on Graphic Reading Most studies on graphic reading are based on Mayer's cognitive theory of multimedia learning (CTML) (Mayer, 2005, 2014) and Schnotz's integrated model of graphic comprehension (Schnotz et al., 2014). According to the dual coding theory (Paivio, 1986), the human cognitive system comprises two independent but interrelated subsystems: the language system and the image system. This theory assumes that visual images and Chinese characters require different processing channels. CTML is based on three basic assumptions. First, human information processing involves two independent and symmetrical channels: auditory and image. Second, memory capacity is limited, with individuals able to remember about 7 ± 2 elements in each system, according to Miller's (1956) and Simon's (1974) Memory Span Test. Third, learners actively construct knowledge by selecting, organizing, and integrating information and prior knowledge to create a coherent mental model. Empirical research supports CTML's assumption that combining pictures and text enhances learning compared to text-only presentation. Mayer suggests that images facilitate high-level learning by helping readers form a mental model of concepts. Since the cognitive system's working memory has limited capacity for retention and operation, presenting visual and textual information together allows for processing and encoding of information through the dual channels of spoken language and images. This approach reduces cognitive load and promotes learning (Wang Zining & Jian Yuqin, 2022).

Current literature on eye movement research in graphic reading can be broadly categorized into two directions. The first direction, business psychology, explores whether people follow specific eye movement patterns when viewing advertisements or browsing the Web, and which objects attract their attention. The second direction, cognitive learning, examines the cognitive processes involved in scientific graphic reading, including the reader's reading path and how information is processed. It also investigates how graphics aid in the integration of graphics and text, such as whether pictures enhance understanding of concepts and whether important information marked on pictures aids in comprehension.

2.2 Eye Tracker

Eyes are crucial for communication with the world, and eye tracking technology can measure gaze points (Cornsweet, 1958). Eye tracking research on reading behavior focuses on fixations, saccades, and regressions (Poole & Ball, 2005). The number of gaze times in a specific area indicates its importance, with longer gaze times suggesting more challenging information extraction or more attractive targets. For informational pictures, more gaze times may indicate complexity and multiple processing cycles. Time spent gazing correlates with gaze length during the gaze period (Just & Carpenter, 1976).

Hyona, Lorch, and Kaakinen (2002) found that personal reading strategies, knowledge, and experience influence gaze time and position when viewing pictures or text. Chen et al. (2014) used eye-tracking to investigate the impact of different presentation modes on college students' performance on computerized assessments, and found that rereading time predicted their evaluation performance. Students with longer gaze times and shorter saccade distances performed better on physical concept tasks, indicating that pictures convey physical concepts faster and more effectively than text. Eye-tracking can provide substantial evidence for understanding students' performance.

2.3 Investigating the Reading Process Using Eye-Tracking Technology

Eye-tracking technology is used to track the movement of the eyeballs and link them with stimuli, allowing researchers to determine where individuals place visual attention on the stimulus, how long they sustain their focus, and the sequence of their gaze patterns (Holmqvist et al., 2011). Metrics include fixation, saccade, and gaze. Gaze refers to almost stationary eyes that retrieve information from the stimulus. Position and duration of gaze indicate which information the individual has accessed and how deeply they have processed it. The fixation point duration usually ranges from 100 to 500 ms (Rayner, 1998). Nearly 100 indicators, such as AOI and POI, including observation time, are used for detailed analysis of eyeball trajectories (Chen Xuezhi et al., 2010).

It is suggested that the use of graphic and text materials in learning has a better effect than using only text materials (Zhang & Gu, 2011). Eye-tracking metrics have been used to explore the impact of news and text layout on attention distribution, and changes in the relative positions of pictures and text have been found to affect attention distribution (Tang & Zhuang, 2005). Design principles of teaching materials have been classified based on six eye movement studies related to graphics and text (Mayer, 2010).

Eye-tracking technology has been used to record students' eye movement behavior and provide important indicators such as gaze time, back gaze, and gaze shift, which reflect attention distribution and cognitive processes (Henderson & Hollingworth, 1999; Josephson & Holmes, 2002; Rayner & Pollatsek, 1987; Yang & McConkie, 1999).

3 Research Methods

3.1 Research Participants

This eye movement experiment was conducted during data mining. The research participants were self-selected students, with an effective sample of 44 participants, including 16 females and 28 males. Each subject completed the eye movement experiment and then answered a viewing questionnaire.

3.2 Research Tools

Eye-Tracking Measurement Instrument

This study utilized Real-eye Web eye tracking software, where participants only needed a network camera to collect data. The software performed eyeball correction before data collection. The data collected included hotspots and eye movement tracks, which recorded the time and frequency of participants' gaze and scanning behavior.

3.3 Defining the Area of Interest (AOI) in Images

Research materials were taken from online coffee resources, and subjects were placed in a cafe environment through text descriptions. Different environmental elements and coffee cup colors were used to assess how they affect mood and coffee perception. Figures 1 and 2 presented two different atmospheres, cool and warm, and identified areas of interest including coffee, coffee pots, tablets, books, and coffee cups of different sizes, blue cups, water cups, and desserts (Figs. 3 and 4).

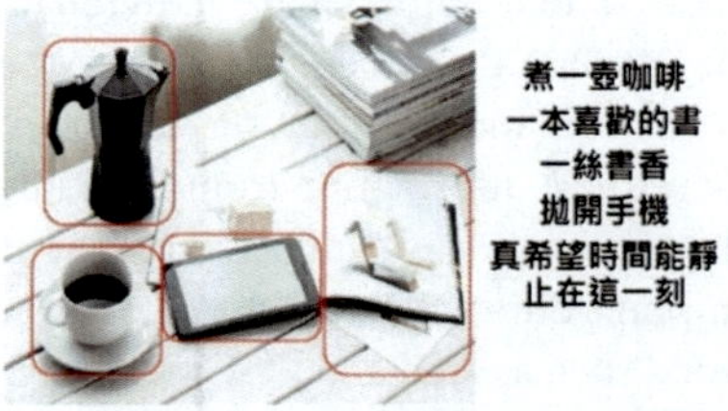

Fig. 1. Display of coffee book and tablet in black and white color style.

Fig. 2. Display coffee, water glasses and bread in a wood tone color scheme.

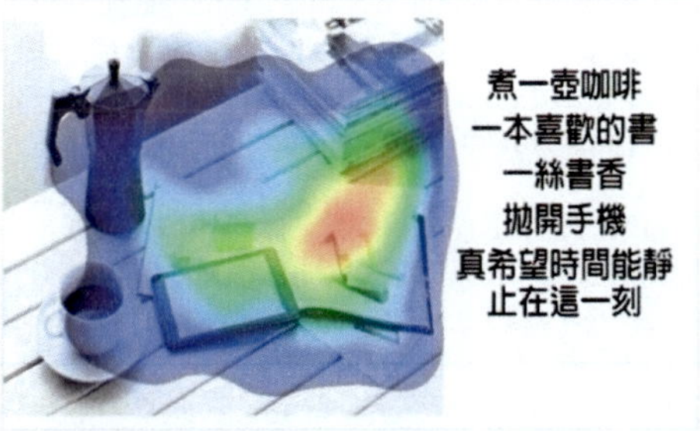

Fig. 3. The heat map of eye movement analysis in Fig. 1.

Fig. 4. The heat map of eye movement analysis in Fig. 2.

4 Discussion and Suggestion

This study uses eye-tracking to study how graphic designs affect cognition, emotion, and perception. Colors have strong emotional impacts, such as red for excitement, orange for stimulation, and blue for comfort and safety (Ballast, 2002; Wexner, 1982). Studies show that people of all ages, races, and education levels prefer culturally associated colors (Adams & Osgood, 1973; Eysenck, 1941).

The study found that subjects spend less time looking at cool-colored designs, feeling calm and relaxed, and perceiving coffee as bitter. On the other hand, warm-colored designs lead to a happy and relaxed emotional state and a perception of coffee as sweet. The hot zones in the coffee cups were also more appealing to the participants. These findings can be used as a reference for designing teaching materials with consideration of color configuration to enhance learners' interest.

References

Jian, Y.C., Wu, C.-J.: The effect of arrows in an illustration when reading scientific text: evidence from eye movements and reading tests. Chin. J. Psychol. **54**, 385–402 (2012). https://doi.org/10.6129/cjp.2012.5403.07

Adams, F.M., Osgood, C.E.: A cross-cultural study of the affective meanings of color. J. Cross Cult. Psychol. **4**(2), 135–156 (1973)

Ballast, D.: Interior Design Reference Manual. Professional Publication Inc., Belmont (2002)

Cohen, J., Cohen, P., West, S.G., Aiken, L.S.: Applied Multiple Regression/Correlation Analysis for the Behavioral Sciences, 3rd edn. Erlbaum, Mahwah (2003)

Eysenck, H.J.: A critical and experimental study of colour preferences. Am. J. Psychol. 385–394 (1941)

Guo, D., Zhang, S., Wright, K.L., McTigue, E.M.: Do you get the picture? A meta-analysis of the effect of graphics on reading comprehension. AERA Open **6**(1) (2020). Advance Online Publication. https://doi.org/10.1177/2332858420901696

Hegarty, M., Canham, M.S., Fabrikant, S.I.: Thinking about the weather: how salience and knowledge affect performance in a graphic inference task. J. Exp. Psychol. Learn. Mem. Cogn. **36**(1), 37–53 (2010). https://doi.org/10.1037/a0017683

Janisse, M.P., Peavler, W.S.: Pupillary research today: emotion in the eye. Psychol. Today **7**, 60–63 (1974)

Lai, M.-L., Tsai, M.-J., Yang, F.-Y., Hsu, C.-Y., Liu, T.-C., Lee, S.W.-Y., et al.: A review of using eye-tracking technology in exploring learning from 2000 to 2012. Educ. Res. Rev. **10**, 90–115 (2013). https://doi.org/10.1016/j.edurev.2013.10.001

McTigue, E.M.: Does multimedia learning theory extend to middle-school students? Contemp. Educ. Psychol. **34**, 143–153 (2009). https://doi.org/10.1016/j.cedpsych.2008.12.003

Mayer, R.E.: Cognitive theory of multimedia learning. In: Mayer, R.E. (ed.) The Cambridge Handbook of Multimedia Learning, pp. 31–48. Cambridge University Press (2005)

Mayer, R.E.: The Cambridge Handbook of Multimedia Learning, 2nd edn. Cambridge University Press (2014)

Paivio, A.: Mental Representations: A Dual-coding Approach. Oxford University Press (1986)

Segers, E., Verhoeven, L., Hulstijn-Hendrikse, N.: Cognitive processes in children's multimedia text learning. Appl. Cogn. Psychol. **22**, 3375–3387 (2008). https://doi.org/10.1002/acp.1413

Schnotz, W.: Integrated model of text and picture comprehension. In: Mayer, R.E. (ed.) The Cambridge Handbook of Multimedia Learning, 2nd edn., pp. 72–103. Cambridge University Press (2014)

Wexner, L.B.: The degree to which colors (hues) are associated with mood-tones. J. Appl. Psychol. **38**(6), 432 (1982)

Design of a Fair Distributed Computing Platform Based on Distributed Ledger Technology and Performance Measurements

Bo-Yan Liao and Jia-Wei Chang(✉)

Department of Computer Science and Information Engineering, National Taichung University of Science and Technology, Taichung, Taiwan
z775357@gmail.com

Abstract. We propose a fair distributed computing platform based on Distributed Ledger Technology (DLT) and performance measurements. The platform integrates DLT and federated learning, enabling users to train machine learning models on their local devices without compromising their privacy by sharing their data with a central server. Instead, only the trained model weights are uploaded to a central server for aggregation. To address privacy concerns associated with federated learning, we integrate various privacy-preserving methods, such as differential privacy, model pruning, and homomorphic encryption, into the platform framework. These techniques help protect user privacy while improving model accuracy. To address the non-IID data problem in federated learning, we use performance measurements to balance the training workload among users, and blacklist malicious users while incentivizing participation. DLT ensures the security and integrity of the platform by validating and recording all data transactions on the ledger. Overall, the proposed platform has the potential to revolutionize machine learning model training by making it more efficient, secure, fair, and transparent.

Keywords: Federated learning · InterPlanetary File System (IPFS) · Blockchain

1 Introduction

As the popularity of smart connected devices such as smartphones, smart homes, and wearables continues to grow, people's lives are increasingly dependent on these intelligent devices. However, the data generated by these devices is mostly stored in various devices, forming multiple Isolated Data Island that cannot be effectively utilized. Moreover, since data involves personal privacy, users do not want to transmit their data to a central server for training, thereby exposing personal privacy information. To solve these problems, the Google AI team proposed the federated learning framework in 2016, which can effectively use distributed data for model training while protecting personal privacy. In federated learning, users can train models on their local devices without transmitting sensitive data directly to the central server. Through federated learning, not only can the problem of Isolated Data Island be solved, but also user privacy can be protected, which has become an attractive research topic for enterprises and researchers in various fields.

J. C. Hung et al. (Eds.): FC 2023, LNEE 1134, pp. 45–50, 2024.
https://doi.org/10.1007/978-981-99-9342-0_5

Despite significant progress in addressing privacy and data ownership issues faced by centralized machine learning, the application of federated learning still faces many challenges. One of the key challenges is how to effectively prevent free-riders or malicious users, and improve the performance and accuracy of federated learning under Non-IID data, by optimizing the order of aggregation weights and using clustering methods. Additionally, as deep learning continues to expand into various fields, the value of model weights cannot be ignored. This paper further explores the fair trade of model weights by establishing a public trading platform based on blockchain and InterPlanetary File System (IPFS) technology. This platform enables each user to safely share their trained model weights with other researchers, while also receiving corresponding virtual currency rewards. Through this approach, not only can the efficiency of model training be effectively improved, but users can also better control their data and model weights, thus protecting personal privacy. Furthermore, the fair trading platform can promote collaboration between different fields to jointly solve various real-world problems.

2 Related Work

2.1 Performance Measurements for AI Applications

In the frameworks of Federated Learning and Swarm Learning, a large number of users participate in the collaboration of the model freely. Participants locally process their own privacy data using technologies such as homomorphic encryption and differential privacy before iterating the model. After completion, the model is uploaded to the server for aggregation, and the participants receive rewards [1]. In this mechanism, participants may modify their local data to obtain more rewards, so that the trained weights can better generalize the model, while attackers may use fake data to maliciously attack the model. Therefore, a fair valuation method is needed to evaluate the quality of the weights provided by the participants. In the figure below, different evaluation methods are used for various fields and tasks in machine learning. Therefore, We combines tools such as classification report, F1 score, and valuation to give a rating of the contribution value to the participants, as shown in Table 1.

2.2 Public and Incorruptible Transaction Platform Using Distributed Ledger Technology (DLT)

To verify the information of participants and record the complete information of model updates for achieving full fairness, the simplest way is to use DLT-related technologies, like Blockchain has the characteristics of immutability and irreversibility, so it can be used to verify whether the relevant information of participants is correct and record the entire process on the chain.

Blockchain

Blockchain is a concept proposed by Satoshi Nakamoto in the Bitcoin white paper in 2008. It heavily utilizes cryptography and consensus mechanisms. Blockchain is composed of blocks linked together, and each block contains an encrypted hash of the previous block, transaction records, and a timestamp. The encrypted hash is calculated

Table 1 .

Domain	Task	Commonly used evaluation metrics
Computer Vision	Object Detection Models	AP, mAP
	Multi-Object Tracking Models	MOTA, MOTP, MT, ML, IDs, FM, IDF1
	Image Data Augmentation	UIQM, PCQI
Natural Language Processing	Machine Translation Models	GLUE, ROUGE, METEOR, CIDEr
	Document Summary Evaluation	METEOR, GLUE, Edmundson, ROUGE
	Code Generation Transformers	BigCloneBench, Defect Detection
Audio	Music source separation	SI-SNRi, SDRi, SDR
	Speaker Diarization	Accuracy, F1 score
	Speech Recognition	SER, S.Corr, WER/CER

by a designated party through a consensus mechanism, which requires a certain amount of resources. Popular consensus algorithms include Proof-of-Work (PoW) and Proof-of-Stake (PoS). This architecture is tamper-resistant. If a transaction record is modified, the content of subsequent blocks should also change accordingly. Therefore, this technology is widely used in various fields to protect data from tampering.

Distributed Ledger
A distributed ledger is a technology that records all transaction contents on the chain through blockchain technology and can verify the authenticity of transactions through miners. Because blockchain has the characteristics of fairness and immutability, it is very suitable for recording transactions.

Consensus Algorithm
In order to verify whether a transaction has been tampered with, miners verify each block on the chain, and the transaction is considered complete and recorded on the blockchain only when most miners reach consensus. Common consensus algorithms include Proof of Work (PoW), Proof of Stake (PoS), Tangle, etc.

Smart Contract
The concept of smart contracts was first proposed by Nick Szabo in 1994, but it wasn't until 2015 that they began to be widely used on the Ethereum platform. Smart contracts are programs that are stored on the blockchain and cannot be tampered with. When certain conditions are met, the program will automatically execute, just like a contract being enforced in a court of law. Smart contracts are very fair and free from tampering concerns, and their execution can be enforced.

(1) IPFS

IPFS is a peer-to-peer (P2P) decentralized file system that can split a file into several file fragments and store them on various nodes. The file fragments are scattered across multiple nodes, and the integrity of the file is ensured by hash values. The location of each node and its hash value are recorded in a Distributed Hash Table (DHT). Since the status of each node cannot be verified during use, each file is typically backed up on multiple nodes to prevent single-point-of-failure issues.

(2) Decentralized identity (Identity verification)

Decentralized identity means that the identity is not verified through a third party. On the blockchain, you can only prove that you are you through a public key, which is usually not linked to the real world, and cannot prove the authenticity of the user. In recent years, with the trend of blockchain, many decentralized applications (DAPPs) have emerged, and many public and private chains have been developed. However, there is no intersection between these chains, so when users use DAPPs, they have to register a new identity on the corresponding chain. In recent years, some banks have gradually introduced FIDO (Fast IDentity Online), a standard released by the FIDO Alliance, which combines public key and biometric technologies to establish identity verification mechanisms.

3 Method

We found that the order of aggregation is critical for the accuracy of the global weights in Non-IID data. Therefore, inspired by clustering methods, a new aggregation method is proposed. In this method, the similarity between users is first calculated, and appropriate clustering and aggregation orders are assigned based on the model weights trained for each user. To implement this aggregation method, we conducted ablation experiments to investigate whether various clustering methods applied to the aggregation of federated learning can increase the accuracy of the globally aggregated model under Non-IID data.

Density-based clustering methods do not require the number of clusters to be specified in advance. We adopted the DBSCAN algorithm to classify each data point as either noise or a member of a cluster. This algorithm accurately classifies clusters of various shapes.

Inspired by the research of Miao, Yinbin, et al. [2], We further compares the weights uploaded by users with the weights calculated by the global model using cosine similarity, and aggregates similar weights from bottom to top. At the same time, the aggregation method is designed to minimize the standard deviation of the similarity between gradients during aggregation and maximize the number of gradient weight models in aggregation. Finally, the accuracy of the weights for the task is evaluated to determine whether to include these weights in the next aggregation. Through multiple bottom-to-top aggregations, weights that are closer in distance will be aggregated, gradually forming larger clusters of positive weights, while malicious weights will be blocked and discarded.

A record should be kept for each upload of weights and aggregation, and the history can be traced at any time. With blockchain technology, each upload of weights or aggregation of the global model by a user will be recorded on the blockchain. The information that should be recorded on the blockchain, as shown in Fig. 1, includes the training mode, time, uploading user, ID of the previous global model, the user who trained the model or which weights were used for aggregation, time spent, and the location where the weights are finally stored in IPFS. With this information, we can always find out on the blockchain which people have operated on the model. If any malicious attacks are detected, the attackers can be blocked or added to a blacklist.

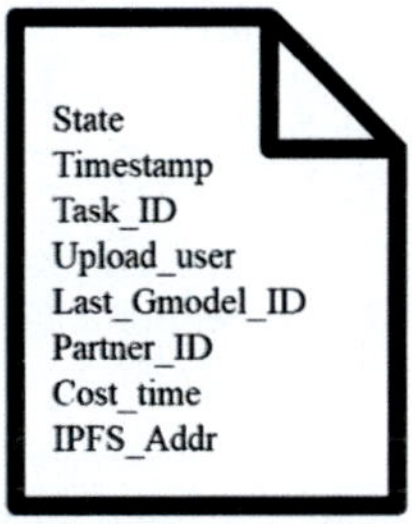

Fig. 1. Transaction Information Structure Recorded on Blockchain

In order to protect the value of the entire global model, We first adds a watermark to the local model weights on individuals' devices, which records the private key of each contributor. When there are doubts about users' usage rights, verification can be performed through the extraction of weights or backdoor attacks. Under the FedIPR framework proposed by Li et al. [3], unless the model is destroyed, the results of the watermark verification cannot be altered, thereby effectively preventing unauthorized commercialization of the trained model by others.

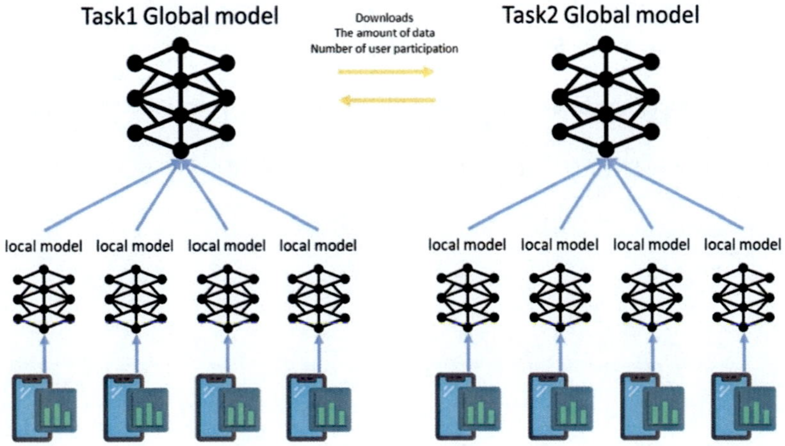

Fig. 2. Fair transaction design for decentralized deep learning task trading platform.

It is difficult for small companies or individuals to have a commercially viable model as it requires high costs for storing data and training the model, which could amount to millions of dollars per year. Additionally, collecting user data for training purposes also incurs significant costs and time. Therefore, we aims to establish a fair trading platform as shown in Fig. 2 where users can freely join to trade or train model weights. Evaluating the value of a model is an important issue. The primary costs of training a model are server costs, the value of user data, and the cost of training on user devices. Therefore, the equivalent relationship between models can be calculated through the download volume of each model, and the real value that a user possesses can be calculated through the value of data that they provide.

4 Results

We discuss the challenges faced by the federated learning framework and propose a fair-trading platform based on blockchain and InterPlanetary File System (IPFS) technology. Our platform allows users to safely share their trained model weights with other researchers and receive virtual currency rewards. We also discuss the need for a fair valuation method to evaluate the quality of the weights provided by participants and present various evaluation metrics for different fields and tasks in machine learning. Lastly, we explore the use of blockchain technology to verify the information of participants and record the complete information of model updates to achieve full fairness.

Acknowledgement. This work was partially supported by the National Science and Technology Council, Taiwan, R.O.C. [grand number NSTC 111-2221-E-025-008].

References

1. Warnat-Herresthal, S., et al.: Swarm learning for decentralized and confidential clinical machine learning. Nature **594**(7862), 265–270 (2021)
2. Miao, Y., Liu, Z., Li, H., Choo, K.K.R., Deng, R.H.: Privacy-preserving byzantine-robust federated learning via blockchain systems. IEEE Trans. Inf. Forens. Secur. **17**, 2848–2861 (2022)
3. Li, B., et al.: Fedipr: ownership verification for federated deep neural network models. IEEE Trans. Pattern Anal. Mach. Intell. (2022)

A Study on the Design of Eye and Eyeball Method Based on MTCNN

Cheng-Yu Hsueh[1], Jason C. Hung[1], Jian-Wei Tzeng[2], Hui-Chun Huang[3], and Chun-Hong Huang[4](✉)

[1] Department of Computer Science and Information Engineering, National Taichung University of Science and Technology, Taichung, Taiwan 404348
{s1811132007,jhung}@nutc.edu.tw
[2] Department of Information Management, National Taichung University of Science and Technology, Taichung, Taiwan 404348
tjw@nutc.edu.tw
[3] Department of Innovation Application and Management, Chang Jung Christian University, Tainan, Taiwan 71101
[4] Department of Computer Information and Network Engineering, Lunghwa University of Science and Technology, Taoyuan, Taiwan 33306
ch.huang@mail.lhu.edu.tw

Abstract. Studies on eye tracking have relied on wearable eye trackers and chin-resting eye trackers, but the high cost of equipment and the need to wear devices during experiments can lead to less natural facial movement. This study collected eye-tracking data by using a raw-video camera without adjusting any parameters. Python was used as the primary programming language. Eye tracking was adjusted through calculations of facial distance, and multitask cascaded convolutional networks were used to collect eye-tracking data. The corrected results were visualized, and linear regression was used to determine correction error. The root mean square error was 221.66, and the mean squared error was 260.48.

Keywords: eye tracker · webcam · multitask cascaded convolutional networks · eye data analysis · eye tracker correction

1 Introduction

Eye-tracking technology has been widely applied in fields such as sports, marketing, and behavior detection. The technology captures eye movements, attention time, and areas of focus for experiments and analysis. Various stimuli are used as experimental variables. Because of the pandemic, students have been unable to take examinations on campus. Eye-tracking technology can be used to detect student behavior and prevent cheating. However, eye tracking requires equipment worn on the face, which may reduce students' willingness to use it. It may also cause cross-infection because of viruses on the equipment. This study designed a contactless eye-tracking method to address this problem. The method is safer and more convenient than chin-rest eye-tracking devices. This study used webcam eye-tracking and increased its accuracy experimentally.

J. C. Hung et al. (Eds.): FC 2023, LNEE 1134, pp. 51–58, 2024.
https://doi.org/10.1007/978-981-99-9342-0_6

1.1 Webcam Eye Tracking

Eye trackers can be chin resting, wearable, and webcam based. The chin-resting and wearable versions cover a large area of the face, limiting the device's ability to analyze eye movement. Neural models cannot be used to analyze facial features and other features, thereby reducing the effectiveness of data augmentation. During experiments, these eye trackers also restrict the movement of the head. Webcam eye trackers enable the collection of motion data through facial recognition and eye detection without any contact between the device and body. This method is both safe and easy to implement and has been increasingly adopted in eye-tracking research (Table 1).

Table 1. Eye tracking device

Eye Tracking Devices	Chin-Resting Eye Tracker	Wearable Eye Tracker	Webcam Eye Tracking
Pictures	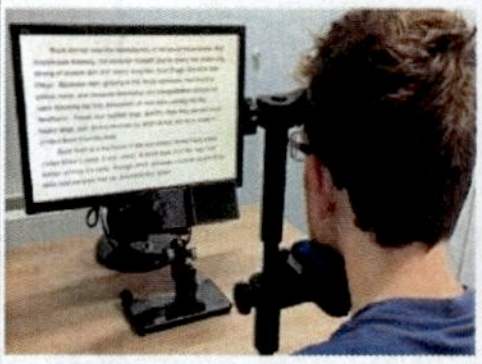		

1.2 Limitations of Eye Trackers in Experiments

Eye tracking must account for facial movement, which can cause calibration error, reduced accuracy, and experimental bias. Chin-resting eye trackers have been widely used because having the face placed on a frame and only focusing on eye tracking prevent errors. However, wearable trackers, which have cameras near the eye, are more widely used and ensure high accuracy by focusing an infrared light on the eye. The largest drawback of wearable devices is that the head cannot move naturally while the device is worn. Webcams have been increasingly used for eye tracking because of their low cost and safety, and they allow for facial movement during experiments (Table 2).

1.3 The Universality of Eye Trackers

Webcam eye tracking is a form of web-based eye tracking. The tracking device can be calibrated by combining mouse and eye movements and then using computer calculations to adjust the position of the eyes. The 9-point calibration method is commonly used in experiments, but the accuracy around the edges of the screen is low (WebGazer) [1]. PyGaze is an eye-tracking module developed using the universal language Python, but it has not been updated since 2018 and is only compatible with Python2, making it prone to error. This study developed Python3.0, a universal version, to increase the accuracy of eye tracking.

Table 2. Advantages and Disadvantages of Eye Tracking Devices

Eye-tracking Device	Chin-resting eye tracker	Wearable eye tracker	Webcam eye tracker
Advantages	High accuracy facilitates analysis of eye movement	Positioned closer to the eye, allowing for accurate collection of data by using infrared technology	Eye tracking with a camera or webcam, enabling large-scale experiments
Disadvantages	The head is fixed on the chin rest and cannot move freely	High cost, making large-scale experiments difficult; prone to masking issues, making analysis challenging	Accuracy may not be as high as that of wearable devices; may require adjustment for high accuracy

2 Literature

2.1 Face Detection

Facial recognition typically involves extracting information regarding facial features such as the eyes, nose, and mouth to identify human faces. Large quantities of data are collected and labelled to increase accuracy. Integral image and feature detection combined with an AdaBoost classifier can be used to quickly identify facial features from a large data set and to remove the background [2]. Dlib facial recognition can also be used to identify faces with obstructed regions, and by extracting feature points and vectors, the accuracy and sensitivity of dynamic image face recognition can be increased [3]. Multitask cascaded convolutional networks (MTCNNs) use a neural network and the P-Net, O-Net, and R-Net methods to segment facial regions into eyes, nose, and mouth corners for facial recognition [4].

2.2 Eye Tracker

When light enters the eye through the pupil, the cornea and lens focus the light onto the retina, and the diameter of the pupil controls the amount of light entering the eye and the resulting image intensity, ensuring clear images [5]. When engaged in cognitive processing, the fovea (the central indentation in the eye) points to the stimulus being processed and denotes the behavior of fixation, but the eye does not continuously fixate on the same stimulus, and saccades occur. Eye tracking can be based on point-of-gaze estimation, three-dimensional (3D) gaze estimation, and pupil center detection. The algorithm based on point-of-gaze estimation [7] uses the vector between the center of the iris and the eye corner as a feature to estimate the gaze point and analyze the positional relationship of eye features in images during head movement. The data collection for the eye tracking method is based on the model approach to the appearance of eye images [8].

Eye tracking involves the use of various mapping functions to calculate gaze fixation. Deep learning techniques yield more accurate results for eye tracking when camera-captured images are used. 3D gaze estimation is based on a novel deep neural network architecture designed specifically for monocular gaze estimation, which simplifies the direct regression of the eye in 3D by regressing two angles for the pitch and yaw of the eyeball [9]. An unsupervised learning-based method is used to estimate the eye gaze in 3D space; geometric spectral photometric consistency constraints and spatial consistency constraints are applied to multiple views in video sequences to refine the initial depth values on the detected iris landmark [10]. Pupil centers are detected by combining the You Only Look Once model with a convolutional neural network (CNN), resulting in an eye-tracking method based on deep learning with a detection accuracy of up to 80% and a recall rate approaching 83% in experimental designs [11]. Deep learning involves training a CNN with eye images as the input, segmenting the pupil area in IR images by using the UNet model, finding the pupil center, and using the pupil center as the regression result [12].

3 Experimental

3.1 Experimental Framework

To address generalizability issues in eye tracking, facial information and distance must be determined. The experimental design is shown as Fig. 1. First, each frame of the video feed is captured, and eye regions are generated through the MTCNN architecture, which captures the position of the eyeballs and then processes the image data. To address facial anomalies in the video in real time, facial distance calculations exclude abnormal values, thereby increasing data accuracy. The collected eye data are compared with the facial distance data to determine eye size and maximum eye movement distance and to ultimately project them onto a screen to visualize the coordinates of the eye's range of sight.

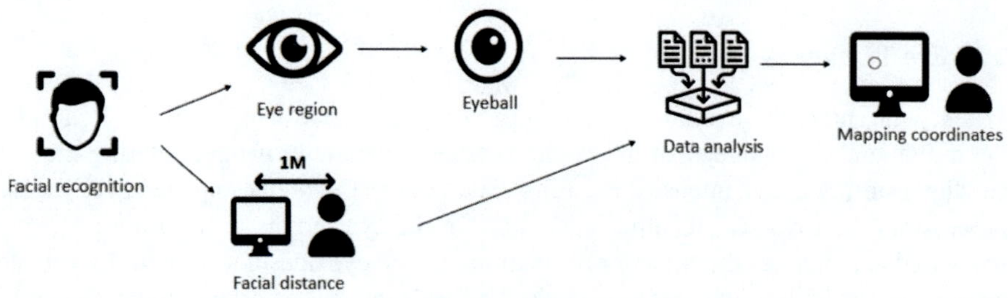

Fig. 1. Experimental Design

3.2 Face Detection

Face detection is a mature technology that enables the extraction of facial information through the collection of facial features. First, an MTCNN model is used to extract facial features, including the eyes, nose, and mouth corners. After the eye area feature is extracted, eye area feature segmentation is combined with HoughCircles to locate the circular shape of the eyeball. The eyeball can be monitored by collecting eye position information, including information regarding both the eye area and the eyeball coordinates (Table 3).

Table 3. Eye Area and Eyeball

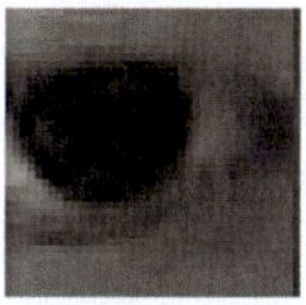 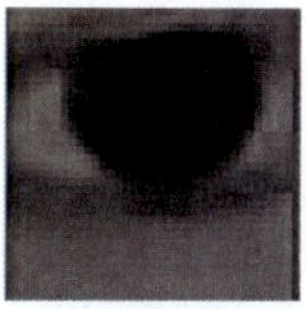	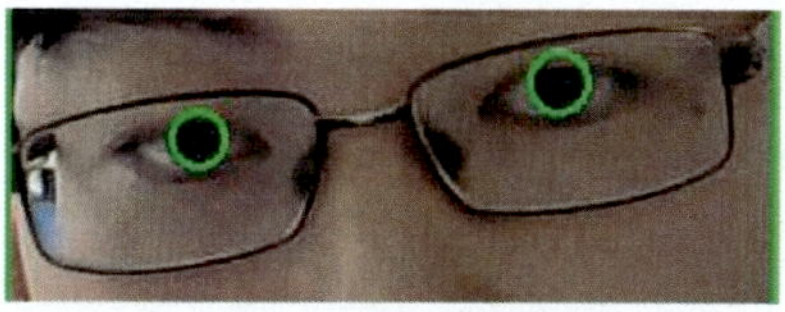
Eye Area	Eye Ball Detection

3.3 Face Distance

$$D(fs, f, as) = (as * f)/f \tag{1}$$

Function $D(fs, f, as)$ represents the distance between the face and the camera, where fs is the size of the face, f is the focal length of the camera, and as is the actual size of the face. When these parameters are input, the distance between the object (with a face as an example) and the camera can be calculated. The principles of geometry and optics are used for this calculation, with the face distance being calculated to identify significant deviations in the face. Large deviations can lead to a decrease in the accuracy of correction, making the collection of facial distance information particularly important.

3.4 Eye and Eyeball

The eye area is usually extracted using face detection, but this method does not perfectly extract eye position. Table 4 indicates the average error of the eye area and eyeball position during calibration. The center of the eye area may not be the true center of the eyeball. When the eyeball and eye area are observed with 13 calibration points, the average error on the x-axis is 2.6 to 10.5, and the average error on the y-axis is 2.7 to 8.5.

Table 4. Average Error Between Eye Area and eyeball

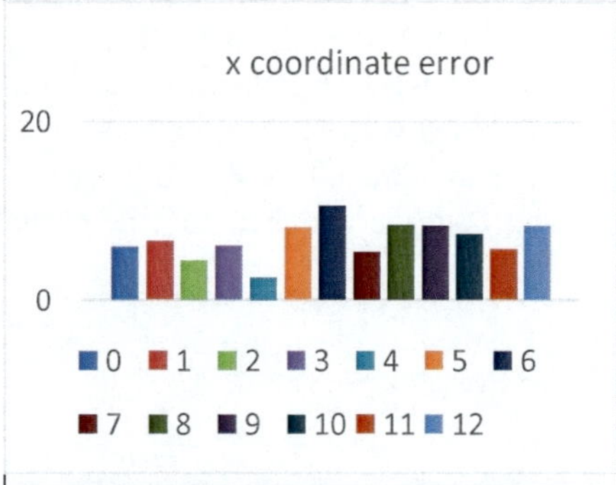	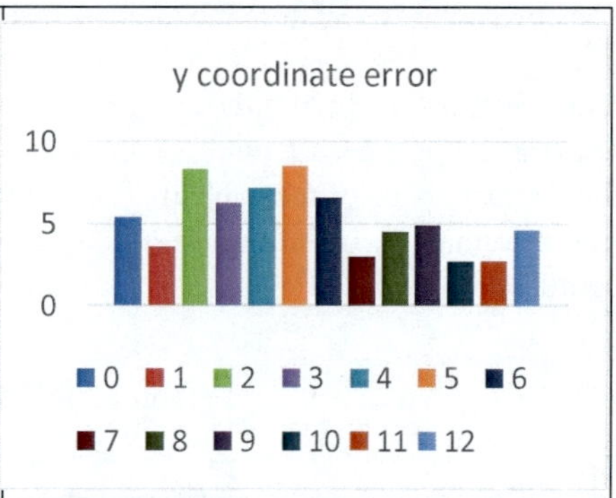
Average Error Between Eyeball and Eye Area's x-Coordinate	Average Error Between Eyeball and Eye Area's y-Coordinate

3.5 Eye Tracker Correction

Eye tracking calibration methods usually rely on 5 points. As shown as Fig. 2. Webgazer's design improves accuracy by using 9 points; the circle turns red after being clicked five times. In this experiment, to further improve accuracy, a 13-point design was used, and calibration and eye-tracking information collection were performed by clicking 15 points in a specified order.

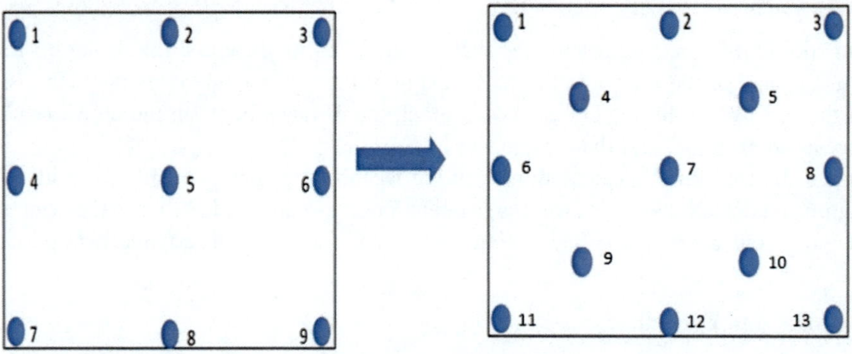

Fig. 2. Calibration point

3.6 Eye Mapping

The mapping method is based on using the maximum number of coordinates to which that the eyeball can move. The maximum distance from the eyeball to the eye corner is projected onto a 1920 × 1080 panel, and the eyeball coordinates (blue dot) of the clicked coordinates are displayed. This ensures that the coordinates correspond to the x–y coordinates of the eyeball when they are viewed on the screen. The average coordinates are marked using a green "x".

4 Conclusions and Directions for Future Research

With the third calibration point in Fig. 3 as an example, most of the values are close to the error range of the linear regression, except for one point in the bottom-left corner that has a larger error. This may have been due to saccades, which caused the point in the bottom-left corner to fall outside the error range of the linear regression. The root mean square error and mean squared error for this point are 221.66 and 260.48, respectively.

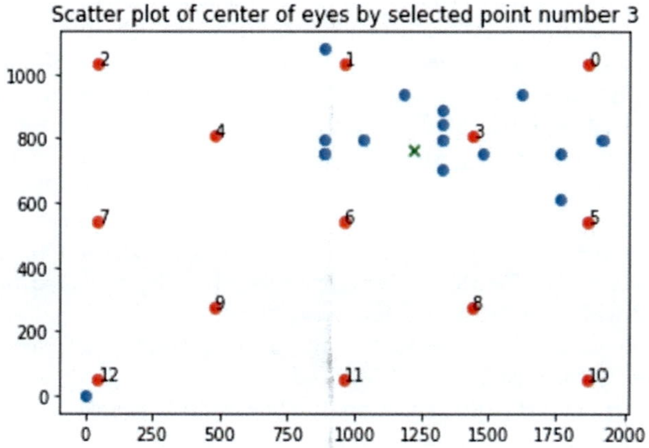

Fig. 3. Accuracy of Left Eye Calibration

The experiment involved capturing facial images using a camera and accurately assessing the movement of the eye-ball from the eye. This process also determined the distance of the face and addressed any abnormalities or abnormal values during calibration. However, real-time video can be affected by environmental factors, leading to issues in face recognition. To overcome this problem and improve accuracy, a solution is proposed. It involves increasing the number of calibration points, carefully evaluating their results, and continuously collecting the eye coordinates of the selected points in real-time. The resulting mapping is then compared to the screen to determine its accuracy. This process aims to establish a straightforward and real-time method for collecting eye data within a Python-based system. By incorporating more calibration points and monitoring the eye coordinates, it seeks to enhance accuracy and address abnormalities during face recognition.

The accuracy is acceptable, but the deviation is uncontrollable when an insufficient number of eye positioning points are collected. The design of calibration points will be based on seconds to ensure more precise calibration points are collected for experimental testing and design and to determine whether the deviation during calibration decreases. In addition to improving accuracy, we will consider facial movement deviation. This study only explored facial deviation and excluded abnormal values during facial recognition. We will consider extending our research to the rotation of the face and parameter design to address the problems in recalibration when the head area moves excessively.

References

1. Papoutsaki, A., Sangkloy, P., Laskey, J., Daskalova, N., Huang, J., Hays, J.: Webgazer: scalable webcam eye tracking using user interactions. In: Proceedings of the Twenty-Fifth International Joint Conference on Artificial Intelligence (IJCAI 2016), pp. 3839–3845. AAAI Press (2016)
2. Viola, P., Jones, M.: Rapid object detection using a boosted cascade of simple features. In: Proceedings of the 2001 IEEE Computer Society Conference on Computer Vision and Pattern Recognition (CVPR 2001), vol. 1, p. I (2001
3. Zhang, D., Li, J., Shan, Z.: Implementation of Dlib deep learning face recognition technology. In: 2020 International Conference on Robots and Intelligent System (ICRIS), pp. 88–91 (2020)
4. Zhang, K., Zhang, Z., Li, Z., Qiao, Y.: Joint face detection and alignment using multitask cascaded convolutional networks. IEEE Signal Process. Lett. **23**(10), 1499–1503 (2016)
5. Wade, N., Tatler, B.W.: The Moving Tablet of the Eye: The Origins of Modern Eye Movement Research. Oxford University Press, New York (2005)
6. Rayner, K., Reingold, E.M.: Evidence for direct cognitive control of fixation durations during reading. Curr. Opin. Behav. Sci. **1**, 107–112 (2015)
7. Hu, D., Qin, H., Liu, H., Zhang, S.: Gaze tracking algorithm based on projective mapping correction and gaze point compensation in natural light. In: 2019 IEEE 15th International Conference on Control and Automation (ICCA), pp. 1150–1155 (2019)
8. Modi, N., Singh, J.: A comparative analysis of deep learning algorithms in eye gaze estimation. In: 2022 International Conference on Data Analytics for Business and Industry (ICDABI), pp. 444–447 (2022)
9. Park, S., Spurr, A., Hilliges, O.: Deep pictorial gaze estimation. In: Proceedings of the European Conference on Computer Vision (ECCV), pp. 721–738 (2018)
10. Lu, Y., Wang, Y., Xin, Y., Wu, D., Lu, G.: Unsupervised gaze: exploration of geometric constraints for 3D gaze estimation. In: Lokoč, J., et al. (eds.) MMM 2021. LNCS, vol. 12573, pp. 121–133. Springer, Cham (2021). https://doi.org/10.1007/978-3-030-67835-7_11
11. Ou, W.L., Kuo, T.L., Chang, C.C., Fan, C.P.: Deep-learning-based pupil center detection and tracking technology for visible-light wearable gaze tracking devices. Appl. Sci. **11**(2), 851 (2021)
12. Han, S.Y., Kwon, H.J., Kim, Y., Cho, N.I.: Noise-robust pupil center detection through CNN-based segmentation with shape-prior loss. IEEE Access **8**, 64739–64749 (2020)

A Comparative Study of GPT-2 and GPT-2 Based On Enhanced Self-attention Mechanism

Wei-Hung Tu[1], Neil Yen[2(✉)], and Yan Pei[2]

[1] Graduate School of Computer Science and Engineering, University of Aizu, Aizuwakamatsu, Fukushima 965-8580, Japan
m5252117@u-aizu.ac.jp

[2] Computer Science Division, University of Aizu, Aizuwakamatsu, Fukushima 965-8580, Japan
{neilyyen,peiyan}@u-aizu.ac.jp

Abstract. In natural language processing, the quality of language models impacts applications such as machine translation and speech recognition. GPT-2, a powerful auto-regressive model with 150 million parameters, performs exceptionally well in various tasks but struggles with computational efficiency for long sequences. We have developed an optimization strategy to mitigate this issue by randomly shortening the auto-regressive length during generation. Our strategy was tested on the GPT-2 medium model using BLEU as the evaluation metric. The results revealed significant improvements in the BLEU scores, with the optimized model outperforming the original. Furthermore, the optimization also improved scores in both the top and bottom 10% of the data. Despite the promising results, there is still room for further exploration and improvement. We are currently investigating adaptive adjustments to the auto-regressive length and applying this strategy to other models, such as GPT-3. In summary, our research proposes a new strategy that enhances GPT-2's efficiency and boosts its performance, as evidenced by the improved BLEU scores. This strategy provides valuable insights for future language model optimization, holding the potential to advance the field of NLP.

Keywords: GPT-2 · Auto regressive · Generative model · Artificial intelligence · Comparative study

1 Introduction

1.1 Introduction to GPT-2

GPT-2, developed by OpenAI in 2019, has made a significant impact in the world of language models. It focuses on generating text that sounds natural by predicting the next word in a sequence. GPT-2 is an upgraded version of GPT, featuring more parameters, a larger model size, and better performance. The primary foundation of GPT-2 is the transformer architecture, which is a successful deep-learning model in natural language processing. One of its notable

J. C. Hung et al. (Eds.): FC 2023, LNEE 1134, pp. 59–65, 2024.
https://doi.org/10.1007/978-981-99-9342-0_7

features is the self-attention mechanism, which enables the model to capture long-distance dependencies in sequences. This capability is crucial for language modeling tasks, as words often depend on each other even when they are far apart. As stated in [1], we propose a new simple network architecture, the transformer, based solely on attention mechanisms, dispensing with recurrence and convolutions entirely.

One of the game-changing aspects of GPT-2 is its use of unsupervised learning for pre-training. During this stage, the model learns language patterns from massive amounts of text data without any manual labeling. As mentioned in [2], language models begin to learn these tasks without any explicit supervision when trained on a new dataset of millions of webpages called WebText. The pre-trained model can then be fine-tuned for specific NLP tasks, such as text generation, sentiment analysis, Q&A, and more. This approach fully leverages the abundance of unlabeled data available and reduces reliance on expensive human-labeled data.

GPT-2 has achieved remarkable success in various natural language processing tasks, surpassing many of the best models available at the time. With its impressive generation capabilities and versatility, it has become a go-to tool for researchers and developers addressing NLP challenges. However, it has also ignited debates surrounding potential risks, such as generating fake news and producing manipulative or unethical content.

Despite its impressive accomplishments, GPT-2 is not without flaws. There are instances where it can generate text that is grammatically correct but lacks coherence or logical consistency, and its support for different languages is not uniformly robust. Additionally, its resource-intensive nature can hinder deployment and scalability. Nevertheless, GPT-2 represents a significant milestone in the field of natural language processing, establishing a foundation for future research and applications. As stated in [2], The capacity of the language model is essential to the success of zero-shot task transfer, and increasing it improves performance in a log-linear fashion across tasks.

Since the introduction of GPT-2, OpenAI has released even more advanced models like GPT-3, which have pushed the boundaries of generation capabilities and versatility even further. However, GPT-2 remains a significant milestone in the history of NLP and continues to hold considerable research value. By improving and optimizing GPT-2, researchers can further explore the potential of self-attention mechanisms and unsupervised learning methods. As stated in [1], experiments on two machine translation tasks show these models to be superior in quality while being more parallelization and requiring significantly less time to train.

Furthermore, researchers can delve into the explainability and robustness of GPT-2 to gain a better understanding of its inner workings and make it more resilient against malicious attacks and manipulations. The success of GPT-2 in the field of natural language processing provides valuable insights and references for the development of subsequent models and technologies.

In summary, GPT-2, as an innovative and large-scale language model, has achieved significant results in the field of natural language processing and has laid a solid foundation for further research. Despite its limitations, GPT-2 continues to hold substantial research value. By improving and optimizing this model, researchers can continue to explore the potential of self-attention mechanisms and unsupervised learning methods. Understanding the background and overview of GPT-2 provides a better context for the comparative research we will discuss later, which focuses on enhancing the self-attention mechanism.

1.2 Research Objective and Motivation

In recent years, the field of natural language processing has made significant progress, largely driven by advancements in deep learning technologies. One standout model in this field is GPT-2, which utilizes self-attention and auto-regressive mechanisms to excel in various NLP tasks. However, despite its impressive performance, there is still room for improvement, particularly in terms of computational efficiency and capturing long-distance dependencies. Thus, the objectives and motivations of our research are as follows.

- Optimizing the self-attention mechanism: Our goal is to optimize the self-attention mechanism to improve computational efficiency and enhance the model's performance in NLP tasks.
- Improving the auto-regressive mechanism: We seek to optimize the auto-regressive mechanism to address error accumulation issues and enhance the grammatical and semantic consistency of the generated text.
- Balancing local and global context information: Striking a balance between local and global context information is crucial in text generation. Overemphasizing local information can result in the loss of global consistency, while excessive attention to global information can lead to overlooked details. We will investigate techniques to achieve a balance between local and global context information within the self-attention and auto-regressive mechanisms.
- Validating the effectiveness of the improvements: Through experiments across different NLP tasks, we will compare the performance of the optimized model with the baseline model to demonstrate the effectiveness and scalability of our proposed improvements.

Our motivation lies in the potential to enhance the performance of the GPT-2 model in NLP tasks while optimizing computational efficiency. The outcomes of our research could significantly contribute to the advancement of the NLP field and offer valuable insights to researchers in related areas. Moreover, the improved model could find practical applications in intelligent dialogues, automatic summarization, knowledge graph construction, and more, providing users with higher-quality NLP services.

Throughout our research, we will employ various optimization strategies and techniques to enhance the self-attention and auto-regressive mechanisms and integrate these improvements into the GPT-2 model. Our experiments will encompass multiple NLP tasks to comprehensively evaluate the performance of the optimized model. By analyzing the experimental results, we will assess the effectiveness and scalability of our improvements.

Finally, we will summarize our achievements, highlight the contributions we have made in optimizing the self-attention and auto-regressive mechanisms in the GPT-2 model, and discuss future research directions and potential applications. Through our research, we aspire to push the development of NLP technologies further, providing more powerful and efficient NLP solutions for real-world applications.

2 Experimental Design and Evaluation Methods

2.1 Implementation Details of the Optimized Model

In this study, our focus is on optimizing the auto-regressive mechanism within the self-attention mechanism of the GPT-2 model. Our specific strategy involves randomly shortening certain auto-regressive lengths during the text generation process. For example, instead of considering all previous 299 words when generating the 300th word, our optimization strategy would only take into account the previous 200 words. The goal of this strategy is to reduce the model's heavy reliance on past contexts, thereby increasing the diversity of generated text and improving generation speed.

The motivation behind this method can be summarized into three main points. Firstly, shortening the auto-regressive length helps to promote diversity in the generated text and prevents the production of excessively repetitive content. Secondly, it reduces computational load, resulting in faster text generation. Lastly, this strategy enhances the model's ability to generalize, enabling it to adapt better to new samples and text styles that may differ from the training data.

In our forthcoming experiments, we will explore the impact of this strategy on the generative capabilities of the GPT-2 model. We will conduct a series of experiments to analyze how different auto-regressive lengths affect the quality of generated content. Additionally, we will evaluate the effectiveness of this method across various natural language processing tasks. Through these experiments, we aim to provide valuable insights into language model optimization and offer guidance for future improvements.

In conclusion, this research focuses on optimizing the auto-regressive mechanism within the self-attention mechanism of the GPT-2 model by randomly shortening auto-regressive lengths during text generation. This approach has the potential to enhance the diversity of generated text, accelerate the generation process, and improve the model's generalization capability. Subsequent experiments will validate the effectiveness and feasibility of this strategy, providing empirical support for further optimization of language models.

2.2 Evaluation Metrics and Experimental Setup

We use BLEU (Bilingual Evaluation Understudy) as our primary evaluation metric in this study. BLEU is commonly used to measure the grammatical and semantic similarity between generated text and manually written reference text. It relies on n-gram precision for scoring. The BLEU score ranges from 0 to 1, with higher values indicating a higher similarity between the model-generated text and the human-written reference text, thus reflecting better model performance.

For our experiments, we selected the GPT-2 medium model configuration, which consists of 1.5 billion parameters. Among different versions of GPT-2, such as GPT-2 small or GPT-2 large, the medium version strikes a suitable balance between computational requirements and performance. It exhibits strong performance across various natural language processing tasks, making it an appropriate baseline model for optimizing the auto-regressive mechanism within the self-attention mechanism.

To accurately evaluate the impact of our optimization, we will compare the optimized GPT-2 medium model with the original GPT-2 medium model. The original model serves as our baseline, and we expect the optimized model to outperform it in terms of BLEU scores.

This experimental setup allows us to precisely measure the effect of optimizing the auto-regressive mechanism within the self-attention mechanism on the generative capabilities of the GPT-2 medium model. By adopting this approach, we aim to enhance the credibility of our results and contribute to the advancements in the field of natural language processing. We are excited about the potential insights and possibilities this research may bring to the self-attention mechanism and its application in auto-regressive models.

3 Experimental Results and Analysis

In terms of BLEU scoring, the optimized GPT-2 model demonstrated significant improvements compared to the original GPT-2 model. The average BLEU score of our optimized model was 2.902508133906813e-06, while the original GPT-2 model had an average score of only 2.803913135029713e-158 (Table 1). This result confirms the effectiveness of our optimization strategy and highlights the superior generative capabilities of our optimized model.

Further analysis of our data revealed that in the top 10% of the generated text in Table 2, the average BLEU score of the optimized model reached 2.9156350551154368e-05, which is significantly higher than the original GPT-2 model's score of 2.81659414920824e-157. This finding indicates that our optimization strategy successfully enhanced the GPT-2 model's ability to generate high-quality text in the top percentile.

However, in the lowest 10% of the generated text, our optimized model's performance was similar to that of the original model, with BLEU scores of 4.797929067427243e-232 and 4.808473266007582e-232, respectively. This suggests that our optimization strategy has a limited impact on improving the lower-end generative capabilities of the model. We hypothesize that this may be because our optimization strategy primarily focuses on enhancing the model's top-end performance while having a limited effect on the lower end.

Table 1. BLEU scores of optimized and original GPT-2 models

Metric	Average
Optimized GPT-2	$2.902508133906813 \times 10^{-6}$
Original GPT-2	$2.803913135029713 \times 10^{-158}$

Table 2. Average BLEU scores (top 10% and bottom 10%) of optimized and original GPT-2 models

Metric	Top 10% Average	Bottom 10% Average
Optimized GPT-2	$2.9156350551154368 \times 10^{-5}$	$4.797929067427243 \times 10^{-232}$
Original GPT-2	$2.81659414920824 \times 10^{-157}$	$4.808473266007582 \times 10^{-232}$

4 Conclusions and Future Work

4.1 Summary of Research Results

Upon reviewing our experimental results, we can conclude that our optimization strategy has indeed improved the generative capabilities of the GPT-2 model to a certain degree. This improvement is particularly notable in the top 10% of the generated text, where our optimized model achieved a significantly higher BLEU score compared to the original model. This outcome provides strong evidence for the effectiveness of our optimization strategy.

However, it is important to acknowledge that our approach showed limited success in enhancing the lower-end performance of the model. This finding highlights the need for further refinement and exploration in our optimization strategy to address this aspect of the model's generative capabilities.

These results emphasize the importance of achieving a balanced performance across both high-quality and low-quality segments of generated text in future optimization efforts. It is crucial to strive for comprehensive robustness in the model's performance across all aspects of text generation, rather than focusing solely on improving the high-quality domain. By addressing the limitations identified in our study, we can pave the way for more robust and reliable language models in the future.

4.2 Future Work

In our future work, our primary focus will be on refining our optimization strategy, specifically targeting the improvement of the model's generative capabilities at the lower-end performance range. We will explore various methods for adjusting the self-attention and autoregressive mechanisms to enhance the model's proficiency in generating lower-quality text segments, while still maintaining the quality of high-end outputs.

Additionally, we aim to incorporate additional evaluation metrics, such as ROUGE and METEOR, to provide a more comprehensive assessment of the performance of our optimized model. By considering multiple evaluation metrics, we can gain a more nuanced understanding of the model's strengths and weaknesses.

Moreover, we have plans to extend our optimization strategy to other generative models, such as GPT-3 and GPT-4, to examine the applicability and effectiveness of our approach across a broader range of models.

In conclusion, our research has successfully optimized the GPT-2 model, offering new perspectives for future investigations. We are excited to continue improving the generative capabilities of the GPT-2 model in our upcoming work and to expand the application of our findings to a wide range of scenarios and domains.

References

1. Vaswani, A., et al.: Attention is all you need. In: Advances in Neural Information Processing Systems, vol. 30 (2017)
2. Radford, A., Wu, J., Child, R., Luan, D., Amodei, D., Sutskever, I., et al.: Language models are unsupervised multitask learners. OpenAI blog **1**(8), 9 (2019)

Case Classification System Based on Taiwanese Civil Summary Court Cases

Ming-Yi Chen[1(✉)], Jia-Wei Chang[1], Hsiao-Chin Lo[2], and Ying-Hung Pu[3]

[1] Department of Computer Science and Information Engineering, National Taichung University of Science and Technology, Taichung, Taiwan 404348
{s1811132015,jwchang}@nutc.edu.tw

[2] Department of Japanese Studies, National Taichung University of Science and Technology, Taichung, Taiwan 404348
hclo@nutc.edu.tw

[3] Department of College of Languages, National Taichung University of Science and Technology, Taichung, Taiwan 404348
yhpu@nutc.edu.tw

Abstract. This experiment classifies cases based on the top 20 most frequently used categories of case reasons found in civil summary court judgments provided by the Judicial Yuan of Taiwan from 2012 to 2022. We built case classifiers using two methods: machine learning with TF-IDF+SVM and deep learning with BERT. We then compared the results of both classifiers. In the classification results using TF-IDF+SVM, an accuracy of 89.3% was achieved, while with BERT, an accuracy of 93.825% was achieved.

Keywords: Natural Language Processing (NLP) · Text Analysis · Machine Learning · Deep Learning · Case Classification

1 Introduction

Natural Language Processing (NLP) is a machine learning technique that allows computers to interpret and understand human language by translating and comprehending concepts similar to humans, using binary code (0 s and 1 s) as a medium. The difference between Chinese and English lies in the fact that Chinese sentences usually do not have any punctuation marks to separate words [1]. When machines process and understand texts, they often start by segmenting long strings of sentences into individual words or tokens, a process known as word segmentation or tokenization. This allows for the extraction of textual features through the process of word segmentation. In many Chinese NLP applications, such as machine translation, text summarization, and others. Chinese word segmentation is often a necessary preprocessing step. This process involves dividing Chinese sentences into individual words or tokens to facilitate further analysis and processing in various NLP tasks. The two major challenges in Chinese word segmentation are ambiguity and unknown words. The issue of ambiguity arises when the same Chinese character sequence may have different word segmentation results in different

J. C. Hung et al. (Eds.): FC 2023, LNEE 1134, pp. 66–76, 2024.
https://doi.org/10.1007/978-981-99-9342-0_8

texts or contexts. Unknown words refer to words that are not included in the Chinese word segmentation dictionary, including names of people, places, organizations, legal terms, and their abbreviations [2]. Legal Artificial Intelligence (LegalAI) refers to the application of artificial intelligence methods to legal tasks, which helps improve the efficiency of legal professionals and provides assistance to individuals who may not have a strong knowledge of the law [3]. In the past few years, deep learning-based methods have achieved significant advancements in text classification tasks. BERT (Bidirectional Encoder Representations from Transformers) [4] is a revolutionary language model that obtains text representations by pre-training on large-scale unlabeled data. It has achieved remarkable performance on various NLP tasks. The introduction of the BERT model has brought new breakthroughs to text classification tasks. Its capabilities surpass traditional feature engineering-based methods, enabling the model to automatically learn key features from raw text. In this experiment, we used publicly available civil summary court judgments from the Taiwan Judicial Yuan for the years 2012 to 2022. We selected the top 20 most frequently occurring case categories as our data and compared the classification results between machine learning and deep learning methods.

2 Related Works

2.1 Legal Artificial Intelligence

Indeed, even before the widespread adoption of artificial intelligence technologies, there were studies that employed statistical methods to analyze legal cases [5, 6]. With advancements in technology, there has been significant research in recent years on applying artificial intelligence to the field of law. Some examples include studies on legal judgment prediction [7–9], reading comprehension [10], and case retrieval [11]. These efforts aim to leverage AI to enhance various aspects of the legal domain. Luo et al. [7] utilized three methods, FastText, TFIDF+SVM, and CNN, to train and test on over 2.6 million criminal cases published by the Supreme People's Court of China, and comparing the results of the three approaches. Xiao et al. [8] proposed a neural network approach based on attention mechanism. They developed a unified framework for jointly modeling the tasks of determining appropriate charges and extracting relevant legal articles for a given criminal case. Zhong et al. [9] mentioned that legal judgments consist of multiple sub-tasks, including decisions of applicable law articles, charges, fines, and the term of penalty. These sub-tasks are considered as a directed acyclic graph (DAG) with dependencies among them. They proposed a spectrum-based multi-task learning framework called TOPJUDGE, which integrates multi-task learning and DAG dependencies into judgment prediction. Duan et al. [10] introduced a dataset for Chinese legal reading comprehension, comprising approximately 10,000 court documents and 50,000 expert-annotated questions with answers. They built two powerful baseline models based on BERT and BiDAF for this task. Shao et al. [11] utilized BERT to capture paragraph-level semantic relations and inferred the relevance between two cases by aggregating paragraph-level interactions.

2.2 Application of NLP in Taiwanese Court Judgments

The current research on the application of NLP in Taiwanese court judgments can be broadly categorized into three types: "Judgment Retrieval Systems", "Case Classification or Clustering", "Judgment Factor Analysis and Prediction of Judgment Outcomes".

"Judgment Retrieval Systems" refer to the development or improvement of systems used for retrieving court judgments, aiming to enhance retrieval efficiency and the accuracy of search results. Hsieh [12] proposed using vocabulary combinations from factual paragraphs in judgment documents to improve retrieval results. In this experiment, they presented methods for extracting Chinese vocabulary from judgment documents, extracting important word phrases from factual paragraphs, and searching for similar cases based on these word phrases. Lin [13] established a factor table for civil judgments related to copyright law and proposed a method to extract relevant factors from judgment documents using regular expressions. This allowed for the analysis of the relationships among various factors.

"Case Classification or Clustering" refers to the process of categorizing and grouping court judgments based on different case types or legal issues using various approaches. The goal is to organize and classify the judgments into meaningful groups, allowing for easier retrieval, analysis, and understanding of the legal content within the judgments. Lia [14] developed a case-based inference system based on gambling and theft cases, combining the system with rules established by domain experts to improve classification performance. Additionally, they proposed a method for automatically annotating semantic information in factual paragraphs of judgment documents to extract the abstract structure of case facts. Ho [15] employed a hierarchical clustering method for grouping civil judgment digests. The study proposed a similarity measurement approach for civil judgment digests and compared the clustering effectiveness of various hierarchical clustering methods in civil judgments. Furthermore, they utilized a method that incorporates weighted legal keywords to enhance the clustering performance.

Research in "Judicial Factor Analysis and Judgment Outcome Prediction" involves extracting judgment factors from court rulings and using them to predict judgment outcomes or analyze the relationship between various judgment factors and outcomes. Huang [16] proposes a method for extracting sentencing and penalty factors from guilty verdicts in criminal cases related to trademark law. The study utilizes regular expressions to extract the paragraphs containing the factors from the judgments and clusters the keywords. By manually labeling the clustering results according to the sentencing standards prescribed by criminal law, specific types of cases can be obtained with their corresponding prosecution and sentencing factors. Chen [17] examines the correlation between the textual consistency of written orders on applications for release from pretrial detention in criminal proceedings and various key influencing factors such as judgment time, court, and the alleged criminal charges. The study also analyzes the relationship between these factors and the judgment outcomes.

3 Experimental

3.1 Dataset

The dataset used in this experiment consists of all the judgments from the civil summary court in Taiwan provided by the Judicial Yuan from 2012 to 2022. The dataset comprises a total of 1,179,705 records, with each judgment stored in JSON format.

Each judgment includes eight label contents: "JID" (file name), "JYEAR" (year), "JCASE" (court of judgment), "JNO" (judgment number), "JDATE" (judgment date), "JTITLE" (judgment case), "JFULL" (full text of the judgment), and "JPDF" (PDF download link for the judgment) (Fig. 1).

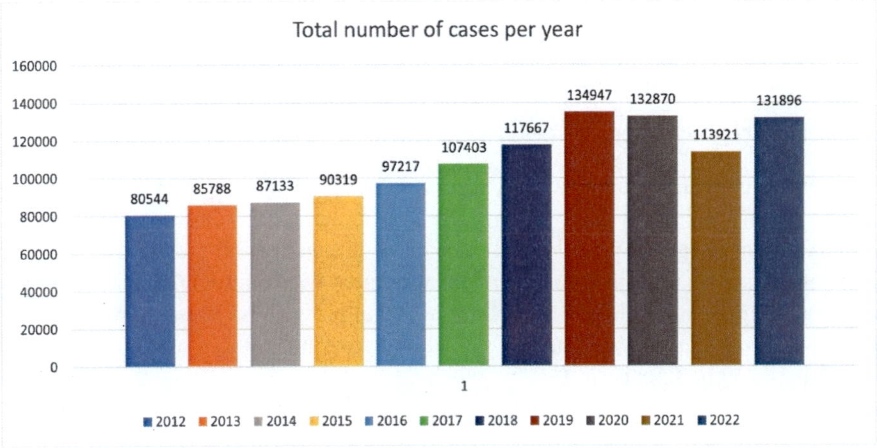

Fig. 1. Number of Cases from 2012 to 2022

First, the total number of judgment categories in the period from 2012 to 2022 was calculated. After the analysis, it was found that there were 9,613 different categories.

In this experiment, the top 20 categories with the highest number of cases were selected for case classification. The top 20 categories of cases are as follows: Damages compensation, Repayment of loans, Payment of credit card consumption expenses, Debt settlement, Damages compensation for tortious acts, Return of loans, Payment of bills, Return of credit card consumption expenses, Confirmation of non-existence of promissory note debt, Repayment of credit card consumption expenses, Transfer of property, Payment of credit card consumption expenses, etc., Payment of telecommunication fees, Payment of credit card debts, Repayment of credit card loans, Return of undue profits, Payment of management fees, Debt settlement for consumer purchases, Division of co-owned property, Lawsuit against debtor's objection. The judgment documents were filtered to include only the top 20 categories of cases, resulting in a total of 85,2168 documents. From each category, a random sample of 1000 documents was selected, resulting in a dataset of 20,000 documents in total (Fig. 2).

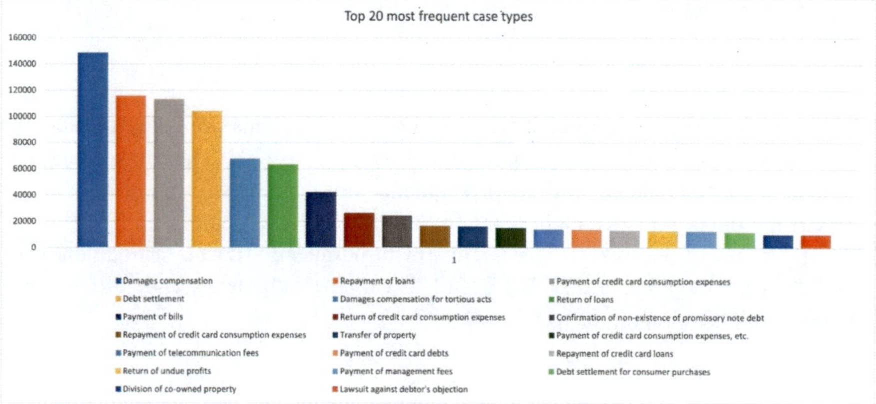

Fig. 2. Number of Top 20 Case Categories from 2012 to 2022

3.2 Data Preprocessing

The content of all the judgments was accessed through web scraping from the Taiwan Judicial Yuan's judgment query system. The system marks the words that appear in the Judicial Yuan's Legal Terminology Dictionary. Additionally, the system lists the legal provisions used in each judgment. In the experiment, web scraping was employed to retrieve the marked words and the referenced legal provisions. These words and legal provisions were then added to the dictionary to improve the accuracy of subsequent word segmentation experiments.

3.3 Word Segmentation

The word segmentation system used in this experiment is Jieba. It was chosen because it allows for the creation of custom word segmentation dictionaries, which helps to achieve more accurate word segmentation results according to our expectations. By adding customized word segmentation dictionaries, the accuracy and integrity of the word segmentation results can be improved. Below is a comparison of the word segmentation results without using a custom dictionary and the results after incorporating the custom dictionary (Fig. 3).

The comparison demonstrates that using a custom word segmentation dictionary improves the accuracy and alignment of the word segmentation results with the expected outcome.

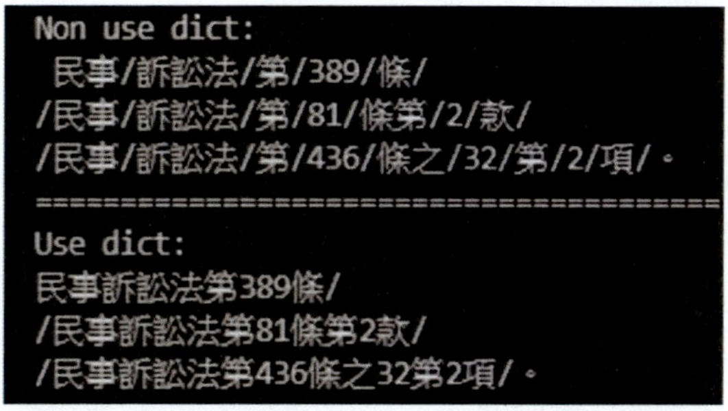

Fig. 3. Comparison of Word Segmentation Results without Using a Custom Dictionary and with Using a Custom Dictionary

In this experiment, regular expressions [18] were used to extract paragraphs from the mention of plaintiffs and defendants in the judgment documents up to the end of the main text. Subsequently, Jieba was used for word segmentation.

3.4 Filtering Non-Chinese Words and Removing Stop Words

Stop words include the most frequently used words in daily life that have high occurrence but little meaningful content, such as "you", "I", "he". In this experiment, not only common language terms but also frequently appearing words in judgment documents were added to the stop word list. These words do not have specific explanations in the Judicial Terminology Dictionary system.

After tokenization, the resulting tokens are further processed using regular expressions to remove non-Chinese characters (such as numbers and English words), followed by the removal of stop words. This preprocessing step helps to eliminate redundant words in the text, reduce computational resources, and improve the efficiency of training the model.

3.5 Term Frequency-Inverse Document Frequency

TF-IDF (Term Frequency-Inverse Document Frequency) is a statistical method for determining the importance of a word in a document. It utilizes two different parameters: term frequency (TF) and inverse document frequency (IDF).

Term frequency refers to the frequency of a term occurring in a document. If a term appears frequently within a document (high term frequency), it is assumed to be important for that particular document.

Inverse document frequency, on the other hand, measures the rarity of a term across the entire document collection. If a term is rare in other documents (high inverse document frequency), it suggests that the term is more significant for the document in question.

By combining term frequency and inverse document frequency, TF-IDF assigns a weight to each term, reflecting its importance within a specific document in the context of the entire document collection. This allows the identification of keywords that are indicative of the content and relevance of a particular document.

If a term appears frequently in a document but also appears frequently in other documents, it should not be considered a keyword for that document. The TF-IDF (Term Frequency-Inverse Document Frequency) algorithm uses two parameters: Term Frequency (TF) and Inverse Document Frequency (IDF), to calculate the importance of a term in a document. The calculation method of TF-IDF(k,f,A) is the multiplication of the term frequency and the inverse document frequency. The formula is shown as Eq. (1):

$$TF - IDF(k, f, A) = TF(k, f) * IDF(k, A) \tag{1}$$

The term frequency TF(k,f) represents the frequency of term k appearing in document f. Assuming the term "artificial intelligence" appears 20 times in a particular document, and the total number of words in that document is 320, the frequency of the term would be 20/320 = 0. 0625. The formula for TF(k,f) is given by Eq. (2):

$$TF(k, f) = \frac{F_f(i)}{max_{w \in f} F_f(w)} \tag{2}$$

The inverse document frequency IDF(k,A) represents the reciprocal of the proportion of documents in a collection that contain the term k. The IDF value decreases as the term appears more frequently across the documents, and vice versa. The formula for IDF(k,A) is given by Eq. (3):

$$IDF(k, A) = ln\left(\frac{|A|}{|\{a \in A : i \in k \in a\}|}\right) \tag{3}$$

Calculating term frequency TF(k,f) alone is insufficient to identify representative keywords for a document because the term may also appear frequently in other documents. Hence, the concept of inverse document frequency IDF(k,A) is introduced to provide a comprehensive evaluation. The combination of TF and IDF yields the final TF-IDF(k,f,A) result, which represents the weight and importance of the term within the document.

In this experiment, we compared the results using two classifiers: TF-IDF with Support Vector Machine (SVM) machine learning classifier and BERT deep learning classifier.

3.6 TF-IDF+SVM

By adding the processed text to the corpus and extracting the judgment categories from each judgment document as labels, we calculated the TF-IDF values for each word in the corpus. Next, the data was split into training and testing sets in a 7:3 ratio. The training set consisted of 14,000 instances, while the testing set had 6,000 instances. The distribution of judgment categories in the training and testing sets is shown in Figs. 4 and 5.

Next, a SVM classifier is constructed and trained on the training set. The trained model is then used to test the performance on the test set, resulting in an accuracy of 89. 3%.

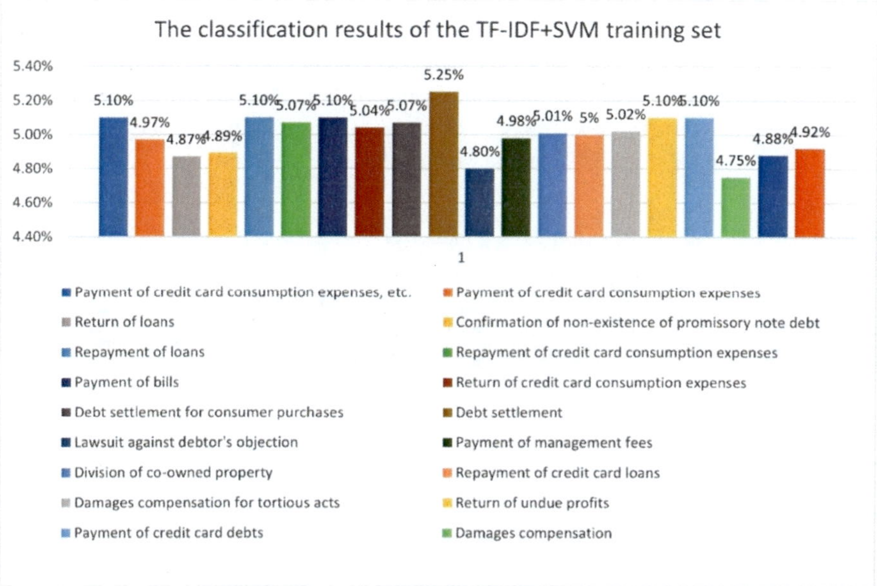

Fig. 4. Proportions of Each Case Category in the Training Set for TF-IDF+SVM

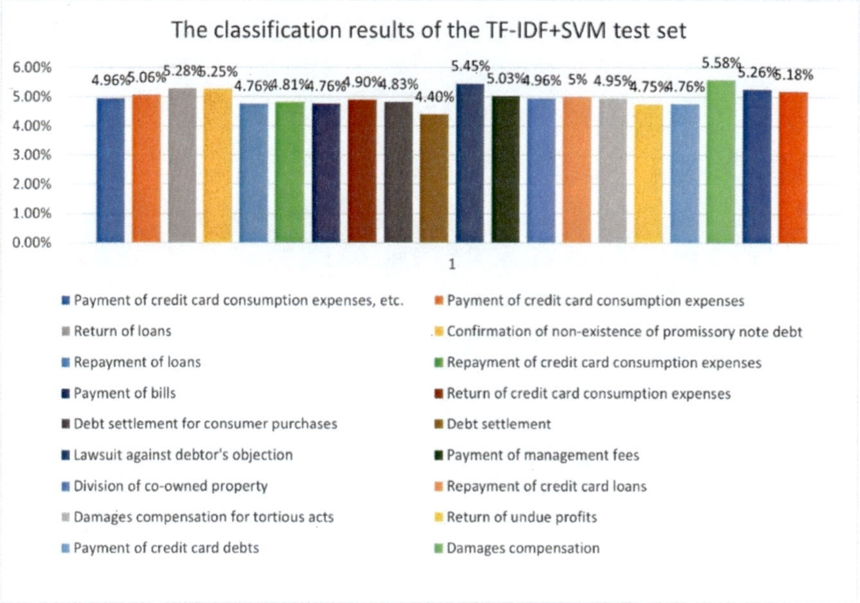

Fig. 5. Proportions of Each Case Category in the test Set for TF-IDF+SVM

3.7 BERT

In the BERT experiment, the pre-trained bert-base-chinese model was used. The dataset was split into training and test sets with a ratio of 8:2. The training set consisted of 16,000 samples, while the test set had 4,000 samples. The distribution of case categories in the training and test sets is shown in Figs. 6 and 7.

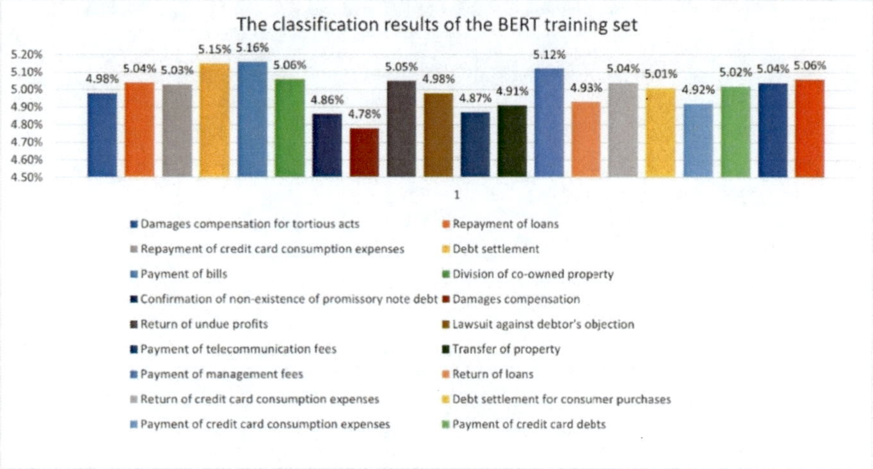

Fig. 6. Proportions of Each Case Category in the training Set for BERT

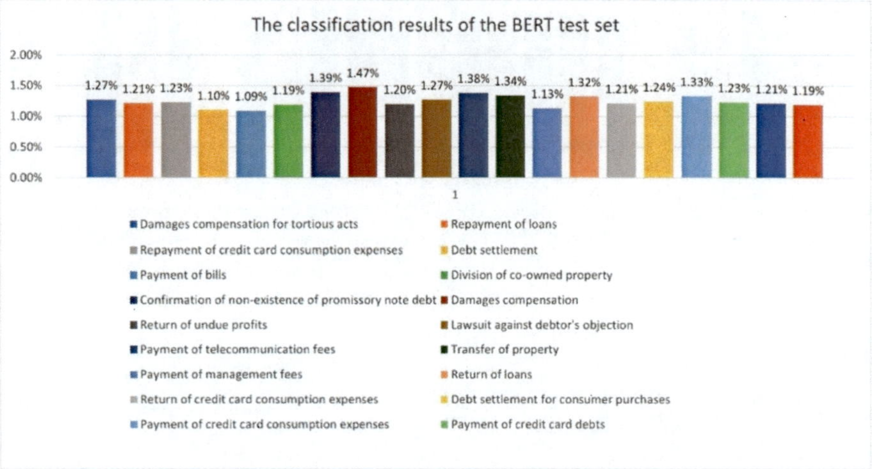

Fig. 7. Proportions of Each Case Category in the test Set for BERT

Next, the batch size was set to 32, and the maximum input length was set to 128. The training process was performed on an Intel(R) Core(TM) i7–10700 CPU@2.90 GHz. The AdamW optimizer was chosen with a learning rate of 1^{e-5}. After training, the model achieved a testing accuracy of 93.825% (Fig. 8).

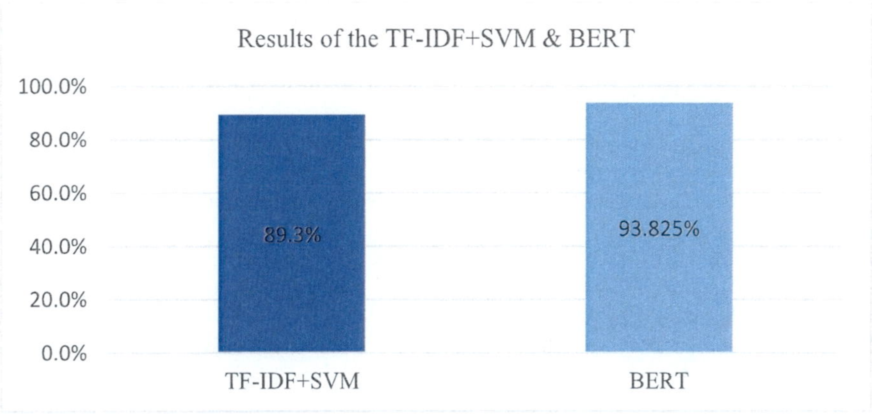

Fig. 8. Comparison of Results between BERT and TF-IDF+SVM

4 Conclusions and Future Directions

In this experiment, two approaches, machine learning and deep learning, were used for the classification of legal case categories based on court judgments. The results of both methods were compared to evaluate their performance.

Currently, only the top 20 most common case categories in civil summary courts are being used, with 1,000 judgment documents extracted for each category. In future experiments, the number of case categories will be expanded, and more judgment documents will be included to allow the model to learn the description patterns and relevant legal provisions associated with different case categories. This will enhance the accuracy of the classification results.

Acknowledgements. This work has received support from the funding provided by the National Science and Technology Council, Project No. 111-2410-H-025-018.

References

1. Ma, W.Y., Chen, K.J.: A bottom-up merging algorithm for Chinese unknown word extraction. In: Proceedings of the Second SIGHAN Workshop on Chinese Language Processing, pp. 31–38 (2003)
2. Lin, Q.X., et al.: A simple and effective closed test for Chinese word segmentation based on sequence labeling. Int. J. Comput. Linguist. Chinese Lang. Process. **15**(3–4) (2010)
3. Zhong, H., Xiao, C., Tu, C., Zhang, T., Liu, Z., Sun, M.: How does NLP benefit legal system: a summary of legal artificial intelligence. arXiv preprint arXiv:2004.12158 (2020). Zhang, D., Li, J., Shan, Z.: Implementation of Dlib deep learning face recognition technology. In: 2020 International Conference on Robots & Intelligent System (ICRIS), pp. 88–91. IEEE (2020)
4. Devlin, J., Chang, M.W., Lee, K., Toutanova, K.: Bert: pre-training of deep bidirectional transformers for language understanding. arXiv preprint arXiv:1810.04805 (2018)

5. Kort, F.: Predicting Supreme Court decisions mathematically: a quantitative analysis of the "right to counsel" cases. Am. Polit. Sci. Rev. **51**(1), 1–12 (1957)
6. Segal, J.A.: Predicting Supreme Court cases probabilistically: the search and seizure cases, 1962–1981. Am. Polit. Sci. Rev. **78**(4), 891–900 (1984)
7. Luo, B., Feng, Y., Xu, J., Zhang, X., Zhao, D.: Learning to predict charges for criminal cases with legal basis. arXiv preprint arXiv:1707.09168 (2017)
8. Xiao, C., et al. Cail2018: a large-scale legal dataset for judgment prediction. arXiv preprint arXiv:1807.02478 (2018)
9. Zhong, H., Guo, Z., Tu, C., Xiao, C., Liu, Z., Sun, M.: Legal judgment prediction via topological learning. In: Proceedings of the 2018 Conference on Empirical Methods in Natural Language Processing, pp. 3540–3549 (2018)
10. Duan, X., et al.: CJRC: a reliable human-annotated benchmark dataset for Chinese judicial reading comprehension. In: Proceedings of the Chinese Computational Linguistics: 18th China National Conference (CCL 2019), Kunming, 18–20 October 2019, vol. 18, pp. 439–451. Springer, Cham (2019)
11. Shao, Y., et al.: BERT-PLI: modeling paragraph-level interactions for legal case retrieval. In: IJCAI, pp. 3501–3507 (2020)
12. Hsieh, C.D.: An exploration of indexing Chinese judicial documents with term (2005)
13. Lin, H.-L.: Implementation of text mining techniques in court decisions: focusing on compensation of copyright infringement (2013)
14. Lia, T.-M.: Classification and discourse analysis of gambling and larceny cases that infringe multiple criminal-law articles (2004)
15. Ho, J.H.: An application of hierarchical clustering of documents for civil judgments (2006)
16. Huang, Y.-T.: Study on the Prosecutorial Sentencing Factors by Text Mining – Focusing on the Intellectual Property Law in Taiwan (2012)
17. Chen, J.-Y.: A study of the consistency of judicial decisions based on text mining: using corpuses from rulings on applications for suspension of detention (2015)
18. Kleene, S.C.: Representation of events in nerve nets and finite automata. Autom. Stud. **34**, 3–41 (1956)

Innovative Interaction Mode in VR Games

Yi-Chun Liao(✉)

Department of Digital Multimedia Design, China University of Technology, Taipei City, Taiwan
ech_liao@cute.edu.tw

Abstract. Virtual reality (VR) games have gained significant popularity in recent years, offering immersive and interactive experiences. The success of VR games relies heavily on innovative interaction modes that enhance player engagement and immersion. This paper explores the concept of innovative interaction modes in VR games and their impact on player experiences. We examine various forms of interaction modes, including gesture-based controls, motion tracking, and haptic feedback, and analyze their effectiveness in enhancing gameplay. Furthermore, we discuss the challenges and opportunities in designing and implementing innovative interaction modes. Through a comprehensive review of existing research and case studies, this paper provides insights into the potential of innovative interaction modes to revolutionize the gaming industry. The findings underscore the importance of continuous innovation and experimentation to create compelling and immersive VR game experiences. By understanding and leveraging these innovative interaction modes, game developers can deliver more engaging and memorable gameplay, transforming the way players interact with virtual worlds. In this study, we conducted a prototype system to evaluate the impact of innovative interaction modes on player engagement and gameplay experiences in VR games. We developed two different interaction modes: gesture-based controls and in-game control tool.

Keywords: interactive tool · VR game · gesture tracking

1 Introduction

1.1 Background

Virtual reality (VR) has emerged as an innovative technology that offers immersive and interactive experiences across various domains, including entertainment, education, training, and healthcare. In recent years, VR gaming has gained significant popularity, captivating players with its ability to transport them into virtual worlds and provide a heightened sense of presence and engagement. Central to the success of VR games is the design and implementation of effective interaction modes that allow players to interact with the virtual environment and game mechanics.

The traditional modes of interaction in video games, such as keyboard and mouse or gamepad controls, may not fully leverage the immersive potential of VR. As a result, game developers have been exploring and implementing innovative interaction modes

J. C. Hung et al. (Eds.): FC 2023, LNEE 1134, pp. 77–86, 2024.
https://doi.org/10.1007/978-981-99-9342-0_9

to enhance the player experience and create more intuitive and immersive gameplay. These interaction modes utilize the unique capabilities of VR devices, such as hand tracking, gesture recognition, motion controllers, and haptic feedback, to enable natural and immersive interactions within the virtual world.

The objective of this research paper is to explore the concept of innovative interaction modes in VR games and examine their impact on player engagement and gameplay experiences. By understanding and evaluating the effectiveness of these modes, game developers can make informed decisions in designing and implementing VR games that provide captivating and immersive experiences for players.

Specifically, this paper aims to:

- Identify and categorize different types of innovative interaction modes used in VR games.
- Assess the impact of these interaction modes on player engagement, presence, and enjoyment.
- Examine the challenges and opportunities in designing and implementing innovative interaction modes in VR games.
- Provide insights and recommendations for game developers to create compelling and immersive VR gaming experiences.

The study of innovative interaction modes in VR games holds significant relevance and implications for the gaming industry, academic research, and the broader field of virtual reality. By investigating and understanding the impact of these modes, this research contributes to the advancement of VR game design and provides valuable insights for game developers, researchers, and practitioners.

Furthermore, this research contributes to the academic discourse on VR gaming and interaction design. By examining the impact of innovative interaction modes, this study expands our understanding of the factors that contribute to player engagement and immersion in virtual environments. The findings can serve as a foundation for further research in this area and stimulate discussions on the design and implementation of interaction modes in VR games. The rest of this paper is organized as follows: In the next section, we review relevant literature on VR gaming and the importance of interaction modes in enhancing player experiences. This is followed by a theoretical framework that conceptualizes innovative interaction modes and discusses their significance in VR game design. The methodology section details the research design, participant recruitment, and data collection procedures. Subsequently, we present the results of our study, followed by a discussion of the findings and their implications. Finally, we conclude the paper by summarizing key findings, highlighting contributions to the field, and providing recommendations for future research and game development practices.

2 Related Work

2.1 Background

Virtual reality gaming has gained significant attention in recent years due to its potential to provide highly immersive and interactive experiences. VR technology offers a simulated environment that enables users to interact with three-dimensional virtual worlds using specialized headsets and controllers. This immersive nature of VR gaming holds promise for creating more engaging and realistic gameplay experiences [1].

Research in the field of VR gaming has focused on exploring various aspects of player experiences, such as presence, immersion, enjoyment [2], and flow. Presence refers to the feeling of being physically present in the virtual environment, while immersion refers to the extent to which the user feels psychologically absorbed in the virtual world. These factors contribute to the overall enjoyment and engagement of players [3–5].

2.2 Interaction Modes in VR Games

Interaction modes play a crucial role in shaping the player's experience in VR games. Traditional input methods, such as keyboard and mouse or gamepad controls, are often inadequate for fully exploiting the immersive potential of VR. As a result, game developers have been exploring innovative interaction modes that leverage the capabilities of VR devices, such as hand tracking, gesture recognition, motion controllers, and haptic feedback [6, 7].

Gesture-based controls allow players to use their hand movements and gestures to interact with objects and navigate within the virtual environment. This mode enables more intuitive and natural interactions, enhancing the sense of presence and immersion. Motion controllers, such as handheld devices with built-in sensors [8, 9], provide precise tracking of the player's hand movements, enabling realistic interactions and object manipulation.

Voice recognition is another innovative interaction mode that allows players to use voice commands to control the game. This mode offers hands-free interaction and can be particularly useful in games that require verbal communication with virtual characters or complex command inputs.

Haptic feedback technology provides physical sensations [17] or vibrations to simulate the sense of touch and enhance the realism of interactions [10]. This mode can add a new dimension to gameplay by providing tactile feedback, such as the sensation of objects or virtual environments.

2.3 Impact of Innovative Interaction Modes on Player Engagement

Several studies have investigated the impact of innovative interaction modes on player engagement and gameplay experiences in VR. Research has shown that these modes can significantly enhance player immersion, presence, and enjoyment compared to traditional input methods.

Gesture-based controls have been found to provide a more intuitive and natural way of interacting with the virtual environment. Players report a greater sense of agency and physical presence [18], leading to increased engagement and enjoyment. The ability to reach out and grab objects or perform gestures that correspond to in-game actions enhances the sense of embodiment within the virtual world.

Voice recognition has shown promise in enabling seamless and immersive interactions. Players find voice commands to be convenient and immersive, as they can communicate with virtual characters or control the game without the need for physical input devices. This mode can facilitate more natural and expressive communication in VR games, enhancing the overall gameplay experience.

Motion controllers, with their precise tracking capabilities, offer realistic hand interactions and object manipulation. Players feel a sense of control and agency as they physically reach out and manipulate virtual objects. This mode enhances the immersion and embodiment within the virtual world, leading to heightened engagement and enjoyment.

Haptic feedback has the potential to enrich gameplay experiences by providing realistic tactile sensations [17]. Players experience a stronger sense of presence and immersion when they can feel the virtual environment or receive feedback through vibrations. Haptic feedback can enhance the realism of interactions and contribute to a more engaging and immersive gameplay experience.

While there has been significant progress in understanding the impact of innovative interaction modes in VR games, there are still gaps and limitations in the existing research that warrant further investigation.

Firstly, many studies have focused on the immediate effects of different interaction modes on player engagement and enjoyment during gameplay sessions. However, there is a need to examine the long-term effects and sustainability of these modes over extended play durations. Understanding how players' experiences evolve and adapt over time can provide insights into the long-term effectiveness and potential challenges associated with innovative interaction modes.

Secondly, most research has primarily examined the subjective experiences and perceptions of players through self-report measures. While self-report measures offer valuable insights, objective measurements and performance metrics can provide a more comprehensive understanding of player engagement and gameplay experiences. Incorporating physiological measures, eye-tracking, or behavioral analysis can offer additional quantitative data to complement the subjective feedback provided by players.

Furthermore, the majority of studies have focused on specific genres or types of VR games, limiting the generalizability of the findings. Exploring a wider range of game genres and contexts can provide a more comprehensive understanding of the impact of innovative interaction modes across different game experiences. Additionally, investigating the preferences and experiences of diverse user groups, including individuals with varying levels of gaming experience or different demographic backgrounds, can provide insights into the effectiveness and accessibility of these modes for a broader audience [17, 18].

Lastly, the design and implementation of innovative interaction modes require careful consideration of technical constraints, such as hardware limitations, software compatibility, and development resources. Future research could explore the technical feasibility and challenges associated with implementing these modes in real-world game development scenarios. Understanding the practical considerations and potential barriers can assist game developers in effectively integrating innovative interaction modes into their projects.

The existing research suggests that these innovative interaction modes can enhance player immersion, presence, and enjoyment in VR games compared to traditional input methods [16]. However, there are gaps and limitations in the current body of research, such as the need for long-term investigations, objective measurements, wider game genre exploration, diverse user group analysis, and consideration of technical constraints.

3 Theoretical Framework

The theoretical framework section aims to provide a conceptual foundation for understanding the role and significance of innovative interaction modes in VR game design. It explores theoretical perspectives and frameworks that inform the design and implementation of these modes, as well as their influence on player engagement and gameplay experiences (Fig. 1).

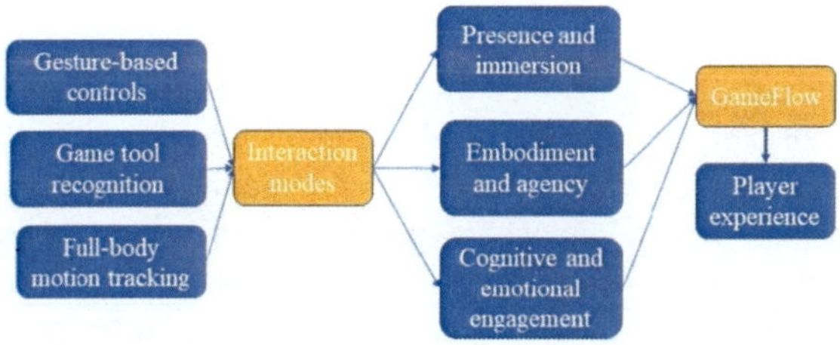

Fig. 1. The theoretical framework

3.1 Presence and Immersion

Presence and immersion are two key concepts that underpin the theoretical framework of this study. Presence refers to the subjective feeling of being physically present in a virtual environment, while immersion refers to the extent to which an individual becomes psychologically absorbed in the virtual world. These concepts are fundamental to understanding the impact of innovative interaction modes on player experiences.

According to the presence theory, a high level of presence is essential for creating a sense of realism and engagement in virtual environments. It is influenced by various factors, including sensory feedback, interactivity, and the extent to which the virtual world aligns with users' expectations and cognitive processes. The use of innovative interaction modes can enhance the sense of presence by providing more natural and intuitive ways of interacting with the virtual environment.

Immersion, on the other hand, is associated with the depth and richness of the virtual experience. It involves the feeling of being fully engrossed in the virtual world and losing awareness of the physical environment. Innovative interaction modes can contribute to immersion by enabling more realistic and embodied interactions, enhancing the player's sense of agency and involvement within the virtual environment.

3.2 Flow Theory

Flow theory, developed by Csikszentmihalyi, provides insights into the psychological state of optimal engagement and immersion in an activity [11]. Flow is characterized by a deep sense of focus, enjoyment, and a feeling of being fully absorbed in the task at hand. It occurs when the challenges presented in the activity match the individual's skills and abilities.

In the context of VR gaming, innovative interaction modes can facilitate the experience of flow by providing a seamless and intuitive interaction process. When the player's actions and inputs align with the virtual environment's feedback and challenges, a state of flow can be achieved. This leads to enhanced enjoyment, sustained engagement, and a sense of accomplishment.

Embodiment and agency are key aspects of player experiences in VR games and are closely related to the effectiveness of innovative interaction modes. Embodiment refers to the feeling of being physically present and connected to a virtual body or avatar. It involves the perception of bodily sensations, movements, and interactions within the virtual environment [12–14].

Innovative interaction modes, such as gesture-based controls and motion tracking, enable a greater sense of embodiment by allowing players to use their own body movements to interact with the virtual world. This embodiment can enhance the player's sense of presence and immersion, as they perceive the virtual body as an extension of their own. Agency, on the other hand, refers to the sense of control and influence over the virtual environment. Innovative interaction modes that offer precise tracking and responsive feedback enable players to have a greater sense of agency, as their actions directly impact the virtual world. This sense of agency contributes to player engagement and satisfaction.

3.3 Cognitive and Emotional Engagement

Cognitive and emotional engagement are important dimensions of player experiences in VR games. Cognitive engagement involves mental processes such as attention, problem-solving, and decision-making. Innovative interaction modes can enhance cognitive engagement by providing more intuitive and immersive interactions that require active mental involvement.

Emotional engagement, on the other hand, refers to the player's emotional responses and attachment to the virtual experience. Innovative interaction modes that elicit emotional reactions, such as joy, excitement, or fear, can enhance the emotional engagement of players. This emotional engagement contributes to the overall enjoyment and memorable experiences in VR games.

4 Proposed Interaction Mode and Prototype System

4.1 Theoretical Framework

The theoretical framework provides a conceptual lens for understanding the impact of innovative interaction modes in VR games. It explores key theoretical perspectives such as presence, immersion, flow theory, embodiment, agency, cognitive engagement, and emotional engagement. These concepts help frame the understanding of how innovative interaction modes influence player experiences in VR games.

The proposed interaction mode has two key parts, all shown in Fig. 2:

- **Gesture-based controls**: offer the benefits of providing a more embodied interaction experience, allowing users to engage in actions and expressions that are closer to real-world counterparts. This control method also offers increased freedom and expressiveness, enabling users to interact with digital content, virtual worlds, or applications in a more natural and intuitive manner. Designing gesture-based controls involves considerations for gesture recognition and interpretation to understand the user's intentions and translate them into appropriate actions. This involve technologies such as machine learning, computer vision, sensor technologies, and algorithms.
- **In-game Control tool**: has implications for gameplay mechanics, player immersion, and overall game design. By accurately recognizing and responding to the player's use of tools, the game can provide a more dynamic and interactive experience, allowing players to fully utilize the available tools and enhance their engagement with the game world. Proposed system implements specific algorithms to handle in-game tool recognition based on the game's design and mechanics. This allows for more precise and context-aware interactions, enabling players to fully leverage the functionalities and possibilities offered by the in-game tools.

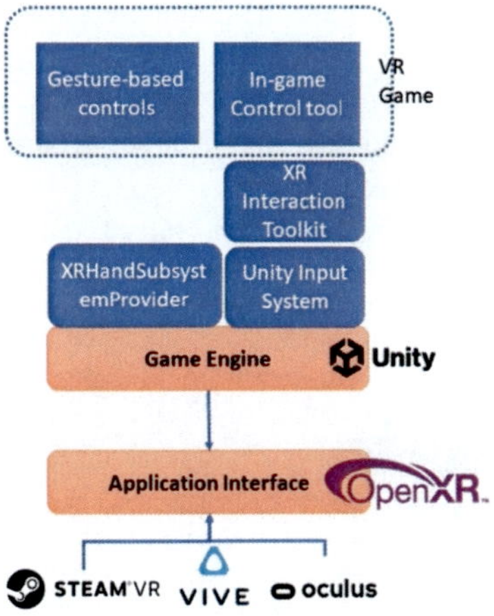

Fig. 2. Game system architecture

4.2 Prototype Implementation

In the hardware and software architecture, it includes a virtual reality (VR) headset, the game program main body, and Arduino sensing devices. Our project implements the following features:

- Arduino Communication Module.
- VR Control Module.
- Character Control Module.

- UI Module.
- Monster AI Module.

Arduino Interfacing Devices: Utilizing Arduino devices to interface with Unity for achieving a multiplayer-operable VR effect. To connect Arduino to Unity, follow these steps:

1. Ensure that your Arduino board is properly connected to your computer and functioning correctly. Make sure you have the correct drivers installed and the Arduino IDE set up.
2. Create a new project or open an existing project in Unity.
3. Download and install the Arduino connection package for Unity. You can find the relevant package from platforms like the Unity Asset Store or GitHub. These packages typically provide APIs and tools for communication with Arduino.
4. Create a script in Unity for serial communication. You can use the C# programming language to write this script. This script will handle the communication with Arduino, such as reading or writing data.
5. Set up and manage the serial communication using the SerialPort class in Unity. This class provides a range of methods and properties to communicate with Arduino through the serial port.
6. Use your script in Unity to configure the parameters for serial communication, such as the serial port number, baud rate, and data format.
7. Use the appropriate methods in your script to send or receive data to/from Arduino. You can send commands to Arduino through the serial communication and read sensor data or other feedback from Arduino.
8. Test and validate the connection. Run your Unity scene or application to ensure that Unity can communicate correctly with Arduino. Check for any communication errors or warnings in Unity's console or logs (Fig. 3).

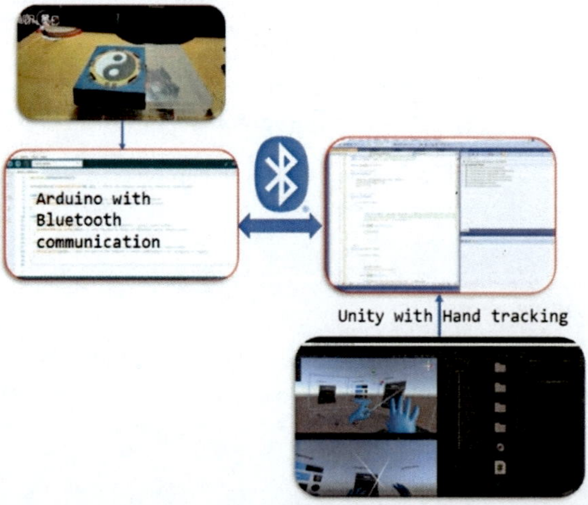

Fig. 3. Proposed communication module.

5 Conclusion

In this study, we explored the impact of innovative interaction modes in VR games, specifically focusing on Gesture-based Controls, In-game Control tool, and Handheld Controllers. The findings of this research contribute to our understanding of player experiences and provide insights for game designers and developers in creating immersive and engaging VR gaming experiences.

In conclusion, this research contributes to the field of VR game design by examining the impact of innovative interaction modes on player engagement, immersion, and presence. By considering the strengths and limitations of different modes, game designers can create more immersive and captivating experiences for players. Future research should continue to explore and refine interaction modes to enhance player experiences and push the boundaries of VR gaming. With the continued advancements in VR technology, there are exciting opportunities to create more immersive and interactive experiences that captivate players and elevate the gaming industry.

References

1. Bowman, D.A., McMahan, R.P.: Virtual reality: how much immersion is enough? Computer **40**(7), 36–43 (2007)
2. Radianti, J., Majchrzak, T.A., Fromm, J., Wohlgenannt, I.: A systematic review of immersive virtual reality applications for higher education: design elements, lessons learned, and research agenda. Comput. Educ. **147**, 103778 (2020)
3. Xi-Ning, W., Gareth William, Y., Adéla, P., Conor, Mc G.: Utilizing virtual reality to assist social competence education and social support for children from under-represented backgrounds. Comput. Educ. **4**, 104815 (2023)
4. Ifanov, J.P., Salim, S., Edo Syahputra, M., Andam, S.P.: A systematic literature review on implementation of virtual reality for learning. Procedia Comput. Sci. **216**, 260–265 (2023)
5. Castelo-Branco, R., Leitão, A.: Algorithmic design in virtual reality. Architecture **2**(1), 31–52 (2022)
6. Vishwakarma, D.K.: Hand gesture recognition using shape and texture evidences in complex background. In: Proceedings of the 2017 International Conference on Inventive Computing and Informatics (ICICI), pp. 278–283, Coimbatore, India (2017)
7. Wu, M.Y., Ting, P.W., Tang, Y.H., Chou, E.T., Fu, L.C.: Hand pose estimation in object-interaction based on deep learning for virtual reality applications. J. Vis. Commun. Image Represent. **70**, 102802 (2022)
8. Baudel T., Beaudouin-Lafon M.: Charade: remote control of objects using free-hand gestures. Commun. ACM **36**(7), 28–35 (1993)
9. Ruiz, J., Li, Y., Lank, E.: User-defined motion gestures for mobile interaction. In: Proceedings of the SIGCHI Conference on Human Factors in Computing Systems, ACM, Vancouver, Canada, pp. 197–206 (2011)
10. Jin, R., Yang J.: Domain adaptive hand pose estimation based on self-looping adversarial training strategy. Sensors **22**(22), 8843 (2022)
11. Csikszentmihalyi, M.: Flow: The Psychology of Optimal Experience. Harper & Row (1990)
12. Sweetser, P., Wyeth, P.: GameFlow: a model for evaluating player enjoyment in games. ACM Comput. Entertainment **3**(3), 1–24 (2005)
13. Hunicke, R., LeBlanc, M., Zubek, R.: MDA: a formal approach to game design and game research. In: Proceedings of the AAAI Workshop on Challenges in Game AI, vol. 4, p. 1722 (2004)

14. Cardona-Rivera, R.E., Zagal, J.P., Debus, M.S.: GFI: a formal approach to narrative design and game research, interactive storytelling. In: 13th International Conference on Interactive Digital Storytelling, ICIDS 2020, pp.133–148 (2020)
15. Reyes-Lecuona, A., Diaz-Estrella, A.: New interaction paradigms in virtual environments. In: Proceeding on IEEE Mediterranean Electrotechnical Conference, pp. 449–452 (2006)
16. Theodoropoulos, A., Stavropoulou, D., Papadopoulos, P., Platis, N., Lepouras, G.: Developing an interactive VR CAVE for immersive shared gaming experiences. Virtual Worlds **2**, 162–181 (2023)
17. Sebastian, V., Panagiotis, K., Ferran, A., Claudio, P., Maud, M.: Design evaluation and calibration of wearable electrotactile interfaces for enhancing contact information in virtual reality. Comput. Graph. **111**, 199–212 (2023)
18. Yukang, Y., Xin, Y., Chun, Y., Yuanchun, S.: Gesture-based target acquisition in virtual and augmented reality. Virtual Reality Intell. Hardware **1**(3), 276–289 (2019)

A Study on the Integration of Worked Examples and Blended Learning in the Curriculum During the COVID-19 Epidemic

Hung Sun(✉) and Shu-Wei Chang

China University of Technology, Taipei City 116, Taiwan (R.O.C.)
arion3d@gm.cute.edu.tw, book@cute.edu.tw

Abstract. Both blended learning and work examples are used to enhance student learning. Due to the temporary suspension of classes due to COVID-19, the face-to-face course was temporarily converted to a full-scale online course. In order to provide a consistent learning foundation for students of all levels, The Next Generation Art course is based on segmented worked examples, supplemented by tutorial videos and supplementary videos. At the end of the semester, a questionnaire based on the TAM scale was administered to three classes in two school districts. Gender, 3D ability, and basic information of topic type were analyzed by Independent Sample t-test. The results showed that females had better 3D ability than males, which was also reflected in the TAM, and females preferred and were willing to apply the technology of the course. As expected, those with lower 3D ability chose 2D format for their topics. The relationship between the number of professional credits and 3D ability was analyzed by ANOVA. Unlike the prediction, it is not the case that more courses equals higher ability, but rather the students with moderate number of 3D courses considered themselves to have the highest 3D ability. It is possible that the Dunning-Kruger effect is caused by the worked example and the blended learning.

Keywords: Technology Acceptance Model · Blended Learning · Worked Example · Next Generation Art

1 Introduction

Since 2019, due to the epidemic of COVID-19, all educators need to combine face-to-face teaching, distance online teaching and online videos for blended teaching. Combining online and offline blended teaching requires careful instructional design and consideration of different strategies. Nowadays, due to external interference such as part-time jobs and mobile phones, students have low self-control, and the interference of COCVID-19 has resulted in generally ineffective learning of 3D courses. Most students fall into a situation of low ability and low achievement, showing partiality Low learning motivation, low self-efficacy and low participation eventually turn into low academic achievement and low satisfaction [1]. Blended learning emerged as a solution with the potential to improve their learning experience and engagement.

J. C. Hung et al. (Eds.): FC 2023, LNEE 1134, pp. 87–95, 2024.
https://doi.org/10.1007/978-981-99-9342-0_10

Worked example learning is a natural learning style in line with human cognition and provides an effective teaching method [4], and both blended learning can be used to enhance student learning outcomes. In Taiwan, the F2F course was temporarily converted to a full-fledged online course in May 2022 due to the mass suspension of classes, and teachers needed to adapt the course content differently from face-to-face classes for this purpose. The course of study teaches Next-Generation Art (NGA) application techniques and is the last course of all 3D technology courses, with students already in the graduation project stage. The course uses working examples to build a consistent foundation for learning, and video recordings are made during face-to-face and distance learning sessions for students to study after the class. A student survey is conducted at the end of the semester to understand students' perceptions of the work examples and blended learning, and use TAM to evaluate the effectiveness of software teaching in NGA courses.

2 Method and Material

2.1 The Technology Acceptance Model

The Technology Acceptance Model (TAM) [2], first proposed by Davis in 1986, has been a favored method for analyzing the use of new technologies or hardware and software, explaining and predicting the usage behavior of information technology, and is used to determine the decision factors or processes of users in accepting or using a new technology. The TAM framework has five main constructs: Perceived Usefulness (PU), Perceived ease of Use (PEU), Attitude toward Use (ATT), Behavioral Intention to Use (BI), and Actual System Use (AU) [3].

The Next-Generation Art (NGA) course focuses on the use of Adobe Substance 3D Painter, and TAM is used to assess students' acceptance of the technology as a reference for course learning outcomes (Fig. 1).

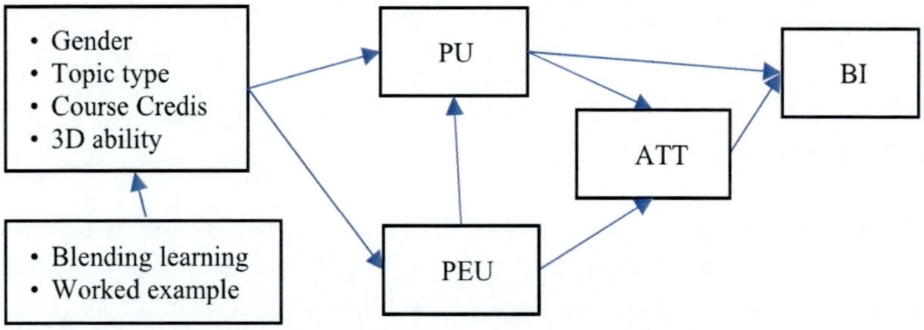

Fig. 1. Technology acceptance model of this research framework

2.2 Segmented Worked Example

The worked example effect is the best known and most widely studied of the cognitive load effect. The field's relative emphasis on this effect is justified. Worked examples provide an effective method of teaching [4]. Worked examples are teaching tools that teach problem-solving skills. They typically involve simulating the problem-solving process in a well-structured design template and presenting the steps and end results of the problem-solving process [5]. In addition, worked examples cannot be adapted to all learning environments and are more suitable for individual learning and not for collaborative learning [6].

Not all students will take all the 3D courses. NGA techniques, as the final integrated application course, requires the most 3D skills as the foundation. Since about half of the digital media students have low achievement in 3D technology and do not have the basic requirements to carry out the course content, in order to make the course progress smoothly, the teachers provide the 3D resources required for the course. However, the study found that effective examples are not always effective. Even when learners are given enough resources to do relevant activities, these activities do not necessarily lead to better learning [7]. Slightly earlier studies also found that giving complete working examples led to misjudgments of students' self-competence [8]. The learning effect involves students' autonomous motivation and ability foundation, so it is changed to provide staged work example files, split the work examples into raw materials, semi-finished products, and finished products, so that students can advance according to the course content on a certain basis, and modify the teacher's own he basic paradigm for the next stage of learning activities.

2.3 Blended Learning

The mix of face-to-face and online components of blended learning provides learners with the opportunity to learn from multiple delivery methods [9]. Implementing blended learning is a challenging process that is influenced by many factors, with the teacher playing the most critical role. Teachers who implement blended learning should possess seven teacher attributes and avoid four attributes that hinder blended learning [10].

Usually, blended learning refers to face-to-face learning and online video learning. There are five ways to enhance the learning effect of video for teaching [11]. The NGA course in this study used OBS software to record videos during both F2F and online learning, and published them in Google Classroom. Three classes had different progress and each had different videos (Table 1).

2.4 Questionnaire

At the end of the semester, during the school closure period, an anonymous online survey was conducted among students in three classes at two campuses. The content of the questionnaire included gender, major credits, topic type, self-rated 3D ability, etc. A five-point Likert-type scale with TAM as the main structure and three sets of qualitative scales were used, totaling 24 items. The collected questionnaires were analyzed and compared with IBM SPSS 26.

Table 1. Principles for improving the effectiveness of video teaching and our corresponding methods

Principles	Correspondence in this study
Dynamic drawing	Real-time drawing and labeling function for emphasis during recording
Gaze guidance	Use of the recording software's highlighting cursor to show mouse clicks and keyboard characters
Generative activity	Not yet included in the course design, as it will increase the load on the course by increasing the teacher's assessment workload
Perspective	Not applicable to software-operated courses
Subtitle	Subtitling of videos requires text correction and verbiage retouching, which is time-consuming and labor-intensive to produce

3 Result and Discussion

3.1 Preliminary Analysis

In two school districts, 76 responses were received from three classes of students, Table 2 shows the basic data list, of which 30 were male and 46 were female, and 55.3% of the graduation projects were full 3D and partial 3D use. Only 18 of the 76 students identified themselves as having 3D ability above the class average, a percentage consistent with the number of people and the number of people who felt they could complete the full NGA work method, and consistent with previous research: Those with a high level of confidence in their abilities are perceived to have a strong sense of efficacy, which leads to better learning outcomes [12].

81.6% of the students needed tutorial videos, while only 4 out of 5.3% felt less needy. As many as 85.5% relied on the videos for learning and review, while 3.9% did not need them or did not need them. The need for videos to complete assignments accounted for 77.7%, indicating the necessity of tutorial videos.

In terms of work examples, about 40.4% of the respondents did not know how to do the work themselves and needed the teacher to provide examples, which is different from the actual situation or the teacher underestimated the students' ability. 72.4% agreed that the work examples were helpful for learning, while 4 people did not think they were very helpful.

After combining the work examples and blended learning, 15.8% felt that they could integrate 3D software, and 23.7% were confident in completing the process of next-generation art techniques, which is roughly similar to the percentage of students with high 3D ability. There were no students who didn't understand it at all, but 9.2% still didn't understand it very well.

Table 2. Basic Information Statistics Table

Basic information	Types	Number of people
gender	Male	30
	Female	46
Campus	Campus A	53
	Campus B	23
Graduation topic form	3D Content	42
	2D Content	34
	< 6 credits	2
Hours of professional courses	6–9 credits	26
	> 9 credits	48
Self-evaluation 3D capability	Good	2
	Not bad	16
	Fair	26
	Not so good	19
	Very bad	13

3.2 Reliability Analysis

A total of 24 items on the scale were analyzed for reliability, and the overall Cronbach's alpha was .860, indicating that the factors in this scale have sufficient reliability and good internal consistency. The results of the reliability analysis for each component are shown in Table 3.

Table 3. Reliability analysis of each facet

Structure	Cronbach's alpha	Cronbach's alpha values based on standardized items	N of Items
PEU	.896	.898	4
PU	.633	.679	4
ATT	.804	.803	4
BI	.788	.798	3
Tutorial Video	.818	.818	3
Worked Example	.387	.430	3
Understand	.747	.742	3

3.3 Analysis of Total Validity and Validity of Various Aspects

For construct validity was examined by factor analysis and the results are shown in Table 4, which can be judged by the KMO quality size according to Kaiser's (1974) selection criteria. The KMO value for this scale is .831, which is greater than the recommended 0.8, indicating that it is suitable for factor analysis. The Bartlett's test value for sphericity is less than 0.05 and therefore significant.

Table 4. KMO and Bartlett test determination.

Kaiser-Meyer-Olkin Sampling Suitability		.831
Bartlett's spherical check	Approximate Cardinality Determination	1080.531
	Degree of freedom	276
	Significance	.000

3.4 Independent Sample t-test

Independent sample t-test was conducted on 3D ability, TAM, teaching video, work example, and comprehension in the form of gender and topic. The results are summarized in Table 5.

Table 5. Results and analysis of the independent sample t-test

Item	Result
Gender - 3D ability	Significance (two-tailed) 0.030 < 0.05, significant, females have stronger 3D ability than males
Gender - TAM	Some of the topics in the ATT and BI sections are significant, and it can be argued that women prefer to use the NGA technique more than men, and are more willing to apply it to topics
Topic - 3D ability	The type of topic and 3D ability were checked, and as expected, those with weak 3D ability chose 2D topic format
Topic - TAM	The students whose topic is in 2D form show significance in some items of PEU and PU and all items of BI, expressing that they feel the ease of use and usefulness of NGA techniques and increase their willingness to use them
Topic - Understanding	Significance (two-tailed) 0.010 < 0.05, indicating that combining work examples with blended learning improved understanding of NGA techniques for students who chose 2D topics (low 3D ability)
Topic - Worked Examples	The students of the 3D topic showed a significant (two-tailed) 0.000 < 0.05 for the item "Using work examples because they can't do it by themselves" It is speculated that those with high 3D ability understand the difficulty of implementation, so their self-assessment is more correct

3.5 Analysis of Variance (ANOVA)

The present study categorized students' enrollment in 3D professional courses into three levels: over 9 credits, 6 to 9 credits, and below 6 credits. The number of credits was used as a factor to analyze the variance in self-evaluated 3D abilities. It was initially hypothesized that students with a higher number of enrolled professional courses would evaluate their 3D abilities more positively. The analysis results are presented in Table 6, and a Scheffe post hoc test was conducted, as shown in Table 7. The results showed that students with an intermediate number of credits evaluated their 3D abilities as the strongest. However, it was found that the NGA technique course, which requires a foundation of all 3D professional course knowledge, adopted a staged working example, which coincidentally made up for the deficiencies of students with an intermediate number of enrolled courses. As a result, they overestimated their abilities. The Dunning-Kruger effect, which describes a tendency for incompetent individuals to overestimate their abilities, cannot be ignored [13]. Although students with the least number of enrolled 3D courses did not exhibit cognitive bias, the impact of the working examples cannot be overlooked.

Table 6. Analysis of variance

	Sum of Squares	Degrees of freedom	Mean Square	F	Significant
Between Groups	10.821	2	5.411	5.200	.008
In Group	75.955	73	1.040		
Total	86.776	75			

Table 7. Scheffe's method

(I) Credit	(J) Credit	Mean Difference (I-J)	Std. Error	Sig	95% confidence interval	
					Lower Bound	Upper Bound
1.00	2.00	.19231	.74850	.968	−1.6781	2.0627
	3.00	.95833	.73615	.433	−.8812	2.7979
2.00	1.00	−.19231	.74850	.968	−2.0627	1.6781
	3.00	.76603*	.24839	.011	.1454	1.3867
3.00	1.00	−.95833	.73615	.433	−2.7979	.8812
	2.00	−.76603*	.24839	.011	−1.3867	−.1454

4 Conclusion

Combining worked examples and blended learning, students with low 3D ability have successfully improved their acceptance of NGA techniques and course understanding. The mixture of face-to-face (physical/distance) and teaching videos in blended learning provides learners with opportunities to learn in a variety of teaching methods and increases the flexibility of students' learning [14]. In addition, the worked example stage is divided into raw materials, semi-finished products, and finished products to reduce the impact of differences in students' 3D abilities. Videos are recorded during lectures to operate the software and explain the functions. The content of the videos is detailed and there are a large number of videos. Students can arrange their own learning or query functions afterwards. In view of the addition of videos that only operate without explanation during the suspension period, it is another reason why students rely on the high proportion of videos. According to research, all detailed explanations should be provided in the initial teaching and minimized when operating with worked examples [4]. Also making videos from material that is taught face-to-face to students in the classroom can lead to content overload, repetition, and is not an effective way to generate blended learning [15]. However, teaching first and then recording two-stage lectures will greatly increase the burden on teachers in teaching, and the school's resource injection is required before it can be considered for implementation.

5 Limitations and Future Prospects

The 3D ability used for comparative evaluation in this study was based on students' self-reported assessments. Self-assessment may be biased because students may not accurately judge their skill level. Therefore, in the future, it is necessary to consider conducting questionnaire surveys in a named manner, and link them with actual classroom grades to make quantitative judgments.

In addition, the work examples are a viable solution for addressing differences in students' basic proficiency levels. However, efforts must be made to exclude students who do not need work examples for learning, as people tend to choose the easiest way to achieve their goals.

References

1. Wang, M.T., Degol, J.: Staying engaged: Knowledge and research needs in student engagement. Child Dev. Perspect. **8**(3), 137–143 (2014)
2. Davis, F.D.: A technology acceptance model for empirically testing new end-user information systems: Theory and results. Diss. Massachusetts Institute of Technology (1985)
3. Davis, F.D., Bagozzi, R.P., Warshaw, P.R.: User acceptance of computer technology: a comparison of two theoretical models. Manag. Sci. **35**(8), 982–1003 (1989)
4. Sweller, J.: The worked example effect and human cognition. Learn. Instr. **16**(2), 165–169 (2006)
5. Renkl, A., et al.: Learning from worked-out examples: the effects of example variability and elicited self-explanations. Contemp. Educ. Psychol. **23**(1), 90–108 (1998)

6. Retnowati, E., Ayres, P., Sweller, J.: Can collaborative learning improve the effectiveness of worked examples in learning mathematics? J. Educ. Psychol. **109**(5), 666 (2017)
7. Moreno, R.: When worked examples don't work: Is cognitive load theory at an impasse? Learn. Instr. **16**(2), 170–181 (2006)
8. Sun, H., Chen, C.C.: Well-designed teaching examples influence the outcome of technology acceptance: the example of next-generation art process learning. Sustainability **13**(23), 13124 (2021)
9. Peimani, N., Kamalipour, H.: The future of design studio education: student experience and perception of blended learning and teaching during the global pandemic. Educ. Sci. **12**(2), 140 (2022)
10. Bruggeman, B., et al.: Experts speaking: crucial teacher attributes for implementing blended learning in higher education. Internet High. Educ. **48**, 100772 (2021)
11. Mayer, R.E., Logan F., Andrew, S.: Five ways to increase the effectiveness of instructional video. Educ. Technol. Res. Dev. **68**(3), 837–852 (2020)
12. Bandura, A.: Self-Efficacy: The Exercise of Control. W.H. Freeman, New York (1997)
13. Kruger, J., Dunning, D.: Unskilled and unaware of it: how difficulties in recognizing one's own incompetence lead to inflated self-assessments. J. Pers. Soc. Psychol. **77**(6), 1121 (1999)
14. Ying, A.N.L., Yang, I.: Academics and learners' perceptions on blended learning as a strategic initiative to improve student learning experience. MATEC Web Conf. **87**, 04005 (2017). EDP Sciences
15. Prifti, R.: Self–efficacy and student satisfaction in the context of blended learning courses. Open learning. J. Open Distance e-Learning **37**(2), 111–125 (2022)

An Intelligent Thermal Compensation System Using Edge Computing for Machine Tools

Lu-Yan Wang[1], Jung-Chun Liu[1], Cheng-Kai Huang[2], Shih-Jie Wei[2], and Chao-Tung Yang[1,3](✉)

[1] Department of Computer Science, Tunghai University, No. 1727, Sec. 4, Taiwan Boulevard, Xitun District, Taichung 407224, Taiwan, ROC
{jcliu,ctyang}@thu.edu.tw

[2] Industrial Technology Research Institute, Sec. 4, Chung Hsing Rd., Chutung, Hsinchu 310401, Taiwan, ROC
{itriA10390,irtiA50317}@itri.org.tw

[3] Research Center for Smart Sustainable Circular Economy, Tunghai University, No. 1727, Sec. 4, Taiwan Boulevard, Xitun District, Taichung 407224, Taiwan, ROC

Abstract. This study explores the application of artificial intelligence in lathe cutting machine tools in smart manufacturing. Long-term processing will cause thermal deformation of the lathe cutting tool machine, which will cause displacement errors of the cutting head and damage to the final product. Using time-series thermal compensation, the research develops a predictive system that can be applied in industry using edge computing technology to predict the thermal displacement of machine tools. The study conducted two experiments to optimize the temperature prediction model and predict the five-axis displacement of the temperature point. Furthermore, a genetic algorithm is used to optimize the LSTM model to predict the thermal displacement of the machine tool. The results show that the GA-LSTM model achieved a thermal displacement prediction accuracy of 0.99, while the average accuracy of the LSTM, GRU, and XGBoost models was 0.97. Based on the analysis of training time and model accuracy, the study recommends using LSTM, GRU, and XGBoost models to design and apply to systems that use edge devices such as Raspberry Pi for thermal compensation.

Keywords: sensor · thermal compensation · time series model · edge computing

1 Introduction

With the rise of Industry 4.0, more and more manufacturing industries are moving towards the era of smart development. The emergence of edge computing (Edge Computing) that is closer to the source of data is mainly aimed at improving the problem of cloud computing, reducing latency, reducing dependence on the network, not excessively concentrating computing resources, and ensuring

J. C. Hung et al. (Eds.): FC 2023, LNEE 1134, pp. 96–102, 2024.
https://doi.org/10.1007/978-981-99-9342-0_11

data privacy sex, and so on [1]. In the error of machine tool processing, thermal error is one of the most influential factors. The traditional way of thermal compensation is to use coolant to cool the inside of the machine, but this method has not completely improved the thermal error of the machine tool [2].

Therefore, the objective of this research was to leverage the power of deep learning models to forecast the axis displacement of a tool machine, and integrate it into a lightweight device for edge computing. This setup offered a direct connection to the tool machine and enabled enhanced machining precision. The study was divided into two parts. Ultimately, a user interface was developed at the Raspberry Pi edge to facilitate the thermal compensation of tool machine data by end-users. The contributions of this article are summarized as follows.

- Two experimental sets were designed to study the time-series data of the turning tool machine.
- A genetic algorithm (GA) was developed to enhance the accuracy of the LSTM model in predicting thermal displacement of the tool machine.
- Multiple time-series models were compared for their effectiveness in compensating for thermal data of the tool machine.
- A system was developed to apply thermal compensation on the tool machine using edge computing with Raspberry Pi.

2 Related Research

2.1 Machine Tool Thermal Compensation

FANUC developed a new AI function that can collect control data of machine tool feed axes and spindles through high-speed sampling [3]. It performs deep learning on the collected data and displays anomaly scores based on the current state of machine components. DMG MORI An adaptive thermal displacement compensation method based on deep learning was developed, and a reliability evaluation method for thermal displacement prediction based on Bayesian dropout was proposed [4]. This method can not only adapt to changes in ambient temperature, but also adapt to cutting heat and working heat generated by spindle rotation or axial movement.

2.2 Time Series Model

Yuan, J et al. [5] proposed a three-stage fault diagnosis method using the Gated Recurrent Unit (GRU) network to carry out intelligent fault diagnosis on large data in the industry, and the results showed very good results. Yangdong He et al. [6]Applying TCN to anomaly detection of time series, they trained TCN on normal sequence and used it to predict the trend of multiple time steps, and the effectiveness was confirmed on three real-world datasets. S.M. Taslim Uddin Raju et al. [7] An ensemble model combining Bagging (Random Forest Regression (RFR)), boosting (Gradient Boosting Regression (GBR) and Extreme Gradient Boosting Regression (XGBR)) and stacking (STACK) to forecast steel industry demand one month in advance.

2.3 Edge Computing for Smart Manufacturing

S. Ren et al. [8]This paper presents the design and implementation of a big data platform for a smart IIoT sensor monitoring system with prior predictions of upcoming errors. This paper proposes a design and implementation scheme of a manufacturing big data platform and an intelligent industrial IoT sensor monitoring system based on edge computing and artificial intelligence.

3 Research Architecture and Related Algorithms

3.1 Research Framework

In this research, the time series model training is first carried out in the Windows environment, and then the selected model is placed in the Raspberry Pi environment to build the human-machine interface system, as shown in Fig. 1 below.

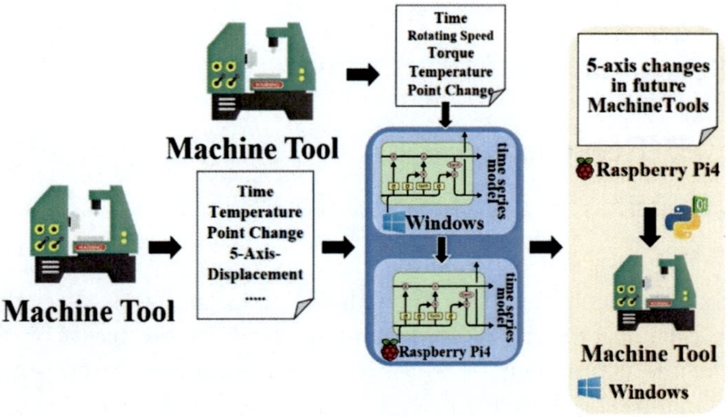

Fig. 1. Research flow chart of intelligent thermal compensation system for machine tools.

3.2 Related Algorithms

GA Optimized LSTM. This study combines GA with LSTM, and uses GA to optimize the parameters in the LSTM training process. The main parameters are the data time step (look_back, lb), the hidden layer of the LSTM model (lstm_nets, ls), the number of LSTM training (epochs, ep), and dropout (dp). These four parameters are optimized through GA. After obtaining the optimal parameter set, the five-axis displacement data set of the machine tool is used as the input data, and the predicted value of the five-axis displacement of the machine tool is used as the output matrix. Adapt to adjust the weight of the model, and finally combine it into a GA-LSTM model, and then train the data set through this model, and finally compare the predicted value with the real value, such as shown in Fig. 2 below.

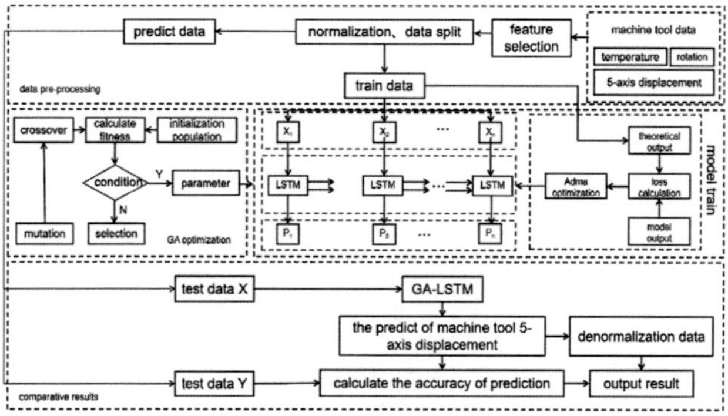

Fig. 2. GA-LSTM Architecture Diagram.

4 Experiment Introduction

4.1 Dataset Introduction

Experiment1. The acquisition of the data set for this research is mainly to collect field data on the machine tool in operation. This research captures the speed and torque of the machine tool in real time and saves it, and uses the infrared temperature sensor to monitor the temperature of the important points of the machine tool. Measure and record, the data set of this study mainly collects 4 temperature points, namely indoor temperature, condensing agent temperature, rotating shaft temperature and motor stator temperature. In this experiment, the changes of the four temperature points are recorded every minute, and the temperature points and the speed and torque of the machine tool are recorded through the production.

Experiment2. This data set is provided by the Industrial Technology Research Institute, mainly for the temperature change and 5-axis displacement of two machine tools under different conditions. Five-axis machining is a processing mode of digital machine tools, using X, Y, Z, For the linear interpolation motion of any 5 coordinates in A, B, and C, the machine tools used in five-axis machining are usually called five-axis machine tools or five-axis machining centers.

As shown in Table 1, the first machine tool mainly provides data sets in three situations, and it has 41 fields in total, which are time, rotational speed, 34 temperature points and displacement changes of 5 axes. The second machine tool provides data sets in four situations, and it has 60 fields in total, including time, rotation speed, 53 temperature points and displacement changes of 5 axes.

Table 1. Description of machine tool operating conditions

Machine Tool	Operating Conditions
Tool1	Spray water to heat 10 degrees
	Spindle 2350 RPM-turn 8 stop 2
	Water spray heating 10 degrees - spindle 2350 RPM - turn 8 stop 2
Tool2	Room temperature plus 15 degrees
	Room temperature plus 15 degrees - spindle 2350 RPM - turn 8 stop 2
	Room temperature plus 15 degrees - water spray heating 10 degrees
	Room temperature 20 degrees - spindle 2350RPM - turn 8 stop 2 - no cooling machine

5 Analysis of Experimental Results

5.1 Experiment 1 Results and Discussion

For the data set of Experiment 1, there are a total of 300 records. In this study, the data set will be divided into training set and test set at a ratio of 8:2 to train the model. This time, RMSE is mainly used to check whether the requirements are met, such as Fig. 3 shows.Both the results of the GRU model and the LSTM model are evaluated at a lower level, which means that the results predicted by these two AI models are more accurate than the actual results.

Although most of the experimental results of the TCN model are also in the case of RMSE less than 1, compared with the LSTM and GRU models, the results are not good, and sometimes the loss value during training will not converge. If this model is used, it will Increased forecast instability. In addition, although the final prediction effect of the Stacking Ensemble Learning algorithm is generally good, this paper finds that if different rotation speeds are given for prediction, its results will vary greatly. Therefore, if the machine tool is working at a fixed speed, you can consider using this AI model, because its training speed and final accuracy are at a better level, but if the machine tool needs a large range of speed changes If so, it is not recommended to use this model.

5.2 Experiment 2 Results and Discussion

For the data set of Experiment 2, there are a total of 6433 data sets for machine tool 1 and 7423 data sets for machine tool 2. This research will divide each data set into a training set and a test set at a ratio of 8:2 to train the model. In this study, the data set of machine tools was screened according to the eight temperature points before the sensitivity analysis, and the same data set was put into the established deep learning model for training. The evaluation index

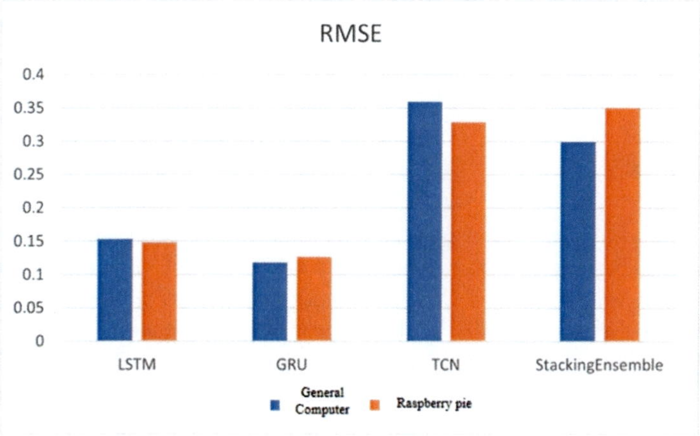

Fig. 3. Histogram of AI model result evaluation in Experiment 1.

of the model prediction results used the coefficient of determination (R2score), in the regression model, this coefficient mainly reflects the accuracy comparison between the predicted value of the model and the actual value. The larger the coefficient, the higher the prediction accuracy of the model, and the value ranges from 0 to 1, the calculation formula is asDownSurface Eq.(2–3).

$$S_n = \frac{X_1 + X_2 + \cdots + X_n}{n} \tag{1}$$

$$S_n = \frac{1}{n} \sum_{i}^{n} X_i \tag{2}$$

As shown in Fig. 4 below, it can be found that under the uniform usage of the same dataset and the same hardware environment, the research results of

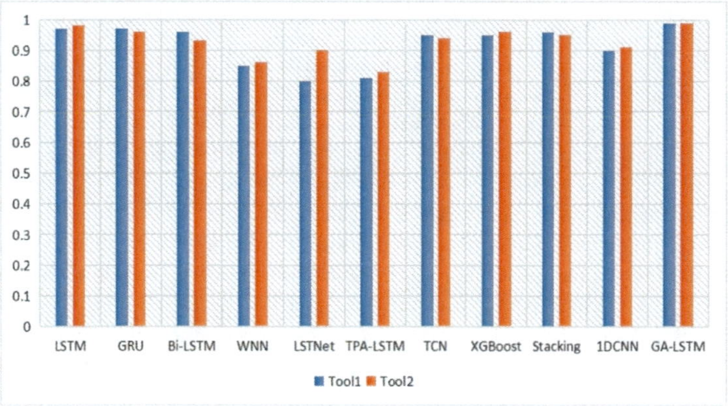

Fig. 4. Histogram of model R2score results of experiment 2.

this experiment can be found that the time series correlation models of this R2 score all reach above 0.8, and the GA optimized LSTM developed in this study also showed good results, with an accuracy of 0.99.

6 Conclusion

Based on the above two experimental studies, it is found that the final prediction results of putting the machine tool data set in different time series models are good, and the GA-optimized LSTM model developed in the second experiment also achieved the best results. In the future, this research will continue to optimize the problem of too long training time of GA-LSTM, and will also analyze the multiple temperature points of the machine tool thermal compensation data set with more algorithms, and hope that the temperature point can be selected in the future A number of recommended features are added to the smart thermal compensation system that has been developed on the Raspberry Pi so far. According to the research results, this study finally selected three AI models to be put into the edge computing terminal for development, and the effect also showed a good state. In addition, we modified the operating system of the system in this study so that it can also be downloaded and used on the machine cloud of the Industrial Technology Research Institute. Finally, the accuracy of the model in this study reached above 0.96, which is in line with the current situation that the manufacturing industry uses deep learning to improve processing thermal errors.

References

1. Chen, B., et al.: Edge computing in IoT-based manufacturing. IEEE Commun. Mag. **56**(9), 103–109 (2018)
2. Ezugwua, E.O., et al.: Int. J. Mach. Tools Manufact. **45**, 1009–1014 (2005)
3. FANUC Corporation,“FANUC's new AI functions that utilize machine learning and deep learning”, https://www.fanuc.co.jp/en.
4. Su, C.: Discussion on DMG Thermal Displacement Compensation Technology. http://www.maonline.com.tw/article_inside.php
5. Yuan, J., Tian, Y.: An intelligent fault diagnosis method using GRU neural network towards sequential data in dynamic processes. Processes **7**(3), 152 (2019)
6. He, Y., Zhao, J.: Temporal convolutional networks for anomaly detection in time series. J. Phys.: Conf. Series. **1213**(4), 042050 IOP Publishing (2019)
7. Raju, S.T.U., et al.: An approach for demand forecasting in steel industries using ensemble learning. Complexity **2022** 1–19 (2022)
8. Song, W., Shigeru, F.: Sensor data prediction in process industry by capturing mixed length of time dependencies. In: 2021 IEEE International Conference on Industrial Engineering and Engineering Management (IEEM). IEEE (2021)

Cluster-Based Blockchain Systems for Multi-access Edge Computing

Chih Peng Lin(✉) and Hui Yu Fan

National Taipei University, No. 151, University Rd., Sanxia Dist., New Taipei City 237303, Taiwan (R.O.C.)
chih.p37@msa.hinet.net
http://www.users/iekeland/web/welcome.html

Abstract. The computing power and storage requirements of the Internet of Things (IoT) are likely to increase substantially in the future years. Because of the rapid development of both machine learning (ML) and the Internet of Things (IoT), vast volumes of data created by edge devices such as smartphones, laptops, and artificial intelligence (AI) speakers have been widely used to train ML models. In this study, we used a cluster-based Blockchain method in the Multi-Access Edge Computing (MEC, also known as Mobile Edge Computing) for markets and technological services. We describe a generalized stochastic block model (SBM) for edge computing applications based on the proposed taxonomy. These mobile edge wireless devices (WD) provide efficient resource allocation in mobile network situations. In our studies, we compared the approximate solutions obtained by the SBM to those generated by the cluster-based Blockchain algorithm. However, the high latency and low scalability of traditional blockchain systems limit mobile transactions on the public blockchain. To reduce the consumption of competitive mobile transactions created by linear sequencing blocks, reconstructed blockchain systems have been developed. This study's use of cluster-based blockchain systems provides speedy confirmation and great scalability without significantly compromising security.

Keywords: Machine learning (ML) · cluster-based Blockchain method · Internet of Things (IoT) · stochastic block model (SBM) · Multi-Access Edge Computing (MEC)

1 Introduction

Cluster-based blockchain edge computing is a cutting-edge technology that merges three developing fields: blockchain, edge computing, and cluster computing. Blockchain is a distributed ledger system that enables numerous parties to share a single source of truth without the need for a central authority. Edge computing is a computing paradigm that brings computation and data storage closer to the devices that generate and consume data in order to reduce latency, bandwidth utilization, and reliance on cloud computing. Cluster computing is a

J. C. Hung et al. (Eds.): FC 2023, LNEE 1134, pp. 103–114, 2024.
https://doi.org/10.1007/978-981-99-9342-0_12

technique that allows multiple computers to work together as a cluster to achieve higher performance, availability, and scalability [15]. Cluster-based blockchain edge computing aims to combine the benefits of these three technologies to create a decentralized and efficient computing infrastructure that can support a wide range of applications, from IoT devices to artificial intelligence (AI) algorithms. One potential application of cluster-based blockchain edge computing is in the field of smart cities, where a large number of IoT devices generate data that needs to be processed in real-time. By using a cluster-based blockchain edge computing architecture, smart cities can create a decentralized and secure infrastructure that allows devices to exchange data and compute tasks without relying on a central authority. Another potential application is in the field of AI, where large-scale machine learning models require massive amounts of data and computation. By using a cluster-based blockchain edge computing architecture, AI algorithms can distribute the computation and storage across multiple nodes in a decentralized and fault-tolerant manner, while preserving data privacy and security. This study adopts cluster-based blockchain edge computing, which has the potential to alter the way we build and implement distributed computing systems by providing a flexible, scalable, and secure architecture that can support a wide range of applications and use cases.

Optimally allocating resources in mobile networks is a complex problem that requires balancing the competing demands of various stakeholders, such as users, operators, and service providers. Traditional approaches to resource allocation have relied on centralized control and decision-making, which can be slow, inefficient, and vulnerable to single points of failure. Cluster-based blockchain technology provides a decentralized and secure architecture for managing mobile network resources, making it a possible alternative to traditional resource allocation methodologies. Cluster-based blockchain networks can ensure that resource allocation decisions are made in a distributed and transparent manner without relying on a central authority by employing a blockchain-based consensus process (build an SBM). Mobile network resources can be allocated in a cluster-based blockchain network utilizing smart contracts, which are self-executing computer programs that can autonomously enforce the rules and norms controlling resource allocation. For example, a smart contract could specify the terms of a mobile data plan, such as the amount of data allocated per user, the price of the plan, and the duration of the plan. Cluster-based blockchain networks can also leverage edge computing resources to optimize resource allocation in mobile networks. By using edge computing resources, such as computing power and storage capacity at the network edge, mobile network operators can reduce latency, improve network performance, and enhance the user experience. Overall, cluster-based blockchain technology offers a promising approach to optimally allocating resources in mobile networks. By providing a decentralized and secure infrastructure for managing network resources, cluster-based blockchain networks can enhance network performance, improve user experience, and increase efficiency and transparency in mobile network operations.

Optimally allocating resources in current mobile networks offers three significant issues. Developing strategies for optimal model-based heterogeneous in a 5G and beyond environment based on limited game-based resource allocation schemes. [1]-[3], as well as effective heuristics [6], Machine learning for wireless communications has progressed rapidly since its introduction. In this article, we'll look at three approaches to tailoring deep learning for mobile network applications: mobile data creation, end-to-end Cloud-Edge wireless communications, and network traffic control that can adapt to changing mobile network environments., [4]-[5]. A primal-dual approach for learning resource allocations in wireless networks via low-dimensional action utilizing a zeroth-order deterministic two-point gradient approximation scheme;see,e. g., [7]-[9] space exploration. We analyze the key concerns, techniques, and various state-of-the-art attempts connected to the offloading and task placement QoS Scheduling challenges from a survey-related study. We use a new characterizing network model to investigate the entire job placement offloading policy from mobile devices to the edge cloud. To meet the requirements of practical applications such as robotics and autonomous vehicles, transportation management systems, healthcare, as well as telepresence, virtual reality (VR), augmented reality (AR), and mixed reality (MR), 6G edge computing mobile networks will require massive internet of things connectivity, ultra-reliability, low latency, and extreme high bandwidth. This Fig. 1 edge cloud or edge computing, on the other hand, is a novel concept and technology that can address current cloud computing problems, such as the time it takes to relay information to a centralized data center, which delays decision making. An edge computing solution involves physically relocating computational resources closer to the source of the data, which is typically an IoT device or sensor application. Edge computing removes the need for large amounts of data to be transmitted between servers, clouds, and devices or edge locations to be processed by processing data at the network's edge. Four types of services are deployed from various state-of-the-art MEC, as follows [2]:

- Infrastructure as a service (IaaS) is a type of cloud computing service that offers pay-as-you-go compute, storage, and networking resources on demand. IaaS is one of four types of cloud services, along with software as a service (SaaS), platform as a service (PaaS), and function-as-a-service (serverless).
- Cloud computing services that provide an on-demand environment for designing, testing, delivering, and maintaining software applications are known as platform as a service. PaaS is designed to help developers build web or mobile apps rapidly without having to worry about setting up or managing the underlying infrastructure of servers, storage, networks, and databases.
- Software as a service (SaaS) is a method of delivering software applications internet on demand, typically by subscription. In the case of SaaS, the cloud server and administration of the software application and supporting infrastructure. These servers are also in charge of maintenance tasks including software upgrades and security fixes. Clients gain access to the program over the internet, generally using a web browser on their phone, tablet, or PC.

- FaaS, or Function-as-a-Service, is a cloud computing service that enables clients to execute code in response to events without having to manage the extensive infrastructure that is generally associated with developing and deploying micro-services applications [3].

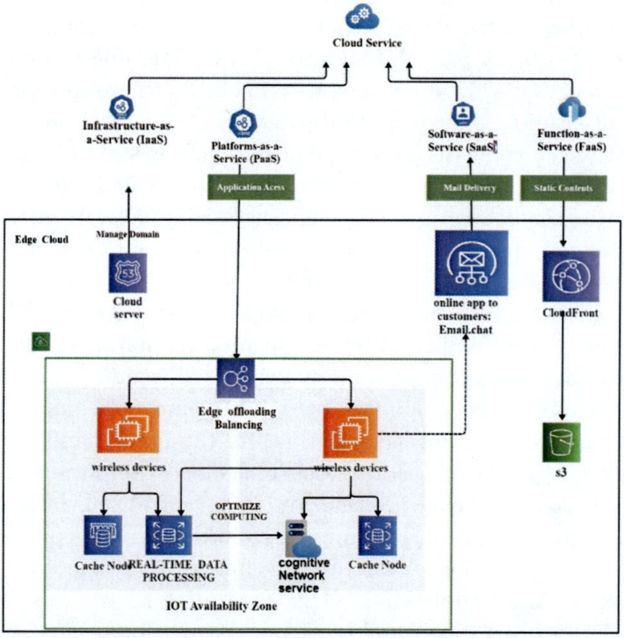

Fig. 1. MEC Infrastructure diagram in 6G network.

The manuscript is organized as follows. Section 2 provides background on MEC optimum resource allocation and the Spectral Graph Theory Concept for Cluster-Based Blockchain Infrastructure. Section 3 presents the design details of Stochastic block model (SBM) for MEC. The prototype implementation and the experimental processing are presented in Sect. 4, as well as the results and data analysis. Our considerations and future works are listed on Sect. 5.

2 Spectral Graph Theory Concept for Cluster-Based Blockchain

In this section, we will show how Cluster-Based Blockchain works with associated matrices such as the adjacency matrix and graph Laplacian. Let $G(V, |E|)$ be a graph. We'll let $n = |V|$ denote the number of vertices/mobile nodes, and $m = |E|$ denote the number of edges. We'll assume that vertices are indexed by $0, \ldots, n-1$, and edges are indexed by $0, \ldots, m-1$. The adjacency matrix** A

is a $n \times n$ matrix with $A_{i,j} = 1$ if $(i, j) \in |E|$ is an edge node, and $A_{i,j} = 0$ if $(i, j) \notin |E|$. If G is an undirected graph, then A is symmetric. If G is directed, then A need not be symmetric. The degree of a node i, $deg(i)$ is the number of neighbors of i, meaning the number of edges which i participates in. You can calculate the vector of degrees (a vector d of length n, where $d_i = deg(i)$), using matrix-vector mulpilication:

Lemma 1. *(Matrix-vector multiplication):*
Given a matrix $A \in d_{i \times j}$
vector of degrees: $A \in d_i$
A and x matrix-vector multiplication is defined as
$d = A\ x$

where x is the vector containing all 1 s of length n. You might alternatively simply add the row entries of all matrix A. We will also use $D = diag(d)$ - a diagonal matrix with $D_{i,i} = d_i$. The incidence matrix B is a $n \times m$ matrix which encodes the relationship between edges and vertices. Let $|E|_k = (i, j)$ be an edge. Then the k-th column of B is all zeros except $B_{i,k} = -1$, and $B_{j,k} = +1$ (for undirected graphs, it doesn't matter which of $B_{i,k}$ and $B_{j,k}$ is $+1$ and which is -1 as long as they have opposite signs). Note that B^T acts as a sort of difference operator on functions of vertices, meaning $B^T f$ is a vector of length m which encodes the difference in function value across all edge nodes. You can check that $B^T x_C = 0$, where x_C is a connected component indicator ($x_C[i] = 1$ if $i \in C$, and $x_C[i] = 0$ otherwise). $C \subseteq V$ is a connected component of the graph if all vertices in C have a path between them, and there are no vertices in V that are connected to C which are not in C. This implies $B^T 1 = 0$. The **graph Laplacian** L is an $n \times n$ matrix $L = D - A = BB^T$. If the graph lies on a regular grid, then $L = -\Delta$ up to scaling by a finite difference width h^2, but the graph Laplacian is defined for all graphs. Note that the null-space of L is the same as the null-space of B^T (the span of indicators on connected components). In Fig. 2, it makes sense to store all these matrices in sparse format. Spectral embeddings are one way of obtaining locations of vertices of a graph for visualization. One way is to pretend that all edges are Hooke's law springs, and to minimize thepotential energy of

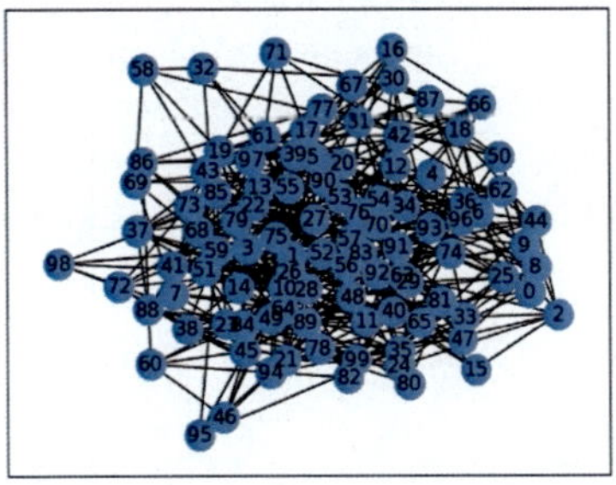

Fig. 2. The graph laplacian of 100 mobile nodes when cluster converge method.

a configuration of vertex locations subject to the constraint that we can't have all points in the same location. In one dimension:

$$\underset{x}{\text{minimize}} \sum_{(i,j)\in|E|} (x_i - x_j)^2$$

$$\text{subject to}$$

$$x^T 1 = 0, \|x\|_2 = 1$$

Note that the objective function is a quadratic form on the embedding vector x:

$$\sum_{(i,j)\in|E|} (x_i - x_j)^2 = x^T BB^T x = x^T Lx \tag{1}$$

Because the vector 1 is in the nullspace of L, this is similar to locating the eigenvector with the *second-smallest* eigenvalue. We can use the eigenvectors for the next-largest eigenvalues for a higher-dimensional embedding. Spectral Graph Theory is the study of graphs, which are mathematical structures used to model relationships between objects. Spectral Graph Theory focuses on the eigenvalues and eigenvectors of the graph's adjacency matrix, which can provide insight into the graph's properties. For example, spectral graph theory can be used to analyze the connectivity and clustering of a graph. Spectral clustering refers to using a spectral embedding to cluster nodes in a graph. Let A, B$\subset$V with $A \cap B = \phi$ We will denote

$$E(A, B) = \{(i, j) \in |E| \mid i \in A, j \in B\} \tag{2}$$

One way to try to find clusters is to attempt to find a set of nodes $S \subset V$ with $\bar{S} = V \setminus S$, so that we minimize the cut objective

$$C(S) = \frac{|E(S, \bar{S})|}{\min\{|S|, |\bar{S}|\}} \tag{3}$$

The inequality bounds the second-smallest eigenvalue of L in terms of the optimal value of $C(S)$. In fact, the way to construct a partition of the graph which is close to the optimal clustering minimizing $C(S)$ is to look at the eigenvector x associated with the second smallest eigenvalue, and let $S = \{i \in V \mid x_i < 0\}$. As Fig. 3, let's look at a graph generated by a stochastic block model with two clusters. The "ground-truth" clusters are the ground-truth communities in the model. As Fig. 4,we obtained. A value of 1 means that we found the true clusters.

3 Stochastic Block Model (SBM) and Cluster-Based Blockchain

This study's SBM structure is a mathematical model used for assessing network structure and community detection, whereas a cluster-based blockchain is a concept that combines clustering with blockchain technology to improve scalability

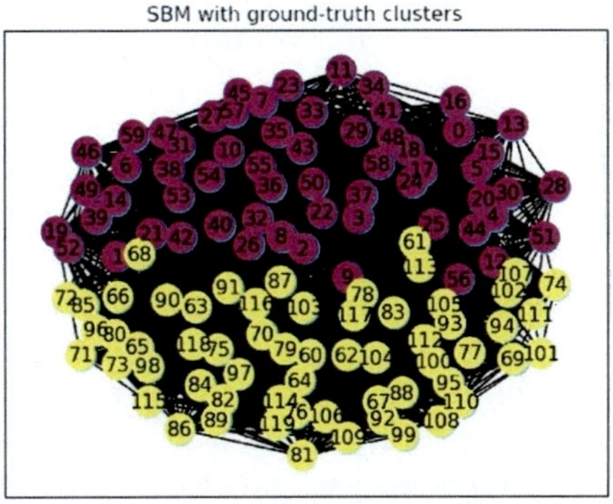

Fig. 3. The graph Laplacian of 100 mobile nodes when spectral clustering to partition into two clusters converge method.

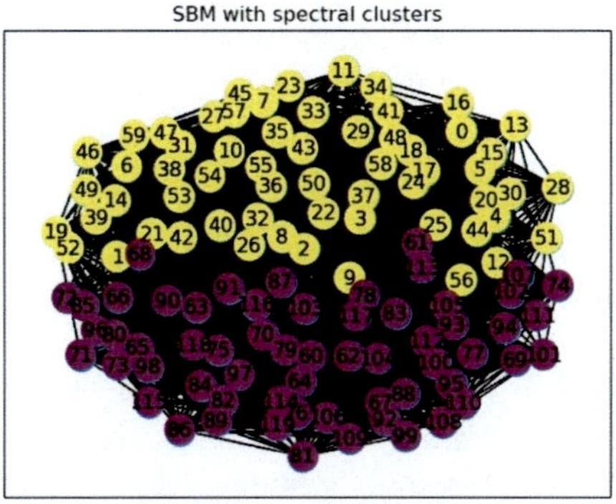

Fig. 4. The graph Laplacian of 100 mobile nodes when the adjusted rand index to measure the quality of the clustering.

and efficiency. It is assumed that nodes in a mobile network are organized into groups or communities, and that edges between nodes are formed based on probabilities that rely on the nodes' community assignments. Assume the following assumptions here: n - the number of mobile nodes in the graph N - $n \times n$ adjacency matrix A - $n \times n$ matrix of probabilities Many statistical network models lie under the umbrella of independent edge random networks, also referred to as the Inhomogeneous Erdos-Renyi (IER) model. The elements of the network's adjacency matrix A are sampled individually from a Bernoulli distribution in this model:

$$A_(i,j) \approx \ Bernoulli(P_{i,j}) \tag{4}$$

If n is the number of mobile nodes, the matrix P is a n*n matrix of probabilities with elements in [0.1] . We can design a variety of specialized models depending on how the matrix P is created. We will now go over a few of these options. It is worth noting that for each model, we assume that there are no loops, or that the diagonal of the matrix P is always set to zero. Each node in the stochastic block model (SBM) is modeled as belonging to a block(sometimes called a community or group). The probability of node i connecting to node j is just a function of the two mobile nodes' block membership. Let n be the number of nodes in the graph, then τ is a length n vector which indicates the block membership of each node in the graph. Let K be the number of blocks, then B is a $K \times K$ matrix of block-block connection probabilities.

$$P_(i,j) = B_{\tau_i \tau_j} \tag{5}$$

In the stochastic block model (abbreviated SBM), we have graphs of the form $G(n, p, q)$. For clarity, consider the following:

Assumption 1 *The class $\mathcal{C}$ is not empty.*
let's assume that n is even and $p > q$

In this paradigm, there are two "communities" of varying sizes $n/2$ so that the probability of an edge existing between any two nodes within a community is p and the probability of an edge between the two communities is q This recovers the communities from a random graph realization $G(V, |E|)$.

The Inhomogeneous Erdos-Renyi model is very simple and lacks many of the properties of networks in real scenarios. It is only a mathematical object with similar phase transition effects. In this study, no communities establish between nodes. An SBM computing for cluster-based blockchain was developed in this study; the majority of these scenarios' MEC models use its variants. Each node in its most mobile nodes belongs to one of C communities, and the occurrence of an edge between two nodes is an event that is independent of the other edges and the probability $\mathbf{Q}_{c_i,c_j}$ (with$\mathbf{Q} \in R^{C\times C}$ definite probability matrix and c_i,c_j node communities i,j respectively).

A graph containing two communities is created by the following cell. Although nodes within the same community have strong connections, nodes within different communities have less connections. Experiment with the two accessible parameters here: 'n' and 'Q'.

This research analyses the qualitative difference between Q with all positive eigenvalues with Q with some negative eigenvalues using two communities to simplify the visualisation. For example, consider the following parameters: 'n=[45, 5, 45, 5]' and $Q = \begin{pmatrix} 0.05 & 0.9 & 0 & 0 \\ 0.9 & 0.8 & 0 & 0.5 \\ 0 & 0 & 0.05 & 0.9 \\ 0 & 0.5 & 0.9 & 0.9 \end{pmatrix}$. How many communities are there about $SBM_n(z, B)$. We know from the graphs that nodes $(0, \frac{n}{2} - 1)$ belong to community A, whereas nodes $(\frac{n}{2} - 1, n)$ belong to community B.

Corollary 1. *Let $p = \alpha \log(n)/n$*
and $q = \beta \log(n)/n$
If:
simulate the probability of exact recovery when

$$\frac{\alpha+\beta}{2} - \sqrt{\alpha\beta} > 1$$

then do the same for

$$\frac{\alpha+\beta}{2} - \sqrt{\alpha\beta} < 1 \qquad (6)$$

4 Results and Discussion

Figure 5 depicts the adjacency matrix for the example, where black and white indicate 1 and 0, respectively. Graphs with binary adjacency matrices are referred to as binary graphs from now on. In the SBM, Fig. 7,n=1000 and each node belongs to one of the $K(< n)$ groups, where K = 2 in the example. Because the groups are unknown before to modelling, a K-vector Z_p is also defined for node $p = 1, 2, \ldots, n$, with all elements 0 except one that takes the value 1 and reflects the group node p belongs to in SBM(z, B). In Fig. 7 network with 'n=[60,60] nodes and block matrix use the following parameters' $Q = \left(\begin{smallmatrix} 0.5 & 0.2 \\ 0.2 & 0.5 \end{smallmatrix}\right)$. In order to describe the generation of the edges of G according to the groups the nodes belong to, a 2×2 cluster block matrix, denoted by C, is introduced. If Fig. 6 G is k-means clustering-based blockchain, for $1 \leq i \leq j \ll K, C_{ij} \epsilon [0, 1]$ and represents the probability of occurrence of an edge between a node in group i and a node in group j. Spectral Graph Theory can be used to analyze the structure of blockchain networks, and to identify nodes that are particularly important for maintaining the network's integrity. Edge computing can be used to improve the performance of blockchain networks by reducing the amount of data that needs to be transmitted over the network. Additionally, edge computing can be used to perform computations related to Spectral Graph Theory, such as the calculation of graph Laplacians, which can be useful for machine learning and other applications. Let p be denote the probability of an edge between nodes in the same cluster, and q denote the probability of an edge between nodes in different clusters. This Study consider Spectral Graph Theory for stochastic block model with $k = 2$ clusters and $n = [60, 60]$ nodes per cluster. Figure 8 Analysis Adjacency spectral embedding when mobile nodes Histogram and Fig. 9 scatter diagram for 2 communities distribution state. Plot a phase diagram of the adjusted rand index (ARI) where p and q are both in the range $[0, 1]$. The Random Dot Product Graph (RDPG) can also be used in blockchain analysis or modelling. The RDPG is utilised in this study to describe relationships or interactions between distinct entities in a blockchain network. A blockchain is a distributed ledger in which transactions are recorded by several nodes or participants. Each transaction may involve a variety of entities, including users, addresses, and smart contracts. The RDPG can help capture and forecast the underlying structure and dynamics of the blockchain network by representing these entities as nodes and their interactions as edges in a graph. To create an RDPG for blockchain analysis, latent vectors or characteristics can be associated with each entity in the blockchain. These latent vectors can reflect a variety of entity traits or properties, such as transaction history, account balances, or network behaviours. In Fig. 10, the dot product of the latent vectors of $k = 5$ clusters under pairwise distance entities can then be utilized to determine the likelihood of an interaction or connection between them. If the dot product, for example, exceeds a specific threshold, an edge can be constructed between the respective nodes in the RDPG. After constructing the RDPG, several graph analysis techniques can be used to acquire insights into the blockchain network. In Fig. 11 $k = 6$ clusters, RDPG can incorporate community discovery, centrality metrics, clustering, and anomaly detection, among other things. Using the RDPG framework, researchers and predictive analysts can investigate the structural characteristics and behaviors of the blockchain network and perhaps find patterns or anomalies that may be useful for understanding its dynamics or detecting fraudulent activity. It is crucial to note that the application of the RDPG to blockchain analysis is still an evolving MEC scenario, and there are numerous computers and methodologies that can be applied depending on the specific cloud edge computing for vehicle and transport.

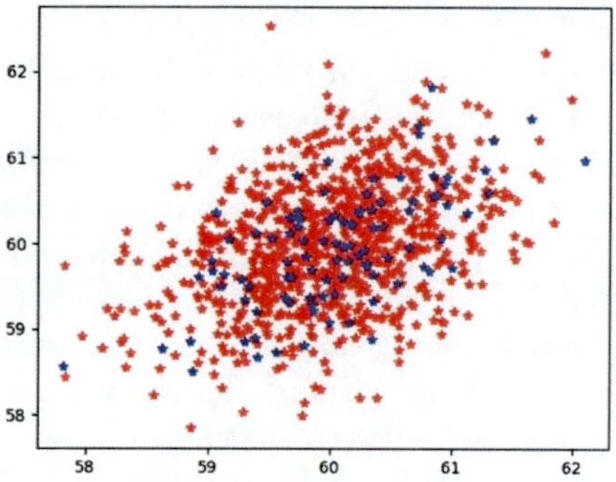

Fig. 5. The graph execution time and energy consumption of each mobile nodes when cluster converge method..

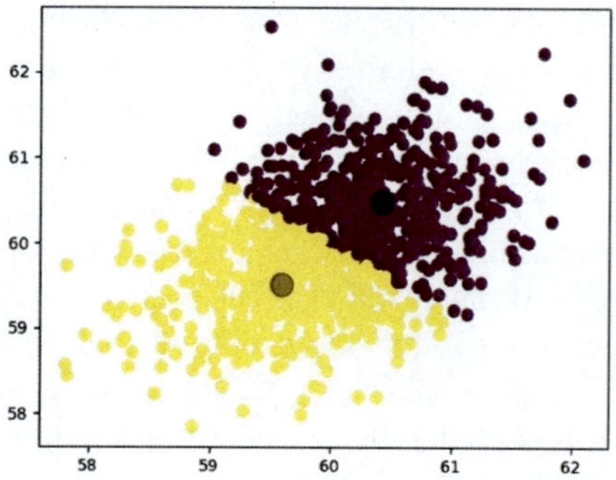

Fig. 6. The graph execution time and energy consumption of each mobile nodes when kmean cluster method..

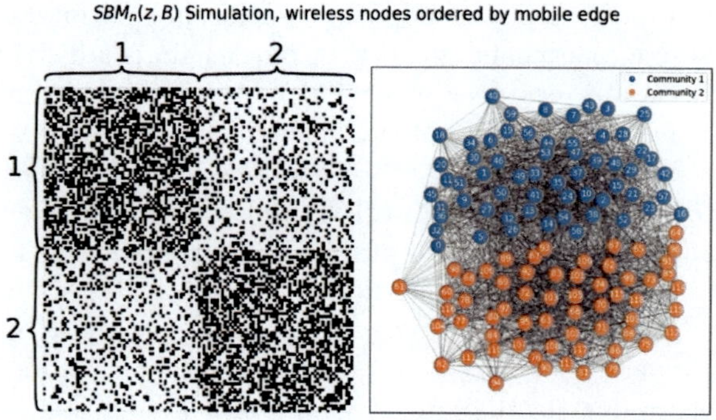

Fig. 7. The graph execution time and energy consumption of each mobile nodes when stochastic Block Model method.

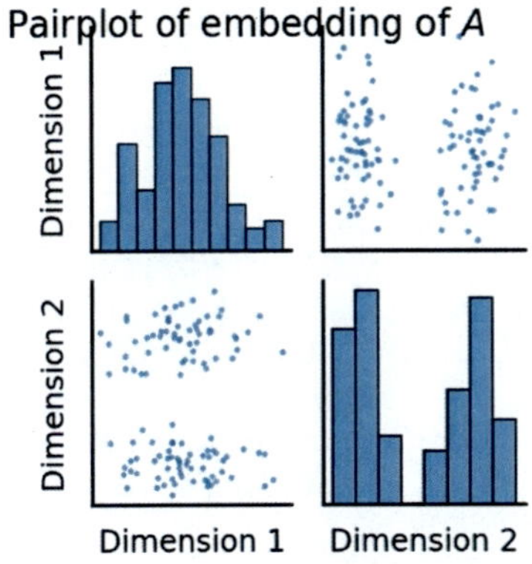

Fig. 8. Analysis Adjacency spectral embedding when mobile nodes Histogram for 2 communities.

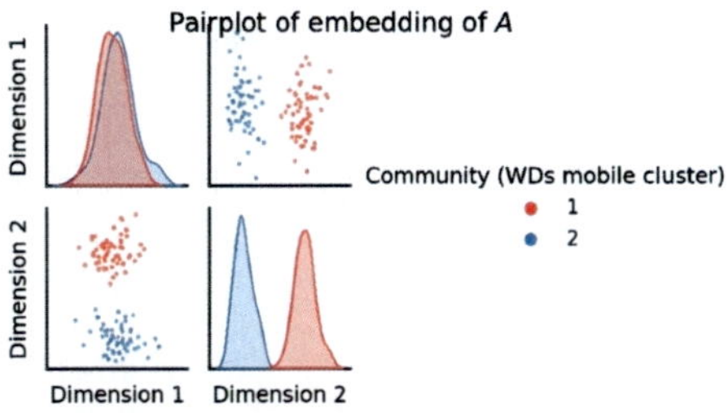

Fig. 9. Analysis Adjacency spectral embedding when mobile nodes scatter diagram for 2 communities.

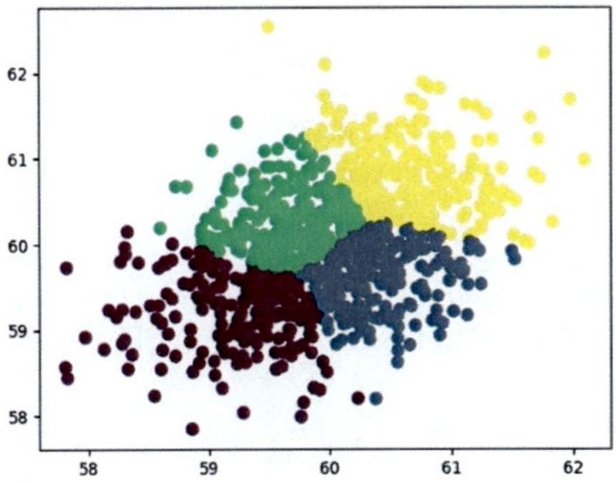

Fig. 10. The graph execution time and energy consumption of each mobile nodes when pairwise distance mode.

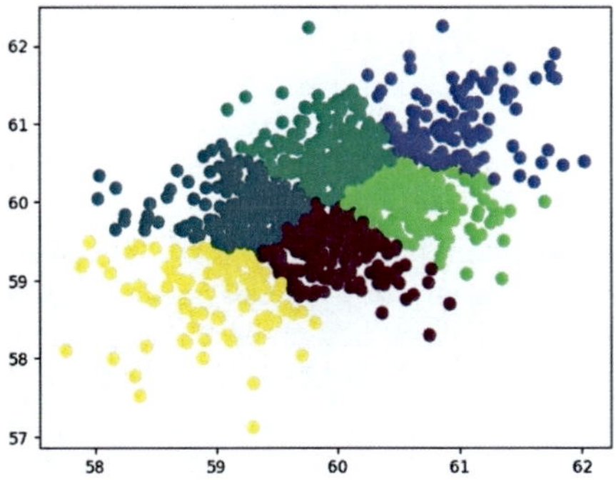

Fig. 11. The graph execution time and energy consumption of each mobile nodes when predict method.

5 Conclusion and Future Work

This investigation Multi-Access edge computing assisted wireless device (IoT) offloading scheme communications is a key component of the future 6G scenario. In this paper, we offer a new SBM-based method for cluster-based Blockchain optimization of transmit reinforcement and resource estimation in a 6G communication system. The technique used by the MEC system while also meeting the greatest transmit power restriction. The simulation results suggest that the proposed offloading technique can reduce the cumulative rate of MEC communication in a short period of task time (CPU time) when compared to the real-world scheme with fixed transmit mobile cloud computing. In the future, we will examine optimal allocation of MEC to IOT using a matching algorithm, as well as deep learning-based design of a 6G cloud integration environment.

References

1. Feng, W., Li, X.: Game-based resource allocation mechanism in B5G HetNets with incomplete information. Appl. Sci. **10**, 1557 (2020)
2. Fossati, F., Hoteit, S., Moretti, S., Secci, S.: Fair resource allocation in systems with complete information sharing. IEEE/ACM Trans. Netw. **26**, 2801–2814 (2018)
3. Xie, R., Wu, J., Wang, R., Huang, T.: a game theoretic approach for hierarchical caching resource sharing in 5G networks with virtualization China Commun. **16**(7), 32–48 (2019)
4. Wu, H., Li, X., Deng, Y.: Deep learning-driven wireless communication for edge-cloud computing: opportunities and challenges. J. Cloud Comput. **9**(1), 1–14 (2020). https://doi.org/10.1186/s13677-020-00168-9
5. Ting, X., Zhao, M., Yao, X., Zhub, Y.: An improved communication resource allocation strategy for wireless networks based on deep reinforcement learning. J. Cloud Comput.: Adv., Syst. Appl. **188**, 90–98 (2022)
6. Poongodi, M., et al.: 5G based blockchain network for authentic and ethical keyword search engine. IET Commun. **16**(1), 1–7 (June 2021)
7. Kalogerias, D.S., Eisen, M., Pappas, G.J., Ribeiro, A.: Model-free learning of optimal ergodic policies in wireless systems. IEEE Trans. Signal Process. **68**, 6272–6286 (2020)
8. Eisen, M., Zhang, C., Chamon, L.F., Lee, D.D., Ribeiro, A.: learning optimal resource allocations in wireless systems. IEEE Trans. Signal Process. **67**(10), 2775–2790 (2019)
9. Hashmi, H., Kalogerias, D.S.: Model-free learning of optimal deterministic resource allocations in wireless systems via action-space exploration. In: IEEE 31st International Workshop on Machine Learning for Signal Processing (MLSP), ISSN:1551-2541, pp. 2775–2790, (Oct 2021)
10. Huang, W., Liu, Y., Chen, Y., Waterman, M.S.: Mixed membership stochastic blockmodels for heterogeneous networks. Bayesian Anal. **15**(3), 711–736 (2020). https://doi.org/10.1214/19-BA1163
11. Zhao, W., Jin, S., Yue, W.: A stochastic model and social optimization of a blockchain system based on a general limited batch service queue. J. Indust. Manage. Optim. AIMS, LLC **17**(4), 1845–1861 (2021). https://doi.org/10.3934/jimo.2020049
12. Maleš, U., Ramljak, D., Krüger, T.J., Davidović, T., Ostojić, D., Haridas, A.: Controlling the difficulty of combinatorial optimization problems for fair proof-of-useful-work-based blockchain consensus protocol. Symmetry **15**(1), 140–172. MDPI, Basel (2023). https://doi.org/10.3390/sym15010140
13. Lekshmi, S.N., Swaminathan, J., Sai Pavan, K.N.: An improved link prediction approach for directed complex networks using stochastic block modeling. Big Data and Cogn. Comput. **7**(1), 31–49. MDPI, Basel (2023). https://doi.org/10.3390/bdcc7010031
14. May, P., Ehrlich, H.-C., Steinke, T.: ZIB structure prediction pipeline: composing a complex biological workflow through web services. In: Nagel, W.E., Walter, W.V., Lehner, W. (eds.) Euro-Par 2006. LNCS, vol. 4128, pp. 1148–1158. Springer, Heidelberg (2006). https://doi.org/10.1007/11823285_121
15. May, P., Ehrlich, H.-C., Steinke, T.: AI-enabled blockchain consensus node selection in cluster-based vehicular networks. IEEE Network. Lett. **5**(2), 115–119. IEEE (2023). https://doi.org/10.1109/LNET.2023.3238964

Scanning QR Codes for Object Detection Based on Yolo-V7 Algorithm and Deblurring Generative Adversarial Network

Huan Chen[1], Hsin-Yao Hsu[1(✉)], Kuan-Ting Lin[1], Jia-You Hsieh[1], Yi-Feng Chang[1], and Bo-Chao Cheng[2]

[1] National Chung Hsing University, Taichung City 40227, Taiwan
huan@nchu.edu.tw, {8109056003,roger.hs}@smail.nchu.edu.tw
[2] National Chung-Cheng University, Chia-Yi 621301, Taiwan
bcheng@ccu.edu.tw

Abstract. Location-based advertising (LBA) has been popular for several years, and the amount of global investment is increasing year by year. Nowadays, in the vigorous development of vehicle vision systems, many recognition tasks can be completed by combining You Only Look Once version 7 (Yolo-v7) object detection algorithms to apply automotive applications, and also involve a QR codes decoding method with deblurring generative adversarial network version 2(DeblurGAN-v2), which can capture the QR codes set on the route in real-time to obtain the LBA placed by the merchant, the results show that the proposed method outperforms the other object detection model and deblurring model, it obtains more efficient for scanning QR codes.

Keywords: Location-based advertising · QR codes · Object detection · Deblurring generative adversarial network

1 Introduction

LBA is a targeted advertising approach that delivers ads based on local cultural practices, ranging from static advertising signs on the roadside to mobile devices. According to [1], it shows that LBA traffic has a higher value than the application filed and investment amounts are expanding year by year. Moreover, the global information market research [2] shows that the market size of LBA is based on geographic positioning.

Nowadays, the value of LBA reveals in the area of mobile service. Yu et al. [3] revealed that the roadside for mobile servers from providers by placing targeted advertisements along the itinerary between the point of passenger boarding and their destination, advertisers can increase their revenue by leveraging local cultural characteristics. However, the service model requires pre-setting the driving route and placing advertisements along the road, which limits the advertisement placement to the designated route.

The demand for systems that support artificial intelligence (AI) increases, it revealed the advanced driver assistance systems (ADAS), which are very popular and the trend

J. C. Hung et al. (Eds.): FC 2023, LNEE 1134, pp. 115–126, 2024.
https://doi.org/10.1007/978-981-99-9342-0_13

of the future. The report shows in [4] from the market size was growing at a compound annual growth rate (CAGR) of 13.8%. The literature in [5, 6] used computer vision for deep learning architecture for the competition of driving needs.

The problem in ADAS system for LBA communication technology, including global positioning system (GPS), Wi-Fi, cellular tower pings, QR code, and radio frequency identification (RFID), it is important that the driving assistance for received the advertisement to consider the detection of higher accuracy, signal strength, and cost, especially using the QR code scanning are confirm to real conditions. The motivation of this research is the challenge of QR code, which is revealed to get more clearly for image collection, rapid for scanning, and error correction [7]. Li et al. [8] points out that the recognition of QR codes may be affected by some motion or focus blur, which makes the uneven road surface may cause horizontal and vertical motion blur within different speeds and road conditions during the driving process.

To address this problem, this study used a vehicle called a donkey car and combine it with raspberry pi and a high-speed camera to capture video on its own. The process of recording during filming is to capture video footage that involves the automotive application, applying different Yolo-v7 algorithms [9] to split training and testing set, which is compared with various speeds, sizes, and angles. Furthermore, the restoring of real pictures for scanning indeed, using end-to-end generative adversarial network (GAN) [10] for single image motion deblurring, named DeblurGAN-v2 [11] to restore the blurred image, and evaluate the proportion of successful scans that can be restored, and also compares the performance to different models, so that can be achieved the proposed method outperforms the other methods. The training and testing set in the dataset are characterized by their detailed information. The former was acquired through a self-propelled device indoors, while the latter was obtained outdoors using the same device.

The remainder of this paper is organized as follows. Section 2 discusses the related work on deep learning for QR codes scanning based on driving assistance and deblur methods. Section 3 describes the device and the proposed deep learning approaches based on architectures. Section 4 presents the experimental results and discussion for evaluation and comparison with different methods. Finally, Sect. 5 provides the conclusions and future work.

2 Related Work

To improve the QR code scanning rate, the QR code for object detection is blurred to increase the reading rate for more discussion. Yuan et al. [12] proposed that neural networks for training such as linear motion, defocus, and Gaussian blur can be distinguished after training. Schuler et al. [13] demonstrated that the use of a neural network-based approach is superior to traditional methods for non-blind image deblurring, particularly in cases of motion blur. Nah et al. [14] proposed Deep Deblur, which is a multi-scale convolutional neural network (MCNN) that cancels the kernel. Mechanism to remove the limitation by the blur kernel, thereby avoiding artifacts. Inspired by Goodfellow et al. [10] proposed GAN image-to-image translation, Kupyn et al. [15] regarded the blurring problem as an image-to-image translation task, and used conditional GAN (cGAN) network structure and loss function to evaluate the generated clear image. The gap between the image and the ground truth, DeblurGAN was proposed, and the best deblurring effect was obtained. To obtain better image quality, which uses the Feature Pyramid Network (FPN) network for feature fusion, and the discriminator uses the loss function of Least Squares GANs (LSGAN) for the overall training process to get more stable.

To remove noise and perform binarization before it is used as input, which has a square shape and functional symbols to capture obvious features of QR code, [16, 17, 18] believe that the task of identifying QR code has a great relationship with the location of Finder Pattern (FIP). Blanger et al. [18] adjusted the Single Shot MultiBox Detector (SSD) architecture in training, and added FIP as a sub-part feature for training to affect the output results, to improve the accuracy of QR code detection. Wang et al. [19] and others proposed a deblurring method under the condition of motion blur. The GAN-based method was used to obtain the deblurred two-dimensional code image, and compared with the traditional method, it was proved that the recognition accuracy and speed of the convolutional neural network (CNN) method were outperformed.

3 Methodology

3.1 Data Collection

In this study, Raspberry Pi was utilized to install vehicle modules for collecting driving QR code images. The following four criteria were used as the basis for collecting and designing the dataset. The first criterion for collecting, labeling, and designing the dataset is the size of the QR code, which determines the amount of space it occupies on the roadside sign. Moreover, it affects the resolution of the in-vehicle camera, especially for smaller QR code sizes, where a higher-resolution lens is required for capturing. The second criterion for collecting and designing the dataset is the driving speed, which can cause shaking and blurring. In this study, the urban speed limit of 50 km/hr was used, and the scale was adjusted based on the size of the QR code. Specifically, for a road surface with a length and width of 1.2 m, and a display driving QR code size of 12cm, the driving speed was reduced to 5 km/hr to minimize shaking and blurring. The third criterion is the distance between the QR code and the vehicle. As the camera on the vehicle may have height limitations, images in the inner lane may be blocked by vehicles in the outer lane. To address this, the environment assumes that the vehicle is driving in the outer lane or a single lane, and the position of the QR code is scaled according to the actual road shoulder width plus the width of the sidewalk. The fourth criterion pertains to the shooting distance, as it is crucial to capture a clear and detailed image of the QR Code. During the data collection process, the initial detection of the QR Code is set at 10 m from the vehicle on the road surface. However, driving speed becomes faster, the QR Code will be early detected at 20 m. The zoom distance of the vehicle is adjusted based on the scale of the QR code. The following are designed is shown as in Fig. 1.

The data collection was in both indoor and outdoor locations, with images captured every 2 s. To analyze the impact of blur restoration at different speeds, it is range from 15 km/hr to 50 km/hr based on the hourly speed.

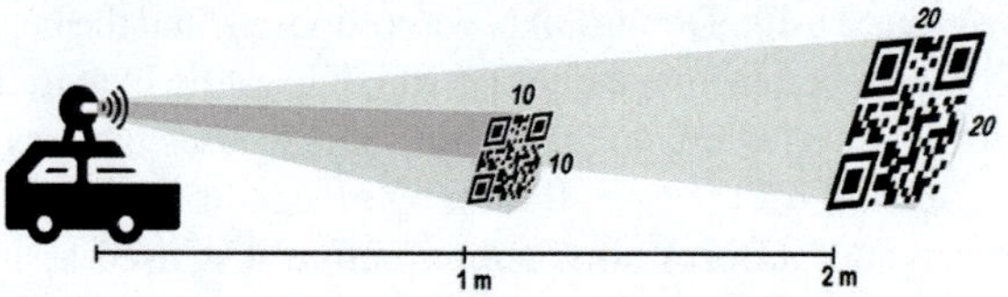

Fig. 1. An illustration of QR code size and distance ratio

3.2 Yolo-V7 Architecture

Real-time object detection requires a faster training speed. To meet this requirement, it is necessary to adopt a model that can achieve a frame rate of over 30 frames per second (fps). The YOLO [20] series offers several versions of object detection algorithms that have demonstrated high execution speeds and accuracy. YOLOv7 [9] employs advanced optimization methods to enhance the model architecture. It combines the original VoVNet [21] architecture with Cross Stage Partial Network (CSPNet) and also improves the gradient path to CSPVoVNet [22] so that the model can learn the weights of different layers more effectively, speed up training and improve accuracy. The model has improved and extended based on Extended Efficient Layer Aggregation Networks (E-ELAN) [23], which stabilizes learning and convergence through methods such as shuffling cardinality, expand cardinality, and merge cardinality, and avoids excessive computational blocks leading to unstable states. A structure is shown in Fig. 2.

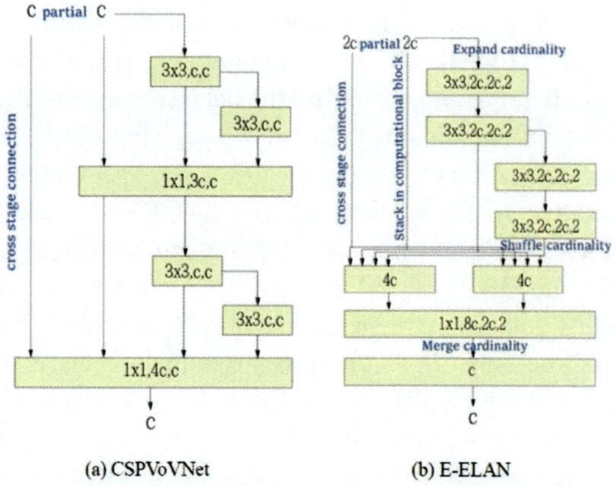

Fig. 2. Extended efficient layer aggregation networks [9]

To address the issue of transition layer width in deep neural networks is through a compound model scaling method. The approach involves scaling the concatenation-based model while maintaining the original nature of the model when the depth of the computational block is scaled [24]. The scaling process also takes into consideration the scaling of the transition layer, which ensures that the width of the transition layer is adjusted proportionally to the scaling of the computational block. The depth of the computational block should be scaled, while the corresponding width can be adjusted through the transition. It allows to maintain the inherent characteristics of the initial model during the design phase and preserve the optimal structure, while also avoiding the issue of reduced computational utilization. Overall, this method aims to optimize the performance of deep neural networks by preserving their original architecture and characteristics while scaling them to handle more complex tasks. A structure is shown in Fig. 3.

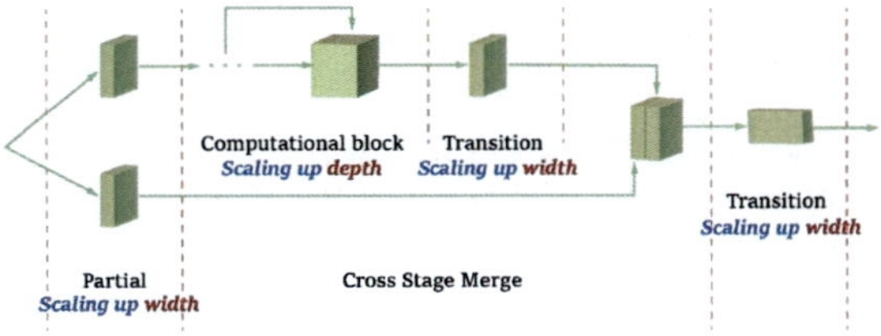

Fig. 3. A compound scaling up depth and width for concatenation-based model [24]

A structural re-parameterization technique with Visual Geometry Group called RepVGG [25], optimizes various indicators such as floating point operations per second (FLOPS), accuracy, and speed through re-parameterization. The RepVGG technique destroys the residual in Residual Neural Networks (ResNet) [26] and connection in Densely Connected Convolutional Networks (DenseNet) [27], to improve their performance of more diversity of gradients for different feature maps. The RepVGG identity connection (ResConvN)is combined with the ResNet shortcut connection. As a solution, YOLOv7 proposes to use RepVGG in ResNet (RepResNet) to remove the identity connection and improve the new architecture. Ultimately, the architecture aims to optimize the accuracy of deep neural networks while reducing the complexity of the architecture [9], it used the planned re-parameterized model and auxiliary head with independent label assignment strategy.

3.3 DeblurGAN-V2 Architecture

GAN [10] is a type of deep learning architecture that consists of two models: a generator and a discriminator. The generator produces fake samples, which are then compared to real input samples during the training process. The discriminator is trained to distinguish between the real and fake samples and adjust its weights, the difference is used to update the weights of the discriminator and improve its ability to distinguish.

During the training process, the generator is used to create synthetic data, and the discriminator is used to identify whether the generated data is real or fake. The process is designed to optimize the generator's ability to create realistic data that can fool the discriminator.

In the training process of a GAN model, two models are used for evaluation. The first is the generator network, which takes in noise data represented as z and generates synthesized data represented as $G(z)$, through the training process. The second model is the discriminator network, which evaluates the difference between the synthesized data and the real data represented as x and computes a probability score. The function

$V(D, G)$ [10] is used to model the interaction between the generator and discriminator networks and is defined in the formula as follows:

$$\min_{G}\max_{D} V(D, G) = E_{x \sim p_{data}(x)}[logD(x)] + E_{x \sim p_z(x)}[log(1 - D(G(z)))] \quad (1)$$

Assume p_z be the data distribution of noise input z and p_{data} be the data distribution of real samples. The function E represents the empirical estimation of the joint probability distribution. In the decision process of the discriminative network, the maximum value of $E_{x \sim p_{data}(x)}[logD(x)]$ to get the fake data $G(z)$ and the expected probability of output is $p_z(x)$ close to the real sample x, which is obtained by $D(G(z))$. Then, the minimization of the fake data probability is obtained by $E_{x \sim_z(x)}[log(1 - D(G(z)))$.

To fix the vanishing gradients and stabilize the training, using the L2-regularized called Least Squares GANs discriminator (LSGAN) [28] to introduce a loss function can fix the vanishing gradients and stabilize training, and provide smoother and unsaturated gradients. The further away the fake samples are from the boundary, receive the greater penalties. By minimizing the Pearson $\mathcal{X}^2$ divergence in the loss function that leads to the better training stability can be achieved. The formula is shown as follows:

$$\min_{D} V(D) = \frac{1}{2}E_{x \sim p_{data}(x)}[(D(x) - 1)^2] + \frac{1}{2}E_{x \sim p_z(x)}[D(G(x))^2] \quad (2)$$

$$\min_{G} V(G) = \frac{1}{2}E_{x \sim p_z(x)}[D(G(x) - 1)^2] \quad (3)$$

The relevant GAN called Double-Scale Relativistic GAN Least Square (RaGAN-LS) for upgrades in DeblurGAN-v2, will be getting the image to become more qualifier and clear, it adopted the relativistic wrapping [29] on the LSGAN cost function, the formula is shown as follows:

$$\begin{aligned} L_D^{RaLSGAN} &= E_{x \sim p_{data}(x)}\left[D(x) - E_{z \sim p_z(x)}D(G(z)) - 1)^2\right] \\ &+ E_{z \sim p_z(x)}\left[D(G(z)) - E_{x \sim p_{data}(x)}D(x) + 1)^2\right] \end{aligned} \quad (4)$$

A relativistic discriminator to estimate the probability that a given real data is more realistic than randomly sampled fake data shows more stable and computationally efficient training.

4 Experimental Results and Discussion

4.1 Experimental Setup

The experimental setup for training was conducted on Google Colab, utilizing the NVIDIA T4 Tensor Core GPU as the primary computing device. In the software implementation, the Python programming language was utilized, with the deep learning framework being implemented through the PyTorch package, and also involve the Pyzbar package to make a barcode reader for decoding the QR codes. In addition, the captured QR codes images were recorded based on the homemade donkey car [30] with data collector and raspberry pi 4 Model B as shown in Fig. 4. A Raspberry Pi Camera Module v2 is used in the camera module part in the hardware implementation.

Fig. 4. Homemade donkey car with data collector and Raspberry Pi

The dataset used in this research is derived from the QR codes image dataset, as described in Sect. 3. It comprises a total of 1099 QR codes being annotated and captured by mounted cameras on donkey carts, which was divided into 779 samples of training set and 320 samples of testing set. The shooting scenes for the study were captured in both indoor and outdoor settings. The indoor shots were taken in a room, while the outdoor shots were taken beside the bike lane in Dahan River Riverside Park, as shown in Fig. 5, with QR codes being available in three sizes: 10 cm × 10 cm, 15 cm × 15 cm, and 20 cm × 20 cm.

(a) Indoors scenario (b) Outdoors scenario

Fig.5. Indoors and outdoors scenario

4.2 Parameters Settings and Evaluation Metrics

The input of each YOLO model for parameters settings are shown in Table1. Backpropagation algorithm was employed in conjunction with Adam optimizer [34] for gradient descent during the training process. The models were trained for a total of 50 epochs, with a batch size of 8 samples used for each epoch. The learning rate was set to 0.01 to facilitate the convergence of the training process. The only exception is for the You Only Look Once version 4 tiny (YOLOv4-tiny) [31] model, which was trained for 60 epochs to ensure optimal performance. Some of the models used Complete-IOU [35], and the others used Focal loss [36] for loss function settings. This study utilizes an evaluation framework to assess the performance of the model, which was divided into two stages.

Table 1. Parameters settings for each Yolo model

Model	Epochs	Optimizer	Learning rate	Batch size	Loss function
YOLOv4	50	Adam	0.01	8	Complete-IoU
YOLOv4-tiny	60	Adam	0.01	8	Complete-IoU
YOLOv7	50	Adam	0.01	8	Complete-IoU
YOLOv7-tiny	50	Adam	0.01	8	Focal loss
YOLOv7-W6	50	Adam	0.01	8	Focal loss

The first stage involves QR code label tracking, with evaluation metrics including Precision, Recall, F1 score, and Intersection over Union (IOU). The second stage consists in deblurring QR code labels, which used the pretrained model of DeblurGAN-v2 with evaluation including success and failure for scanning the number of QR code images.

$$Precision = \frac{TP}{TP + FP} \tag{5}$$

$$Recall = \frac{TP}{TP + FN} \tag{6}$$

$$F1 = \frac{2 * precision * recall}{precision + recall} \tag{7}$$

$$IOU = \frac{\text{Object} \cap \text{Detected box}}{\text{Object} \cup \text{Detected box}} \tag{8}$$

where TP = True Positive, TN = True Negative, FP = False Positive, FN = False Negative, and the object refers to the area of the actual object, while the detected box refers to the predicted candidate area by dividing the overlap area by the union area [32].

4.3 Compared Approaches

In this study, several metrics were compared to the results obtained that the training model and verify the training dataset on indoor data. The data collected from outdoor scenes were used as a testing dataset to evaluate the performance of various YOLO models as shown in Table 2, including YOLOv4, YOLOv4-tiny, YOLOv7, YOLOv7-tiny [9], and a larger YOLOv7 model called YOLOv7-W6 [9, 33] were evaluated. The main comparison criteria were the detection accuracy and efficiency of models. The results showed that the YOLOv4 model exhibited good detection performance. The YOLOv4-tiny model decreased the Precision by 0.03, F1-score by 0.02 and IOU by 0.021. The YOLOv7 model shows the results that the performance has a significant increase, which is better than YOLOv4 and YOLOv4-tiny. YOLOv7-tiny model obtains a decrease by only Precision by 0.01 and IOU by 0.021. The YOLOv7-W6 model achieved the best performance among all models, with a high value for a Precision of 0.97 and an IOU of 0.8611. Additionally, all properties of different YOLOv7 models are better than different YOLOv4 models, and the value for F1-score is almost 0.99.

The experiment results for the deblurring models which are pre-trained models for prediction of testing set, and scanning from a barcode reader with comparing as shown in Table 3, the donkey car for driving is about 0–25 km/h speed range, 212 images can be compared with different models, including without deblurring, Deep Deblur [14], DeblurGAN [15], DeblurGAN-v2 with the plugin of sophisticated backbones are MobileNet and Inception-ResNet-v2 [11]. It is shown that the QR code image without deblurring has a higher rate of failed scanning, it shows that more blurring image it cannot scan more clearly, and the use Deep Deblur model shows that the performance of successful scanning has more QR code image than without scanning. Unfortunately, it has failed to scan and it is worst. Moreover, the DeblurGAN model shows that it is equal to successful and fails to scan to Deep Deblur. Otherwise, the DeblurGAN-v2(Inception-ResNet-v2) returns to the source and read QR code successful scanning is grow up to 89 images, it has a significant increase and is better than others. Nevertheless, it has to be improved for more successful scanning and less failure.

Table 2. A comparison of QR code detection for the results.

Model	Precision	Recall	F1-score	IOU
YOLOv4	0.95	1	0.93	0.836
YOLOv4-tiny	0.92	1	0.91	0.815
YOLOv7	0.95	1	**0.99**	0.857
YOLOv7-tiny	0.94	1	**0.99**	0.836
YOLOv7-W6	**0.97**	1	**0.99**	**0.861**

Table 3. A comparison of QR code deblurring and scanning for the results.

Model	Success	Fail
None	10	202
Deep Deblur	21	191
DeblurGAN	21	191
DeblurGAN-v2 (MobileNet)	31	181
DeblurGAN-v2 (Inception-ResNet-v2)	**89**	**123**

5 Conclusion

This research presents a base for an in-depth learning method combined with Donkey Car for high-speed vehicle reading QR codes efficient structure. The purpose of the research using the model for object detection relies on the detection of QR codes that are strategically placed in low-density areas of the image. These codes are then processed using an image enhancement technique to reduce reconciliation errors and missing QR code images. As a result, the QR code reading success rate is significantly increased, leading to more accurate object detection. In the limitation of this research have the challenge of collecting data for each size, lightness, number of roadside and roadside width, etc. In future works, it can be tried out to add data argumentation and attention mechanism into the training process of detecting QR codes and develop corresponding image processing techniques to enhance the scanning rate of these codes. By addressing these challenges, the accuracy and efficiency of the object detection model can be improved.

Acknowledgements. This work was supported in part by the National Science and Technology Council (NSTC) of Taiwan, R.O.C., under Contract NSTC-110-2221-E-005-032-MY3, NSTC-111-2218-E-005-007-MBK and Qualcomm's UR program support.

References

1. Dhar, S., Varshney, U.: Challenges and business models for mobile location-based services and advertising. Commun. ACM **54**(5), 121–128 (2011)
2. Global location-based advertising market 2021–2025. https://www.grandviewresearch.com/industry-analysis/laboratory-informatics-market. Accessed 08 Jun 2023
3. Liu, C., Wang, Z.: The research on advertising model of self-driving car platform. In: 2017 IEEE 3rd Information Technology and Mechatronics Engineering Conference (ITOEC), IEEE, pp. 95–99 (2017)
4. Advanced driver assistance systems (ADAS) market - global industry analysis, Size, Share, Growth, Trends, Regional Outlook, and Forecast from 2023 to 2032. https://www.precedenceresearch.com/advanced-driver-assistance-systems-market. Accessed 08 Jun 2023
5. Boukerche, A., Hou, Z.: Object detection using deep learning methods in traffic scenarios. ACM Comput. Surv. (CSUR) **54**(2), 1–35 (2021)
6. Arcos-García, Á., Álvarez-García, J.A., Soria-Morillo, L.M.: Evaluation of deep neural networks for traffic sign detection systems. Neurocomputing **316**, 332–344 (2018)
7. Hakimpour, F., Zare Zardiny, A.: Location based service in indoor environment using quick response code technology. Int. Arch. Photogrammetry Remote Sens. Spatial Inf. Sci. **40**(2), 137 (2014)
8. Li, J., et al.: A motion blur QR code identification algorithm based on feature extracting and improved adaptive thresholding. Neurocomputing **493**, 351–361 (2022)
9. Wang, C.-Y., Bochkovskiy, A., Liao, H.-Y.M.: YOLOv7: trainable bag-of-freebies sets new state-of-the-art for real-time object detectors. In: Proceedings of the IEEE/CVF Conference on Computer Vision and Pattern Recognition, pp. 7464–7475 (2023)
10. Goodfellow, I., et al.: Generative adversarial networks. Commun. ACM **63**(11), 139–144 (2020)
11. Kupyn, O., et al.: Deblurgan-v2: Deblurring (orders-of-magnitude) faster and better. In: Proceedings of the IEEE/CVF International Conference on Computer Vision, pp. 8878–8887 (2019)
12. Yuan, Q., et al.: Blind motion deblurring with cycle generative adversarial networks. Vis. Comput. **36**, 1591–1601 (2020)
13. Schuler, C.J., et al.: Learning to deblur. IEEE Trans. Pattern Anal. Mach. Intell. **38**(7), 1439–1451 (2015)

14. Nah, S., Hyun Kim, T., Mu Lee, K.: Deep multi-scale convolutional neural network for dynamic scene deblurring. In: Proceedings of the IEEE Conference on Computer Vision and Pattern Recognition, pp. 3883–3891 (2017)
15. Kupyn, O., et al.: Deblurgan: blind motion deblurring using conditional adversarial networks. In: Proceedings of the IEEE Conference on Computer Vision and Pattern Recognition, pp. 8183–8192 (2018)
16. Hussain, N., Finelli, C.: KP-YOLO: a modification of YOLO algorithm for the keypoint-based detection of QR codes. In: Schilling, FP., Stadelmann, T. (eds.) Artificial Neural Networks in Pattern Recognition. ANNPR 2020. Lecture Notes in Computer Science, vol. 12294, pp. 211–222. Springer, Cham (2020). https://doi.org/10.1007/978-3-030-58309-5_17
17. Peng, J., Yuan, S., Yuan, X.: QR code detection with faster-RCNN based on FPN. In: Sun, X., Wang, J., Bertino, E. (eds.) Artificial Intelligence and Security. ICAIS 2020. LNCS, vol. 12239, pp. 434–443. Springer, Cham (2020). https://doi.org/10.1007/978-3-030-57884-8_38
18. Blanger, L., Hirata, N.S.T.: An evaluation of deep learning techniques for QR code detection. In: 2019 IEEE International Conference on Image Processing (ICIP), pp. 1625–1629. IEEE (2019)
19. Wang, B., et al. Motion deblur of QR code based on generative adversative network. In: Proceedings of the 2019 2nd International Conference on Algorithms, Computing and Artificial Intelligence, pp. 166–170 (2019)
20. Redmon, J., et al.: You only look once: unified, real-time object detection. In: Proceedings of the IEEE Conference on Computer Vision and Pattern Recognition, pp. 779–788 (2016)
21. Lee, Y., et al.: An energy and GPU-computation efficient backbone network for real-time object detection. In: Proceedings of the IEEE/CVF Conference on Computer Vision and Pattern Recognition Workshops (2019)
22. Wang, C.-Y., Bochkovskiy, A., Liao, H.-Y.M.: Scaled-yolov4: scaling cross stage partial network. In: Proceedings of the IEEE/CVF Conference on Computer Vision and Pattern Recognition, pp. 13029–13038 (2021)
23. Wang, C.-Y., Mark Liao, H.-Y., Yeh, I.-H.: Designing network design strategies through gradient path analysis. arXiv preprint arXiv:2211.04800 (2022)
24. Wang, C.-Y, et al.: CSPNet: a new backbone that can enhance learning capability of CNN. In: Proceedings of the IEEE/CVF Conference on Computer Vision and Pattern Recognition Workshops, pp. 390–391 (2020)
25. Ding, X., et al.: Repvgg: making vgg-style convnets great again. In: Proceedings of the IEEE/CVF Conference on Computer Vision and Pattern Recognition, pp. 13733–13742 (2021)
26. He, K., et al.: Deep residual learning for image recognition. In: Proceedings of the IEEE Conference on Computer Vision and Pattern Recognition, pp. 770–778 (2016)
27. Huang, G, et al.: Densely connected convolutional networks. In: Proceedings of the IEEE Conference on Computer Vision and Pattern Recognition, pp. 4700–4708 (2017)
28. Mao, X., et al.: Least squares generative adversarial networks. In: Proceedings of the IEEE International Conference on Computer Vision, pp. 2794–2802 (2017)
29. Jolicoeur-Martineau, A.: The relativistic discriminator: a key element missing from standard GAN. arXiv preprint arXiv:1807.00734 (2018)
30. Donkeycar: a python self-driving library. https://github.com/topics/donkeycar. Accessed 11 May 2023
31. Jiang, Z., et al.: Real-time object detection method based on improved YOLOv4-tiny. arXiv preprint arXiv:2011.04244 (2020)
32. Kocakanat, K., Serif, T.: Turkish traffic sign recognition: comparison of training step numbers and lighting conditions. Avrupa Bilim ve Teknoloji Dergisi **28**, 1469–1475 (2021)
33. Jernbäcker, A.: Kalman filters as an enhancement to object tracking using YOLOv7 (2022)

34. Kingma, D.P., Ba, J.: Adam: a method for stochastic optimization. arXiv preprint arXiv:1412.6980 (2014)
35. Zheng, Z., et al. Distance-IoU loss: faster and better learning for bounding box regression. In: Proceedings of the AAAI Conference on Artificial Intelligence, pp. 12993–13000 (2020)
36. Lin, T.-Y., et al.: Focal loss for dense object detection. In: Proceedings of the IEEE International Conference on Computer Vision, pp. 2980–2988 (2017)

Positive-Unlabeled Learning with Field of View Consistency for Histology Image Segmentation

Xiaoqi Jia[1,3], Chong Fu[1,2(✉)], Jiaxin Hou[3,4], and Wenjian Qin[3]

[1] School of Computer Science and Engineering, Northeastern University, Shenyang 110819, China
[2] Key Laboratory of Intelligent Computing in Medical Image, Ministry of Education, Northeastern University, Shenyang 110819, China
fuchong@mail.neu.edu.cn
[3] Shenzhen Institute of Advanced Technology, Chinese Academy of Sciences, Shenzhen 518055, China
[4] Shenzhen College of Advanced Technology, University of Chinese Academy of Sciences, Shenzhen 518055, China

Abstract. Histology image annotation is costly and time-consuming. Utilizing Positive and Unlabeled (PU) data for model training offers a more resource-efficient alternative. However, previous methods for PU learning suffer from the noise arising from label assignment to unlabeled data. We observe that predictions on noisy data lack consistency under data augmentation. In this paper, we present Field of View (FoV) consistency regularization for PU segmentation in histology images, which effectively reduces the noise influence by promoting consistent predictions across varying FoVs. Using only 20% of positive labels on the Glas Dataset, our approach outperforms previous methods, achieving a Dice score of 90.69%-almost reaching the fully supervised result of 93.30%. Source code is available at: https://github.com/lzaya/PU_with_FoV.

Keywords: Histology Image Segmentation · PU Learning · FoV Consistency Regularization

1 Introduction

The advancement of digital pathology in clinical diagnostics has led to an increasing demand for histology image analysis. Deep learning-based segmentation algorithms, such as fully convolutional networks, have achieved remarkable accuracy and efficiency in histology image analysis, fostering progress in disease diagnosis, treatment, and research [18]. However, model training depends on extensive fully annotated data, and the high costs associated with medical image labeling pose challenges in this field. Positive-Unlabeled (PU) data, a weakly labeled dataset type, can help alleviate annotation efforts in histology image segmentation scenarios. It consists of a subset of positive examples (i.e., instances of the class of

J. C. Hung et al. (Eds.): FC 2023, LNEE 1134, pp. 127–136, 2024.
https://doi.org/10.1007/978-981-99-9342-0_14

interest) and unlabeled examples (i.e., instances not labeled as either positive or negative). Due to the large size and heterogeneity of histology images, identifying specific structures and abnormal tissues may be prone to omissions [3], leading to a situation where only a portion of positive examples is labeled, and the problem can be addressed as a PU learning problem.

Two main approaches exist for handling positive and labeled data [1]. The first identifies reliable positive or negative instances within unlabeled data to expand the labeled set. The second treats all unlabeled data as weighted negative samples. Both methods, however, cannot prevent introducing noise, such as misidentification in the first approach or positive cases within unlabeled data in the second. This noise can lead to performance decline when the model overfits to these inaccurate data points [19].

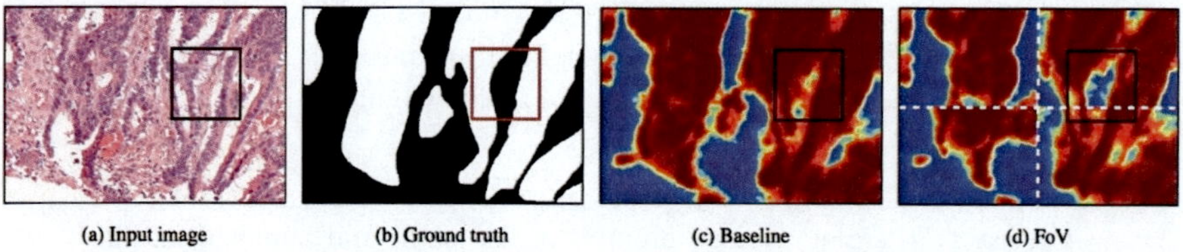

Fig. 1. Comparison of prediction results with and without data augmentation, using FoV disturbance as the augmentation method. (a) The original input image. (b) The ground truth. (c) The prediction without data augmentation. (d) The prediction with data augmentation. The inconsistency between (c) and (d) indicates areas where the model fails to accurately predict.

We observe that predictions of noisy data points are inconsistent under data augmentation, as illustrated in Fig. 1. To address this issue, we apply consistency regularization to PU learning, a strategy commonly employed in weakly-supervised learning [2,9,12,22]. Moreover, we consider the importance of Field of View (FoV) in histology image analysis. Small FoV patches provide cellular-level details, while large FoV images offer more global information. Many multi-magnification studies [8,17] combine pixel-aligned feature maps across different FoVs to enrich representation. The underlying principle of these studies is that semantic information should be consistent under different FoVs. For example, when zoomed in (i.e., small FoV) or out (i.e., large FoV), the same object retains the same essential information. However, this aspect is not explicitly addressed in previous work. Therefore, we introduce FoV consistency regularization to enhance the performance of PU learning for histology image semantic segmentation. Specifically, we divide the entire input image (i.e., large FoV) into several patches (i.e., small FoV) and regularize the prediction of the whole input and the reassembled prediction of these patches. To evaluate our method, we conduct experiments on the Glas dataset and consider a setting where the ratio of unlabeled/all positives is controllable. We average metrics over the last 10 epochs to assess our method since a clean validation set is unavailable in PU learning. Full labels are not accessible during training but are available for performance verification. Our results demonstrate that consistency regularization can mitigate noise impact. For example, when the ratio of unlabeled/all positives is 80%, adding consistency loss to nnPU [10] can improve the Dice score by 1.70% to 6.58%, and our FoV consistency regularization outperforms other popular regularization methods based on data augmentation.

Our contributions are summarized as follows: **1)** First, we empirically demonstrate that consistency regularization can reduce the inevitable noise impact in PU learning. **2)** Second, We introduce a new regularization method called FoV consistency to improve the performance of PU learning in histology image segmentation, leveraging the semantic invariance across different FoVs of the same input. **3)** Finally, on the Glas dataset, even with only 20% positive labels, our method achieves a competitive result with a Dice score of 90.69%, approaching the fully supervised result of 93.30%.

2 Related Work

2.1 PU Learning

PU learning aims to train a binary classifier using a dataset containing only positive and unlabeled (PU) samples, without labeled negatives. The primary challenge in PU learning is handling the unlabeled data. Two main approaches address this issue. The first approach, known as the two-step technique [7,14], seeks to expand the labeled set by identifying unlabeled data points likely to be negative or positive and using them to train the model. However, incorrect identification in this approach can cause the model to overfit on noisy data points, leading to performance degradation [13,19]. The second approach [5,10, 21] treats all unlabeled data as negative samples and accounts for the presence of noise (i.e., positive data) within them. Nonetheless, this approach is also vulnerable to noisy data and may result in suboptimal performance.

2.2 Consistency Regularization

Consistency regularization enforces consistent predictive results under various disturbances, improving the network's generalization capability. It can be applied as a form of supervision without requiring additional manual annotations. Previous studies have demonstrated the effectiveness of consistency regularization in scenarios with limited labeled data, such as weakly-supervised learning. For example, [12] enforces output consistency across multi-scales to achieve more accurate predictions under noisy labels. [22] utilizes cutout consistency to penalize inconsistent segmentation results. Recently, [9] proposes Puzzle-CAM to identify the most integrated pseudo-labels in weakly-supervised semantic segmentation by encouraging consistency between features from separate local patches and the entire image. In contrast, our FoV consistency regularization technique aims to prevent PU learning models from overfitting to noise that arises when assigning labels to unlabeled data.

3 Method

3.1 Review of PU Segmentation

Our method focuses on Positive-Unlabeled learning for binary segmentation tasks (PU segmentation). In a PU dataset $D = \{(x_i, s_i, y_i)\}_{i=1}^n$, $x_i \in R^{H \times W \times 3}$ represents the i-th image instance, $s_i \in \{0,1\}^{H \times W}$ indicates the positive pixels selected for labeling in x_i, and $y_i \in \{0,1\}^{H \times W}$ is the true label, which is unavailable. In the binary mask s_i, s_i^j represents the labeling status of a pixel p_j in x_i. Specifically, $s_i^j = 1$ indicates p_j is a positive pixel, while $s_i^j = 0$ signifies p_j is an unlabeled pixel that could belong to either class.

PU segmentation aims to learn a mapping function $f(\cdot;\theta)$ from PU data by minimizing empirical risk $\min_{\theta} R(f) = E_D[L(f(x;\theta), y)]$, where L is a loss function. Since only part of the positive examples is known, computing empirical risk is challenging. One solution is to relabel the unlabeled data, thereby constructing negative sample sets D_n and positive sample sets D_p. Subsequently, we minimize the approximated empirical risk on the relabeled data, given by:

$$\min_{\theta} E_{D_n}(L(f(x;\theta), 0)) + E_{D_p}(L(f(x;\theta), 1)). \tag{1}$$

However, this approach overlooks noise in D_n and D_p due to incorrect relabeling, potentially causing overfitting and reduced performance. We note that predictions on noisy data points often lack consistency when subjected to data augmentation. To address this, we add a consistency regularization to the optimization function. Consequently, our empirical risk can be expressed as:

$$\min_{\theta} E_{D_n}(L(f(x;\theta), 0)) + E_{D_p}(L(f(x;\theta), 1)) + \lambda \Omega_D(x;\theta), \tag{2}$$

where $\Omega_D(x;\theta)$ is the regularization term, and λ is a weighting factor controlling regularization strength. By enforcing consistency on predictions for noisy data points, we enhance the model's robustness and generalization performance.

3.2 Field of View Consistency

The concept of Field of View (FoV) consistency is inspired by multi-magnification research in histology image analysis. Whole Slide Images (WSIs) are typically too large for direct GPU processing and must be divided into smaller patches for training Convolutional Neural Networks (CNNs). Basic patch-based methods utilize only a single FoV of the input image, whereas multi-magnification approaches integrate representations from multiple FoVs. This process mirrors the way pathologists examine WSIs by zooming in and out to study tissues at various magnifications. They observe details of individual cells at smaller FoVs with high magnification and their surroundings at larger FoVs with low magnification. By incorporating representations from different FoVs, multi-magnification studies enhance input features and yield improved results. The underlying principle is that semantic information remains invariant across various FoVs. In this paper, we explicitly regularize this invariance by designing a consistency regularization term. This loss can be flexibly applied to histology images and other domains that necessitate multi-scale processing.

To implement this consistency regularization term, we divide the input images into smaller patches and feed each patch into the network separately. Then, we merge the results back together to obtain the reassembled prediction at its original size. In our experiment, we divide the image x_i into 4 non-overlapping patches $\{x_i^k\}_{k=1}^4, x_i^k \in R^{\frac{H}{2} \times \frac{W}{2} \times 3}$. The consistency loss can be formulated as:

$$\Omega(x;\theta) = \sum_{i=0}^{n} \|f(x_i;\theta) - merge\{f(x_i^k;\theta)\}_{k=1}^4\|_1, \tag{3}$$

where $f(x_i)$ is the prediction of the full image and $f(x_i^k)$ is the prediction of the j-th patch. The merge operation combines the predictions of the four patches to reconstruct the full image prediction, as shown in (d) of Fig. 1. The consistency loss encourages the predictions of both patches and the full image to be consistent with each other. This helps to ensure that the semantic information is invariant across different FoVs.

3.3 A Simple Two-Step Method

To assess the effectiveness of the FoV consistency regularization, we design a simple two-step method for PU learning, which we refer to as probability thresholding.

In the first step, our objective is to augment the labeled dataset by choosing reliable negative and positive samples based on the model's output. A probability thresholding strategy is employed to ensure the reliability of unlabeled data. Specifically, we classify unlabeled examples as reliable negatives or positives according to their sigmoid probabilities. Examples with probabilities less than 0.5 are considered reliable negatives, while those greater than 0.8 are deemed reliable positives. This simple and intuitive step selects reliable examples for the second step.

During the second step, we employ the relabeled data and labeled positive data to refine the model using the sigmoid activation function and binary cross-entropy loss function.

4 Experiment

4.1 Experimental Setup

Dataset: We perform PU segmentation experiments on the Gland Segmentation in Colon Histology Images (GlaS) Dataset [15], containing 165 images from 16 H&E stained T3/T4 colorectal adenocarcinoma histological sections. The dataset is split into 85 training images (37 benign and 48 malignant cases) and 80 test images (37 benign and 43 malignant cases), following the protocol of [15]. We manually corrupt the original GlaS dataset labels to create a PU dataset by randomly assigning connected positive annotation components to the unlabeled class. We also evaluate our method's performance by varying the ratio (η) of unlabeled/all positives to 20%, 50%, and 80%.

Baseline: We use three PU learning methods as baselines, including uPU [5], nnPU [10], and probability thresholding. Both uPU and nnPU treat unlabeled examples as negative and assume the positive class prior probability π_p is known. In the GlaS dataset, $\pi_p = 0.45$.

We compare FoV with three data augmentation techniques for consistency regularization, including:

- **Scale**: Adjust the input image size by downscaling by 0.5 or upscaling by 1.5.
- **Cutout** [4]: Randomly erase a square region of each input image during training, with an area not exceeding 256×256 pixels.
- **Rotation**: Rotate the input image by angles of $\gamma \cdot 90$, with $\gamma \in 1, 2, 3$.

Evaluation Metrics: We evaluate segmentation performance using Intersection over Union (IoU) and Dice coefficient (Dice), both averaged over the last 10 epochs, following the evaluation methods in [16]. Prior research [6] has shown that training with corrupted labels can cause unstable training, leading to considerable performance fluctuations. Thus, we calculate the average of the segmentation metrics across the final 10 epochs for a more reliable performance estimate.

Implementation Details: In our experiments, we preprocess each image by resizing it to 512×512 pixels. We use the state-of-the-art transformer-based segmentation algorithm Segformer [20] as our backbone model and initialize it with pre-trained weights from the CityScapes dataset. The model is trained using the AdamW optimizer, with a learning rate of 1e-4. Following [11], we set λ as a Gaussian ramp-up curve. Each experiment has a batch size of 8 and runs for 100 epochs. We conduct our experiments on a single NVIDIA Quadro A8000 GPU using the PyTorch library.

4.2 Segmentation Performance

The results of our experimental comparisons are presented in Table 1, which demonstrate the performance of three PU learning methods as well as the upper bound (i.e., the model trained with clean labels). We conduct a comprehensive comparison between our method and other popular consistency regularization methods based on disturbances. Our results indicate that integrating consistency regularization leads to improved performance relative to the baseline. Moreover, the results suggest that the FoV consistency regularization outperforms other methods, particularly when the ratio η of unlabeled/all positives is high. For instance, when η is 80%, our FoV method achieves a Dice score of 90.69%, which closely approaches the upper bound result of 93.32%. To further illustrate the

Table 1. Segmentation performance of PU learning methods with different consistency regularization techniques for different η. The best result is indicated with **bold text**, while the second-best result is underlined. N/A denotes Not Applicable, indicating that the baseline method is employed without any consistency regularization technique.

Baseline method	Consistency method	η-80%		η-50%		η-20%	
		Dice (%)	IoU (%)	Dice (%)	IoU (%)	Dice (%)	IoU (%)
uPU (known π_p)	N/A	43.37 ± 8.24	32.47 ± 7.17	74.12 ± 2.30	61.37 ± 2.96	89.78 ± 1.27	82.49 ± 1.87
	Scale	42.55 ± 9.23	34.47 ± 8.23	85.15 ± 1.75	76.08 ± 2.25	92.00 ± 0.40	85.94 ± 0.61
	Cutout	39.94 ± 6.23	29.74 ± 4.97	75.88 ± 2.76	65.34 ± 3.27	**92.32 ± 0.79**	**86.50 ± 1.19**
	Rotation	**54.46 ± 10.84**	**44.27 ± 10.68**	**85.65 ± 2.69**	**76.69 ± 3.46**	91.92 ± 0.77	85.83 ± 1.20
	FoV	42.06 ± 6.62	33.59 ± 5.81	80.70 ± 4.25	71.48 ± 4.81	92.04 ± 0.82	86.15 ± 1.20
nnPU (known π_p)	N/A	82.94 ± 3.12	72.52 ± 3.82	87.62 ± 1.78	78.75 ± 2.59	85.80 ± 1.00	75.77 ± 1.47
	Scale	86.96 ± 2.43	78.40 ± 2.92	90.32 ± 0.44	82.89 ± 0.74	86.40 ± 1.06	76.53 ± 1.64
	Cutout	84.64 ± 1.70	75.91 ± 1.97	**90.97 ± 1.35**	**84.09 ± 1.93**	**89.27 ± 0.40**	**81.15 ± 0.64**
	Rotation	88.29 ± 0.74	79.87 ± 1.15	88.04 ± 0.51	80.23 ± 0.84	86.52 ± 0.30	76.73 ± 0.47
	FoV	**89.52 ± 0.67**	**81.81 ± 1.07**	90.38 ± 0.52	83.02 ± 0.84	87.84 ± 0.56	78.81 ± 0.90
Probability thresholding (unknown π_p)	N/A	44.24 ± 9.95	33.92 ± 8.59	87.84 ± 2.20	79.66 ± 3.05	91.15 ± 0.54	84.45 ± 0.84
	Scale	89.55 ± 0.60	81.98 ± 0.90	90.61 ± 0.56	83.44 ± 0.85	89.95 ± 0.46	82.30 ± 0.77
	Cutout	87.17 ± 0.99	79.24 ± 1.42	**92.56 ± 0.21**	**86.61 ± 0.36**	**91.87 ± 0.18**	**85.46 ± 0.31**
	Rotation	90.26 ± 0.98	83.15 ± 1.30	91.55 ± 0.32	84.97 ± 0.53	88.94 ± 0.26	81.74 ± 0.43
	FoV	**90.69 ± 0.64**	**83.86 ± 0.96**	89.44 ± 0.36	82.64 ± 0.61	91.29 ± 0.53	84.58 ± 0.72
Upper bound		Dice = 93.32 ± 0.17 %; IoU = 88.03 ± 0.26 %					

superiority of the FoV method, we provide representative segmentation results in Fig. 2 using the probability thresholding baseline. These results confirm the superior qualitative performance of the FoV consistency regularization. Additionally, Fig. 3 shows the test Dice score curves for each regularization method across all epochs. Although the numerical differences among the methods are relatively small, the curve graphs reveal that our method offers more stable performance. The consistent performance of our method throughout the training process emphasizes the effectiveness of FoV consistency regularization in handling noise and achieving robust, accurate segmentation results.

However, we emphasize that the performance of PU segmentation is strongly influenced by the choice of consistency regularization technique. While the FoV method outperforms other techniques in some cases, there are scenarios in which Cutout or Rotation yield better results, such as when η is 50%. This suggests the importance of selecting the most appropriate consistency regularization technique, tailored to the specific dataset and task at hand.

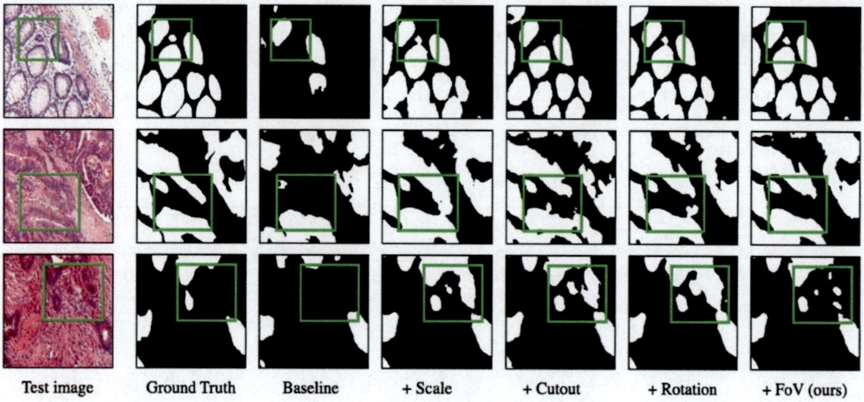

Fig. 2. Comparison of segmentation results when the ratio η of unlabeled positives is 80%.

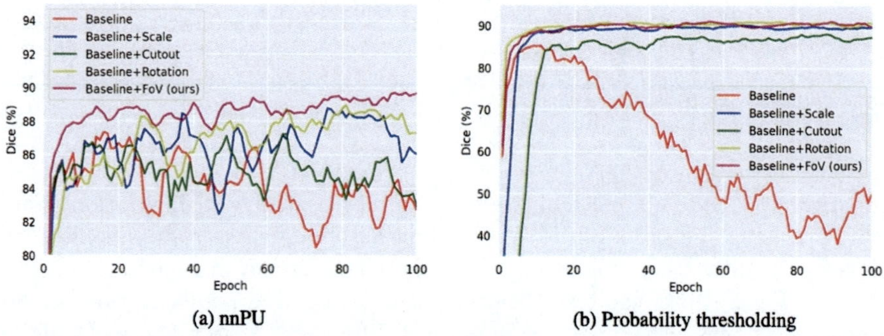

Fig. 3. Test dice score curves for each method across all epochs when the ratio η of unlabeled positives is 80%. The curves demonstrate the stability of each method's performance over time.

5 Conclusion

In this paper, our study investigate the effectiveness of consistency regularization for PU segmentation in histology images. Our experiments demonstrate that the choice of consistency regularization technique strongly influences the performance of PU segmentation, and the most suitable method should be tailored to the specific dataset and task. Additionally, our proposed Field of View (FoV) consistency regularization achieve the best results in scenarios with high ratio of unlabeled positives. The results also demonstrate the stability of our method's performance, indicating its ability to handle noisy data and achieve robust and accurate segmentation results. Our findings provide insights into the selection of appropriate consistency regularization techniques for PU learning in histology image segmentation, which could have significant implications for improving the performance of segmentation models in real-world clinical applications.

Acknowledgements. This work was supported by National Natural Science Foundation of China (No.62271475).

References

1. Bekker, J., Davis, J.: Learning from positive and unlabeled data: a survey. Mach. Learn. **109**, 719–760 (2020)
2. Chen, X., et al.: Self-pu: Self boosted and calibrated positive-unlabeled training. In: International Conference on Machine Learning, pp. 1510–1519. PMLR (2020)
3. Cheng, H.-T., et al.: Self-similarity student for partial label histopathology image segmentation. In: Vedaldi, A., Bischof, H., Brox, T., Frahm, J.-M. (eds.) Computer Vision – ECCV 2020: 16th European Conference, Glasgow, UK, August 23–28, 2020, Proceedings, Part XXV, pp. 117–132. Springer International Publishing, Cham (2020). https://doi.org/10.1007/978-3-030-58595-2_8
4. DeVries, T., Taylor, G.W.: Improved regularization of convolutional neural networks with cutout. arXiv preprint arXiv:1708.04552 (2017)
5. Du Plessis, M.C., Niu, G., Sugiyama, M.: Analysis of learning from positive and unlabeled data. In: Advances in Neural Information Processing Systems 27 (2014)
6. Guo, X., Yuan, Y.: Joint class-affinity loss correction for robust medical image segmentation with noisy labels. In: Wang, L., Dou, Q., Fletcher, P.T., Speidel, S., Li, S. (eds.) Medical Image Computing and Computer Assisted Intervention – MICCAI 2022: 25th International Conference, Singapore, September 18–22, 2022, Proceedings, Part IV, pp. 588–598. Springer Nature Switzerland, Cham (2022). https://doi.org/10.1007/978-3-031-16440-8_56
7. He, F., Liu, T., Webb, G.I., Tao, D.: Instance-dependent pu learning by bayesian optimal relabeling. arXiv preprint arXiv:1808.02180 (2018)
8. Ho, D.J., et al.: Deep multi-magnification networks for multi-class breast cancer image segmentation. Comput. Med. Imaging Graph. **88**, 101866 (2021)
9. Jo, S., Yu, I.J.: Puzzle-cam: improved localization via matching partial and full features. In: 2021 IEEE International Conference on Image Processing (ICIP), pp. 639–643. IEEE (2021)
10. Kiryo, R., Niu, G., Du Plessis, M.C., Sugiyama, M.: Positive-unlabeled learning with non-negative risk estimator. In: Advances in Neural Information Processing Systems 30 (2017)
11. Li, X., Yu, L., Chen, H., Fu, C.W., Xing, L., Heng, P.A.: Transformation-consistent self-ensembling model for semisupervised medical image segmentation. IEEE Trans. Neural Netw. Learn. Syst. **32**(2), 523–534 (2020)
12. Liu, S., Liu, K., Zhu, W., Shen, Y., Fernandez-Granda, C.: Adaptive early-learning correction for segmentation from noisy annotations. In: Proceedings of the IEEE/CVF Conference on Computer Vision and Pattern Recognition, pp. 2606–2616 (2022)
13. Liu, S., Niles-Weed, J., Razavian, N., Fernandez-Granda, C.: Early-learning regularization prevents memorization of noisy labels. Adv. Neural. Inf. Process. Syst. **33**, 20331–20342 (2020)
14. Nguyen, N.V., Rigaud, C., Revel, A., Burie, J.C.: A learning approach with incomplete pixel-level labels for deep neural networks. Neural Netw. **130**, 111–125 (2020)
15. Srinukunwattana, K., et al.: Gland segmentation in colon histology images: the glas challenge contest. Med. Image Anal. **35**, 489–502 (2017)

16. Tan, C., Xia, J., Wu, L., Li, S.Z.: Co-learning: learning from noisy labels with self-supervision. In: Proceedings of the 29th ACM International Conference on Multimedia, pp. 1405–1413 (2021)
17. Van Rijthoven, M., Balkenhol, M., Siliņa, K., Van Der Laak, J., Ciompi, F.: Hooknet: Multi-resolution convolutional neural networks for semantic segmentation in histopathology whole-slide images. Medical Image Anal. **68**, 101890 (2021)
18. Wang, S., Yang, D.M., Rong, R., Zhan, X., Xiao, G.: Pathology image analysis using segmentation deep learning algorithms. Am. J. Pathol. **189**(9), 1686–1698 (2019)
19. Wang, Z., Jiang, J., Long, G.: Positive unlabeled learning by semi-supervised learning. In: 2022 IEEE International Conference on Image Processing (ICIP), pp. 2976–2980. IEEE (2022)
20. Xie, E., Wang, W., Yu, Z., Anandkumar, A., Alvarez, J.M., Luo, P.: Segformer: simple and efficient design for semantic segmentation with transformers. Adv. Neural. Inf. Process. Syst. **34**, 12077–12090 (2021)
21. Xu, Y., Gong, M., Chen, J., Chen, Z., Batmanghelich, K.: 3d-boxsup: positive-unlabeled learning of brain tumor segmentation networks from 3d bounding boxes. Front. Neurosci. **14**, 350 (2020)
22. Zhang, K., Zhuang, X.: ShapePU: a new PU learning framework regularized by global consistency for scribble supervised cardiac segmentation. In: Wang, L., Dou, Q., Fletcher, P.T., Speidel, S., Li, S. (eds.) Medical Image Computing and Computer Assisted Intervention – MICCAI 2022: 25th International Conference, Singapore, September 18–22, 2022, Proceedings, Part VIII, pp. 162–172. Springer Nature Switzerland, Cham (2022). https://doi.org/10.1007/978-3-031-16452-1_16

Check for updates

Cryptanalysis and Improvement to Two Key-Policy Attribute-Based Encryption Schemes for Weighted Threshold Gates

Yi-Fan Tseng(✉) and Pin-Hao Chen

Department of Computer Science, National Chengchi University, Taipei, Taiwan
yftseng@cs.nccu.edu.tw

Abstract. Attribute-based encryption is one of the most suitable access control mechanism for modern data sharing models. To provide better performance, lots of attribute-based encryption schemes are constructed over without pairings. However, these schemes are either with no security proofs or broken. In this manuscript, we give the cryptanalysis of two key-policy attribute-based encryption schemes for weighted threshold gates. We propose two attack methods, the first one is able to generate valid private keys without the master secret keys, and the second one is able to recover the master secret key when an attacker gathers enough number of private keys. Moreover, an improved schemes is given in this manuscript. We also present a security analysis to show that our improved scheme fix the security flaws with only one pairing added.

Keywords: attribute-based encryption · weighted threshold gates · cryptanalysis · access control · collusion attack

1 Introduction

In the era of cloud computing, multi-user scenarios have become increasingly common, and traditional one-to-one encryption mechanisms, such as RSA [13] and ElGamal encryption [3], are no longer suitable for applications nowadays. As a result, many cryptographers are turning to attribute-based encryption (ABE) [4,20] as a solution.

ABE is a type of encryption that enables access control based on attributes, rather than specific identities. This makes it ideal for multi-user scenarios where different users may have varying access rights based on their attributes. For example, in a financial setting, employees with different roles may require different levels of access to sensitive data.

While ABE has many advantages over traditional encryption methods, one of the main challenges is reducing the computational complexity involved in the encryption and decryption process. In response to this challenge, many pairing-free ABE schemes [1,2,8,9,11,12,14–16,19,21], i.e. schemes built over elliptic curves, have been proposed to simplify the process. Unfortunately, these schemes

J. C. Hung et al. (Eds.): FC 2023, LNEE 1134, pp. 137–146, 2024.
https://doi.org/10.1007/978-981-99-9342-0_15

have all been shown to be insecure. In 2017, Herranz [6] broke the schemes of [11,12]. In 2020, Tseng and Huang demonstrated a collusion attack to [2,19], and Herranz [7] further give cryptanalysis to [2,8,9,15,16,21]. Later in 2021, Tseng [17] give a attack method to [15] so that in [15] a ciphertext can be decrypted by an unauthorized user.

In this manuscript, we further show the cryptanalysis to two pairing-free ABE schemes, [5,10]. Both these two schemes are in key-policy setting, i.e., an access structure is associated with the private key, and an attribute set is related to the ciphertext. The access structures supported by both the two schemes are weighted threshold gates $(\mathbb{A}^{\mathsf{WT}}_{k,n}, \mathsf{S}_K)$, which can be satisfied by a set of weighted attributes if the summation of the weight is greater then a pre-defined threshold value k . Unfortunately, we found that both [5,10] are insecure. In this manuscript, we propose two attack methods, which can be applied to these two ABE schemes, due to the structural similarity between [5,10]. Our first attack allows a malicious user with a private key for $(\mathbb{A}^{\mathsf{WT}}_{1,n}, \mathsf{S}_K)$ to compute a private key for $(\mathbb{A}^{\mathsf{WT}}_{k,n}, \mathsf{S}_K)$ without the knowledge of the master secret key. Furthermore, our second attack method allows an attacker colluding with several users to recover the master secret key. Moreover, an improved scheme to fix the security flaws is also given in this manuscript.

1.1 Organization

The rest of the manuscript is organized as follows. In Sect. 2, we introduce the preliminaries for our work, including notations, complexity assumption, definition for ABE, etc. In Sect. 3, we briefly review on the scheme of [5], and show our proposed two attacks to [5]. An improved scheme is demonstrated in Sect. 4. For [10], we only give the high-level description for the scheme and the cryptanalysis, in order to avoid the unnecessary duplication. Finally, we conclude our work in Sect. 6.

2 Preliminaries

In this section, we give the notation used in this manuscript, and the definition of key-policy attribute-based encryption for weighted threshold gate (KP-ABE-WT).

2.1 Notations

The notations used in this manuscript are listed as follows.

- For a set S, by "$x \xleftarrow{\$} S$" we mean uniformly randomly choose an element x from S.
- For an algorithm A, we denote by "$y \leftarrow A$" that y is the output obtained by running A.
- By PPT we mean "probabilistic polynomial-time".

- By $[n, m]$ for some integers $n \leq m$, we mean $\{n, n+1, \ldots, m\}$.
- A function $f : \mathbb{N} \rightarrow \mathbb{R}$ is said negligible in n, if for every $k \in \mathbb{N}$, there is $n_0 \in \mathbb{N}$ such that for every $n \geq n_0, |f(n)| < \frac{1}{n^k}$.

2.2 Bilinear Maps and Complexity Assumption

Let $\mathbb{G}$ and $\mathbb{G}_T$ be multiplicative groups with prime order p. Let g be a generator of $\mathbb{G}$. A bilinear map e, aka pairing, is defined as $e : \mathbb{G} \times \mathbb{G} \rightarrow \mathbb{G}_T$, where the following properties are satisfied.

1. For all $a, b \in \mathbb{Z}_p, e(g^a, g^b) = e(g, g)^{ab}$.
2. There is an efficient algorithm to compute $e(u, v)$ for all $u, v \in \mathbb{G}$.
3. $e(g, g)$ is not the identity of $\mathbb{G}_T$.

We also give a complexity assumption which the security of our improved scheme bases on.

Definition 1 (Discrete-Log Assumption). *The discrete-log assumption says that, no PPT algorithm is able to compute* $\log_g h$ *from a given* $h \in \mathbb{G}$.

Definition 2 (M-DDH$_{\mathbb{G}_T}$ Assumption [18]). *Let* $a, b \xleftarrow{\$} \mathbb{Z}_p$. *Let* $e(g, g) = \mathfrak{g}$ *The M-DDH*$_{\mathbb{G}_T}$ *assumption states that, there is no PPT algorithm, given* $(g, \mathfrak{g}, g^a, \mathfrak{g}^a, \mathfrak{g}^b)$, *tells the difference between* $\mathfrak{g}^{ab}$ *and an element* $Z \xleftarrow{\$} \mathbb{G}_T$.

2.3 Access Structure

In both [5,10], the authors propose a KP-ABE scheme for weighted threshold gates, which is defined as follows. Let S be a set of attributes. A weighted threshold gate is defined by

$$(\mathbb{A}_{k,n}^{\mathsf{WT}}, \mathsf{S}) = \{(k, n), \{w_x \mid x \in \mathsf{S}\}\},$$

where

- $w_x \geq 1$ is the weight of the attribute x;
- n is the total weight of the attributes in an attribute set S;
- $k \in [1, n]$ is the threshold.

A threshold gate is a special case of weighted threshold gate when $w_x = 1$ for all $x \in \mathsf{S}$. For a set $\mathsf{S}' \subseteq \mathsf{S}$, we say that S' satisfies $(\mathbb{A}_{k,n}^{\mathsf{WT}}, \mathsf{S})$ if

$$\sum_{x \in \mathsf{S}'} w_x \geq k.$$

2.4 Lagrange Polynomial Interpolation

Lagrange polynomial interpolation is an algorithm to compute a polynomial f of $k-1$ degree given k points. More precisely, given k points $(x_1, y_1), \ldots, (x_k, y_k)$, the polynomial f passing the k points can be computed by

$$f(x) = \sum_{i=1}^{k} y_i \Delta_i(x),$$

where $\Delta_i(x) = \prod_{j \in [1,k] \setminus \{i\}} \frac{x - x_j}{x_i - x_j}$.

2.5 Key-Policy Attribute-Based Encryption for Weighted Threshold Gates

A KP-ABE scheme for weighted threshold gates consists of the following four algorithms $\mathsf{Setup}, \mathsf{Encrypt}, \mathsf{KeyGen}, \mathsf{Decrypt}$.

$\mathsf{Setup}(1^\lambda)$. Taking as input the security parameter, the algorithm outputs the system parameter params and the master secret key msk. Note that params will be a implicitly input for the following algorithms.

$\mathsf{Encrypt}(\mathsf{S}, \mathsf{M})$. Taking as inputs an attribute set S and a message M, the algorithm outputs a ciphertext CT.

$\mathsf{KeyGen}(\mathsf{msk}, (\mathbb{A}^{\mathsf{WT}}_{k,n}, \mathsf{S}))$. Taking as inputs the master secret key msk and an access structure $(\mathbb{A}^{\mathsf{WT}}_{k,n}, \mathsf{S})$ described in Sect. 2.3, the algorithm outputs a private key D.

$\mathsf{Decrypt}(\mathsf{CT}, \mathsf{D})$. Taking as inputs a ciphertext CT and a private key D, the algorithm outputs a message.

Correctness. For $\mathsf{CT} \leftarrow \mathsf{Encrypt}(\mathsf{S}_C, \mathsf{M}), \mathsf{D} \leftarrow \mathsf{KeyGen}(\mathsf{msk}, (\mathbb{A}^{\mathsf{WT}}_{k,n}, \mathsf{S}_K))$, we have $\mathsf{M} \leftarrow \mathsf{Decrypt}(\mathsf{CT}, \mathsf{D})$ if S_C satisfies $(\mathbb{A}^{\mathsf{WT}}_{k,n}, \mathsf{S}_K)$, denoted by $\mathsf{S}_C \models (\mathbb{A}^{\mathsf{WT}}_{k,n}, \mathsf{S}_K)$.

3 Review and Cryptanalysis on Gu and Lin's KP-ABE-WT Scheme

In this section, we briefly review on the KP-ABE-WT scheme (named $\mathsf{GL22}$) proposed by Gu and Lin [5] in 2022, and give the attacks to break their scheme.

3.1 Review on GL22

$\mathsf{GL22}$ supports small universe, i.e., the set of all attributes in the system is polynomially large. Let $\mathcal{U}$ be the universe in $\mathsf{GL22}$. We omit the description of $\mathsf{Decrypt}$ algorithm since our attack method does not depend on it.
$\mathsf{Setup}(1^\lambda)$. Taking as input the security parameter, the algorithm performs as follows.

1. Choose a group $\mathbb{G}$ over an elliptic curve. Let g be a generator of $\mathbb{G}$ and p be the prime order of $\mathbb{G}$.
2. Choose $t \xleftarrow{\$} \mathbb{Z}_p$ and choose $t_x \xleftarrow{\$} \mathbb{Z}_p$ for each attribute $x \in \mathcal{U}$.
3. Compute $T = g^t$ and $T_x = g^{t_x}$ for each attribute $x \in \mathcal{U}$.
4. Choose a cryptographic hash function $H : \mathbb{G} \to \mathbb{Z}_p$.
5. Output $\mathsf{params} = (p, g, T, \{T_x\}_{x \in \mathcal{U}}, H)$ and $\mathsf{msk} = (t, \{t_x\}_{x \in \mathcal{U}})$.

$\mathsf{Encrypt}(\mathsf{S}, \mathsf{M})$. Taking as inputs an attribute set S_C and a message $\mathsf{M} \in \mathbb{Z}_p$, the algorithm performs as follows.

1. Choose $s \xleftarrow{\$} \mathbb{Z}_p$.
2. Compute $C = \mathsf{M} \cdot H(T^s), C' = g^s$.
3. Compute $C_x = T_x^s$ for each $x \in \mathsf{S}_C$.
4. Output $\mathsf{CT} = (C, C'\{C_x\}_{x \in \mathsf{S}_C})$.

$\mathsf{KeyGen}(\mathsf{msk}, (\mathbb{A}_{k,n}^{\mathsf{WT}}, \mathsf{S}_K))$. Taking as inputs the master secret key msk and an access structure $(\mathbb{A}_{k,n}^{\mathsf{WT}}, \mathsf{S}_K) = \{(k, n), \{w_x \mid x \in \mathsf{S}_K\}\}$, the algorithm performs as follows.

1. For each attribute $x \in \mathsf{S}_K$ and $y \in [1, w_x]$, choose $r_{x,y} \xleftarrow{\$} \mathbb{Z}_p$. Let $R = \{r_{x,y} \mid x \in \mathsf{S}_C, y \in [1, w_x]\}$.
2. For each $r_{x,y} \in R$, compute the corresponding Lagrange basis polynomial

$$\Delta_{r_{x,y}}(z) = \prod_{r \in R \setminus \{r_{x,y}\}} \frac{z - r}{r_{x,y} - r}.$$

3. Choose a $(k-1)$-degree polynomial q such that $q(0) = t$.
4. For each $r_{x,y} \in R$, compute $q_{x,y} = q(r_{x,y}), D_{x,y} = q_{x,y} + t_x$.
5. Output $\mathsf{D} = ((\mathbb{A}_{k,n}^{\mathsf{WT}}, \mathsf{S}_K), \{D_{x,y}, \Delta_{r_{x,y}}(0)\})$.

3.2 Cryptanalysis on GL22

Our attack algorithms focus on collusion attacks, that is, to recover the master secret msk or generate another private key without knowing msk, given enough amount of private keys D. For simplicity, we will consider access structures for threshold gate, i.e., $w_x = 1$ for all $x \in \mathsf{S}_K$ for describing the intuition of our attack algorithms.

Attack 1. Suppose a user query a private key for the access structure $(\mathbb{A}_{1,2}^{\mathsf{WT}}, \mathsf{S}_K) = \{(1, 2), \{w_x = 1 \mid x \in \mathsf{S}_K\}\}$ and $\mathsf{S}_K = \{A, B\}$ for some attributes $A, B \in \mathcal{U}$. Observe that when $k = 1$, the polynomial chosen in Step 2 of KeyGen algorithm is actually a constant polynomial $q(z) = t$, and hence[1]

$$D_A = q(r_A) + t_A = t + t_A$$
$$D_B = q(r_B) + t_B = t + t_B.$$

[1] We omit the subscript y here since all the weight are 1 and $y \in [1, 1]$.

Then the user is able to generate a private key for $(\mathbb{A}_{2,2}^{\mathsf{WT}}, \mathsf{S}_K) = \{(2,2), \{w_x = 1 \mid x \in \mathsf{S}_K\}\}$ and $\mathsf{S}_K = \{A, B\}$, given the private key $\mathsf{D}' = ((\mathbb{A}_{1,2}^{\mathsf{WT}}, \mathsf{S}_K), \{D_A, \Delta_{r_A}(0), D_B, \Delta_{r_B}(0)\})$. The details are shown as follows.

1. Choose $r_A, r_B \xleftarrow{\$} \mathbb{Z}_p$.
2. Compute $\Delta_{r_A}(z) = \frac{z - r_B}{r_A - r_B}, \Delta_{r_B}(z) = \frac{z - r_A}{r_B - r_A}$.
3. Choose $a \xleftarrow{\$} \mathbb{Z}_p$ and compute $D_A = ar_A + (t + t_A), D_B = ar_B + (t + t_A)$.
4. Output the private key for $(\mathbb{A}_{2,2}^{\mathsf{WT}}, \mathsf{S}_K) = \{(2,2), \{w_x = 1 \mid x \in \mathsf{S}_K\}\}$.

In Step 3, our attack algorithm implicitly set the polynomial $q(z) = az + t$. and no master secret is needed since $(t + t_A, t + t_B)$ has been given to the user in the private key D'. Besides, our attack algorithm can be extended into any general weighted threshold gate $(\mathbb{A}_{k,n}^{\mathsf{WT}}, \mathsf{S}_K)$, given a private key $\mathsf{D}_{\mathcal{U}}$ for $(\mathbb{A}_{1,|\mathcal{U}|}^{\mathsf{WT}}, \mathcal{U}) = \{(1, |\mathcal{U}|), \{w_x = 1 \mid x \in \mathsf{S}_K\}\}$, since

- the computation of $\Delta_{r_x}(0)$ for $x \in \mathsf{S}_K$ depends only on the choice of randomness in Step 1, which is fully controlled by the attack algorithm;
- the computation of $D_x = q(r_x) + t_x = a_{k-1}(r_x)^{k-1} + \cdots + a_1 r_A + (t + t_A)$ can be done given $\mathsf{D}_{\mathcal{U}}$.

Attack 2. Consider a private key[2] $\mathsf{D} = ((\mathbb{A}_{k,n}^{\mathsf{WT}}, \mathsf{S}_K), \{D_x, \Delta_{r_x}(0)\})$ for an access structure $\{(k,n), \{w_x = 1 \mid x \in \mathsf{S}_K\}\}$ and $\mathsf{S}_K \subseteq \mathcal{U}$, where $D_x = q(r_x) + t_x$ for $x \in \mathsf{S}_K$. By the correctness of Lagrange polynomial interpolation, we have that, for any subset $U \subset \mathsf{S}_K$ with $|U| = k$,

$$\sum_{x \in U} q(r_x)\Delta_{r_x}(0) = q(0) = t. \tag{1}$$

Therefore, by Eq. (1), we have

$$\sum_{x \in U} D_x \Delta_{r_x}(0) = \sum_{x \in U} (q(r_x) + t_x)\Delta_{r_x}(0) = t + \sum_{x \in U} \Delta_{r_x}(0) \cdot t_x. \tag{2}$$

As $D_x, \Delta_{r_x}(0)$ for $x \in U$ is given in D, there are only $|U| + 1 = k + 1$ unknown variables in Eq. (2), i.e. $t, \{t_x\}_{x \in U}$. Therefore, given private keys $\mathsf{D}^{(1)}, \ldots, \mathsf{D}^{(k+1)}$ for $\{(k,n), \{w_x = 1 \mid x \in \mathsf{S}_K^{(1)}\}\}, \ldots, \{(k,n), \{w_x = 1 \mid x \in \mathsf{S}_K^{(k+1)}\}\}$, respectively, such that $U \subseteq \mathsf{S}_K^{(1)} \cap \cdots \cap \mathsf{S}_K^{(k+1)}$, anyone is able to recover $t, \{t_x\}_{x \in U}$ by solving a linear equation system.

We give the following simple example to illustrate **Attack 2**. Let

$$\begin{aligned}
\mathsf{D}^{(1)} &= ((\mathbb{A}_{2,3}^{\mathsf{WT}(1)}, \mathsf{S}_K^{(1)}), \{D_A^{(1)}, \Delta_{r_A^{(1)}}(0), D_B^{(1)}, \Delta_{r_B^{(1)}}(0), D_C^{(1)}, \Delta_{r_C^{(1)}}(0)\}),\\
\mathsf{D}^{(2)} &= ((\mathbb{A}_{2,3}^{\mathsf{WT}(2)}, \mathsf{S}_K^{(2)}), \{D_A^{(2)}, \Delta_{r_A^{(2)}}(0), D_B^{(2)}, \Delta_{r_B^{(2)}}(0), D_D^{(2)}, \Delta_{r_D^{(2)}}(0)\}),\\
\mathsf{D}^{(3)} &= ((\mathbb{A}_{2,3}^{\mathsf{WT}(3)}, \mathsf{S}_K^{(3)}), \{D_A^{(3)}, \Delta_{r_A^{(3)}}(0), D_B^{(3)}, \Delta_{r_B^{(3)}}(0), D_E^{(3)}, \Delta_{r_E^{(3)}}(0)\}),
\end{aligned}$$

[2] We again omit the subscript y here since all the weight are 1 and $y \in [1, 1]$.

be the private keys for

$$\mathbb{A}_{2,3}^{\mathsf{WT}^{(1)}} = \{(2,3), \{w_x = 1 \mid x \in \mathsf{S}_K^{(1)}\}\}, \mathsf{S}_K^{(1)} = \{A, B, C\},$$
$$\mathbb{A}_{2,3}^{\mathsf{WT}^{(2)}} = \{(2,3), \{w_x = 1 \mid x \in \mathsf{S}_K^{(2)}\}\}, \mathsf{S}_K^{(2)} = \{A, B, D\},$$
$$\mathbb{A}_{2,3}^{\mathsf{WT}^{(3)}} = \{(2,3), \{w_x = 1 \mid x \in \mathsf{S}_K^{(3)}\}\}, \mathsf{S}_K^{(3)} = \{A, B, E\}.$$

In this example, $U = \{A, B\} \subseteq \mathsf{S}_K^{(1)} \cap \mathsf{S}_K^{(2)} \cap \mathsf{S}_K^{(3)}$. By Eq. (2) we have

$$\begin{cases} D_A^{(1)} \cdot \Delta_{r_A^{(1)}}(0) + D_B^{(1)} \cdot \Delta_{r_B^{(1)}}(0) = t + \Delta_{r_A^{(1)}}(0) \cdot t_A + \Delta_{r_B^{(1)}}(0) \cdot t_B, \\ D_A^{(2)} \cdot \Delta_{r_A^{(2)}}(0) + D_B^{(2)} \cdot \Delta_{r_B^{(2)}}(0) = t + \Delta_{r_A^{(2)}}(0) \cdot t_A + \Delta_{r_B^{(2)}}(0) \cdot t_B, \\ D_A^{(3)} \cdot \Delta_{r_A^{(3)}}(0) + D_B^{(3)} \cdot \Delta_{r_B^{(3)}}(0) = t + \Delta_{r_A^{(3)}}(0) \cdot t_A + \Delta_{r_B^{(3)}}(0) \cdot t_B. \end{cases}$$

Thus (t, t_A, t_B) can be easily recovered by solving the linear equation systems shown above.

4 An Improved Scheme

The main reason causing the security flaws shown in Sect. 3.2 is that, the information of the master secret key has been directly exposed in a private key. Equation (2) shows the linear relation between D and msk. A straightforward way to fix the problem is to raise D to the power of g. However, this method would make the number of pairings be $\mathcal{O}(\mathsf{S}_K)$ in $\mathsf{Decrypt}$ algorithm.

To reduce the number of pairings as possible, we move the most of the computations of $\mathsf{GL22}$ to the group $\mathbb{G}_T$, and randomize the components $D_{x,y}$ in D with a new randomness β. We give our improved version below. Let $\mathfrak{g} = e(g, g)$.

Setup is the same as $\mathsf{GL22}$, except that $T = \mathfrak{g}^t$ and $T_x = \mathfrak{g}^{t_x}$ for $x \in \mathcal{U}$.

$\mathsf{Encrypt}$ is the same as $\mathsf{GL22}$, except that $C' = \mathfrak{g}^s$ and an additional component $C'' = g^s$ is added.

KeyGen is the same as $\mathsf{GL22}$, except that

1. a random number β is chosen from $\mathbb{Z}_p$;
2. $D_{x,y}$ is computed as $q_{x,y} + t_x + \beta$;
3. an additional component $E = g^\beta$ is added.

$\mathsf{Decrypt}(\mathsf{CT}, \mathsf{D})$. Taking as inputs a ciphertext $\mathsf{CT} = (C, C', C''\{C_x\}_{x \in \mathsf{S}_C})$ and a private key $\mathsf{D} = ((\mathbb{A}_{k,n}^{\mathsf{WT}}, \mathsf{S}_K^{(1)}), \{D_{x,y}, \Delta_{r_{x,y}}(0)\}, E)$, the algorithm performs as follows.

1. Compute $F = e(C'', E) = e(g^s, g^\beta) = \mathfrak{g}^{s\beta}$.
2. For $x \in \mathsf{S}_K$ and $y \in [1, w_x]$, compute

$$F_{x,y} = \frac{(C')^{D_{x,y}}}{C_x \cdot F} = \frac{\mathfrak{g}^{s(q_{x,y} + t_x + \beta)}}{\mathfrak{g}^{st_x} \cdot \mathfrak{g}^{s\beta}} = \mathfrak{g}^{sq_{x,y}}.$$

3. Compute

$$T^s = \mathfrak{g}^{st} = \prod_{x \in \mathsf{S}_K} F_{x,y}^{\Delta_{r_x,y}(0)}.$$

4. Recover $\mathsf{M} = C/H(T^s)$.

Correctness. The correctness nearly follows that of GL22, except the difference due the newly-added randomness β. Thus, we cancel the term $\mathfrak{g}^{s\beta}$ in Step 2 of Decrypt algorithm, with the cost of only 1 pairing.

Security Analysis. To see why the attacks shown in Sect. 3.2 do not work in our improved scheme, note that there is a newly-added randomness β is added in KeyGen algorithm. β will be sampled each time KeyGen algorithm is perform. Besides, the information of β is hidden in E, which is impossible to be retrieved due to the discrete-log assumption. Therefore, Eq. (2) shown in Sect. 3.2 will become

$$\sum_{x \in U} D_x \Delta_{r_x}(0) = t + \sum_{x \in U} \Delta_{r_x}(0) \cdot t_x + \beta \cdot \left(\sum_{x \in U} \Delta_{r_x}(0) \right). \tag{3}$$

Thanks to the existence of β, the number of unknown variable now increases with the number of private keys obtained by the attacker, which makes the attacker impossible to recover msk by solving a linear equation system. Furthermore, according to the M-DDH$_{\mathbb{G}_T}$ assumption, even with the knowledge of $(g, \mathfrak{g}, C'' = g^s, C' = \mathfrak{g}^s, T = \mathfrak{g}^t)$, no PPT algorithm distinguishes $T^s = \mathfrak{g}^{st}$ from an uniformly random element in $\mathbb{G}_T$. This fact implies that the information of M is hidden from the attacker's view, and thus guarantees the security of our improved scheme.

5 Cryptanalysis on Lin *et al.*'s KP-ABE-WT Scheme

In this section, we show the insecurity of the KP-ABE-WT scheme (named LHXS17) proposed by Lin *et al.* [10] in 2017. Due to the conceptual similarity of GL22 and LHXS17, we only give the high-level description for LHXS17 to avoid the unnecessary duplication, and show the intuition for the corresponding cryptanalysis.

LHXS17 is almost identical to GL22, except that, in GL22 the Langrange coefficients $\Delta_{r_x,y}(0)$ is included as a part of the private key D, while in LHXS17 $\Delta_{r_x,y}(0)$ is computed in Decrypt algorithm. By this operation, GL22 has lower computation cost in Decrypt algorithm than LHXS17, with the cost of doubling the private key size. Besides, since $\Delta_{r_x,y}(0)$ needs to be computed by user, in LHXS17 the randomness r_x used in KeyGen algorithm is set to be some public indices instead of fresh random numbers, which allows anyone to compute $\Delta_{r_x,y}(0)$ for any user. Therefore, our attack methods shown in Sect. 3.2 work well for LHXS17.

6 Conclusion

With the raise of cloud computing, ABE has become one of the most suitable cryptographic primitives for multi-user scenario. In order to reduce the computation cost, lots of ABE schemes are designed without using pairings. However, all of these schemes are either flawed or lacking of security proofs. In this manuscript, we find out the security issues of [5,10] by giving two attack methods. Our attack methods are generate private keys without msk, and even recover msk. Moreover, an improved scheme have been given to fix the security problem of [5,10]. Our improved scheme requires only one pairing, which may be an optimal result when constructing ABE in pairing groups. In the future, we will prove the security of the improved scheme, and attempt to further improve the efficiency and the expressiveness of the proposed scheme.

Acknowledgment. This work was partially supported by the National Science and Technology Council of Taiwan, under grants 111-2221-E-004-005-, 111-2218-E-004-001-MBK.

References

1. Cheng, R., Wu, K., Su, Y., Li, W., Cui, W., Tong, J.: An efficient ECC-based CP-ABE scheme for power IoT. Processes **9**(7) (2021). https://doi.org/10.3390/pr9071176. https://www.mdpi.com/2227-9717/9/7/1176
2. Ding, S., Li, C., Li, H.: A novel efficient pairing-free CP-ABE based on elliptic curve cryptography for IoT. IEEE Access **6**, 27336–27345 (2018). https://doi.org/10.1109/ACCESS.2018.2836350
3. ElGamal, T.: A public key cryptosystem and a signature scheme based on discrete logarithms. In: Blakley, G.R., Chaum, D. (eds.) Advances in Cryptology, pp. 10–18. Springer, Berlin Heidelberg, Berlin, Heidelberg (1985)
4. Goyal, V., Pandey, O., Sahai, A., Waters, B.: Attribute-based encryption for fine-grained access control of encrypted data. In: Proceedings of the ACM Conference on Computer and Communications Security, pp. 89–98 (2006). https://doi.org/10.1145/1180405.1180418
5. Gu, Z., Lin, G.: A pairing-free key policy weighted attributed-based encryption. Available at SSRN 4173677 (2022)
6. Herranz, J.: Attribute-based encryption implies identity-based encryption. IET Inf. Secur. **11**(6), 332–337 (2017)
7. Herranz, J.: Attacking pairing-free attribute-based encryption schemes. IEEE Access 8, 222,226–222,232 (2020). https://doi.org/10.1109/ACCESS.2020.3044143
8. Karati, A., Amin, R., Biswas, G.P.: Provably secure threshold-based ABE scheme without bilinear map. Arab. J. Sci. Eng. **41**, 3201–3213 (2016)
9. Khandla, D., Shahy, H., Bz, M.K., Pais, A.R., Raj, N.: Expressive CP-ABE scheme satisfying constant-size keys and ciphertexts. Cryptology ePrint Archive, Report 2019/1257 (2019).https://ia.cr/2019/1257
10. Lin, G., Hong, H., Xia, Y., Sun, Z.: An expressive, lightweight and secure construction of key policy attribute-based cloud data sharing access control. J. Phys.: Conf. Series **910**(1), 012,010 (2017)

11. Odelu, V., Das, A.K.: Design of a new cp-abe with constant-size secret keys for lightweight devices using elliptic curve cryptography. Security Commun. Netw. **9**(17), 4048–4059 (2016). https://doi.org/10.1002/sec.1587, https://onlinelibrary.wiley.com/doi/abs/10.1002/sec.1587
12. Odelu, V., Das, A.K., Khurram Khan, M., Choo, K.R., Jo, M.: Expressive CP-ABE scheme for mobile devices in IoT satisfying constant-size keys and ciphertexts. IEEE Access **5**, 3273–3283 (2017)
13. Rivest, R.L., Shamir, A., Adleman, L.: A method for obtaining digital signatures and public-key cryptosystems. Commun. ACM **21**(2), 120–126 (1978)
14. Sowjanya, K., Dasgupta, M., Ray, S.: A lightweight key management scheme for key-escrow-free ecc-based CP-ABE for IoT healthcare systems. J. Syst. Architect. **117**, 102,108 (2021). https://doi.org/10.1016/j.sysarc.2021.102108, https://www.sciencedirect.com/science/article/pii/S1383762121000849
15. Sowjanya, K., Dasgupta, M., Ray, S., Obaidat, M.S.: An efficient elliptic curve cryptography-based without pairing KPABE for internet of things. IEEE Syst. J. **14**(2), 2154–2163 (2020). https://doi.org/10.1109/JSYST.2019.2944240
16. Tan, S.Y., Yeow, K.W., Hwang, S.O.: Enhancement of a lightweight attribute-based encryption scheme for the internet of things. IEEE Internet Things J. **6**(4), 6384–6395 (2019). https://doi.org/10.1109/JIOT.2019.2900631
17. Tseng, Y.-F.: Cryptanaylsis to Sowjanya et al.'s ABEs from ECC. In: Tsihrintzis, G.A., Wang, S.-J., Lin, I.-C. (eds.) 2021 International Conference on Security and Information Technologies with AI, Internet Computing and Big-data Applications, pp. 287–294. Springer International Publishing, Cham (2023). https://doi.org/10.1007/978-3-031-05491-4_29
18. Tseng, Y.F., Liu, Z.Y., Tso, R.: Practical inner product encryption with constant private key. Appl. Sci. **10**(23) (2020)
19. Wang, Y., Chen, B., Li, L., Ma, Q., Li, H., He, D.: Efficient and secure ciphertext-policy attribute-based encryption without pairing for cloud-assisted smart grid. IEEE Access **8**, 40704–40713 (2020). https://doi.org/10.1109/ACCESS.2020.2976746
20. Waters, B.: Ciphertext-policy attribute-based encryption: An expressive, efficient, and provably secure realization. In: Catalano, D., Fazio, N., Gennaro, R., Nicolosi, A. (eds.) Public Key Cryptography - PKC 2011, pp. 53–70. Springer, Berlin Heidelberg, Berlin, Heidelberg (2011)
21. Yao, X., Chen, Z., Tian, Y.: A lightweight attribute-based encryption scheme for the internet of things. Future Gen. Comput. Syst. **49**, 104–112 (2015). https://doi.org/10.1016/j.future.2014.10.010, https://www.sciencedirect.com/science/article/pii/S0167739X14002039

Applying Virtual Reality to Teaching the Law of Conservation of Energy in Physics

Tung-Hua Yang(✉), Yi-Ru Yang, and Ching-Chi Huang

Department of Digital Multimedia Design, China University of Technology, Taiwan, Republic of China

dhyang@gm.cute.edu.tw

Abstract. This paper explores the use of virtual reality in developing educational materials for the conservation of energy law in the field of natural sciences, specifically focusing on gravitational potential energy and elastic potential energy. Through immersive experiences in virtual reality, students are provided with an enjoyable learning opportunity. The visual experience in virtual reality is designed to simulate scenarios involving three different gravitational fields of planets, allowing learners to break free from the constraints of reality and experience the conversion between potential energy and kinetic energy within the context of energy conservation. Students are immersed in an engaging learning environment, where they can truly grasp the essence of Newtonian mechanics.

Keywords: Virtual Reality · The Law of Conservation of Energy · K12 Education

1 Literature Discussion

For the teaching physics in the natural field of primary and secondary schools, gamification teaching can provide students with experience in learning the basics of science. Even Jean Piaget, the founder of cognitive psychology, discussed the effectiveness and importance of learning through games [1]. With the gamification of educational methods, teachers should use their imagination and create solutions across disciplines. Intensify knowledge learning by developing cognitive processes (perception, attention, memory, thinking skills) [2]. Virtual reality technology has been proven in K12 science classrooms to enhance the learning experience, thereby increasing achievement and motivation. This is the teaching method of Inquiry-Based Learning (IBL) [3].

In this paper, virtual reality technology is applied to the teaching of the law of energy conservation in physics, and game-based interactive content is designed to achieve innovative basic education content in physics and astronomy.

2 Methods and Steps

In the field of science and technology education, Burke (2014) revised the 5E teaching circle and proposed the 6E teaching mode, which is student-centered, and aims to strengthen the design and inquiry ability in STEM education [4]. The six processes

J. C. Hung et al. (Eds.): FC 2023, LNEE 1134, pp. 147–151, 2024.
https://doi.org/10.1007/978-981-99-9342-0_16

include: 1. Engage 2. Explore 3. Explain 4. Extend/Elaborate 5. Enrich 6. Evaluate. Each cycle of the 6E teaching mode is a process that represents a complete unit. Because students need to keep thinking during the process, it will be a teaching process that is quite suitable for STEM teaching. The project is designed with the teaching steps of 6E, as follows:

1. Engage
   Based on the physical mechanism of the roller coaster and bungee jumping, the design students use the first-person viewpoint inducing think about physical phenomena in the immersive experience.
2. Explore
   The content design students have interactive experiences in three different gravitational fields of the earth, the moon and Mars, and guide them to explore and compare the differences.
3. Explain
   In VR content, the left-hand controller is designed as a tablet to assist students who need to understand the principles in depth. While playing, they can turn on the tablet at any time to read more detailed explanation and theories on learning.
4. Elaborate
   During the VR playing, the user interface displays the current altitude, speed, kinetic energy and potential energy values. Learners can compare differences and deepen their experience with physics formulas.
5. Enrich
   The content design can adjust the height of roller coaster or displacement of the bungee jumping spring for interaction, that students can deepen their understanding of exploring the principles of physical mechanics and guide they are interested in natural science.
6. Evaluate
   Compare pre- and post-teaching examination to check whether students have achieved the learning goals (Fig. 1).

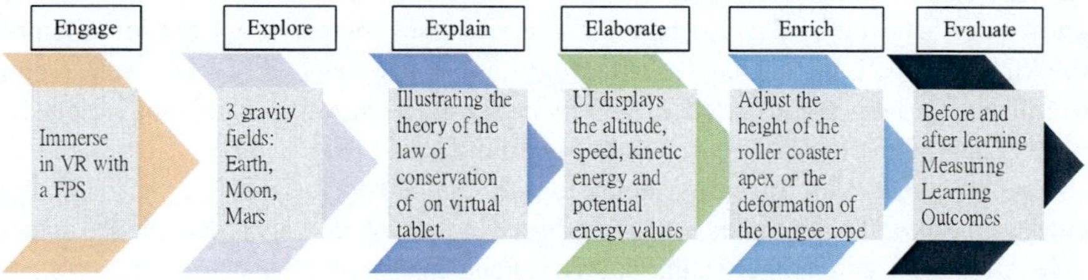

Fig. 1. 6E Design flow chart

2.1 Energy Conservation Law—Gravitational Potential Energy

The content is designed for K6 to K9 students. The objective is teaching as energy conversion between gravitational potential energy and kinetic energy. The gravitational potential energy is proportional to the gravitational field and height. The course design takes the roller coaster as an example and uses virtual reality to design three different gravitational fields with the earth's gravitational acceleration of 9.81 m/s^2, the moon's gravitational acceleration of about 1.625 m/s^2, and Mars' gravitational acceleration of about 3.724 m/s^2.

The roller coaster is gradually pulled to a certain height and rising potential energy at this time. When the car slides down the slope, the height gradually decreases, and the speed gradually increases. The value of energy conversion can be clearly expressed through virtual reality simulation, which is helpful for students to observe and explore. Under the same height and roller coaster track conditions, student can intuitively compare the difference in the change of the gravity field and improve there's understanding of this phenomenon (Figs. 2 and 3).

Fig. 2. The roller coaster rail track.

Fig. 3. The UI and Interaction of roller coaster.

The content is designed to interact with grabbing props, coins or bombs. The speed of grabbing objects corresponds to the speed of the car, to deepen the feeling of the speed of the car.

2.2 Energy Conservation Law—Elastic Potential Energy

The objective is teaching as elastic potential energy and kinetic energy conversion. The elastic potential energy is the potential to the elastic body due to deformation, which is proportional to the square of the spring coefficient k and the deformation.

Elastic potential energy is

$$\frac{k\Delta^2}{2} \tag{1}$$

Kinetic energy is

$$\frac{mv^2}{2} \tag{2}$$

Those two-energy store in different ways. In virtual reality simulates the first-person point of view to feel the change of speed in the jumping cockpit, and the UI is designed to display the values of elastic potential energy and dynamic potential energy in real time, convenient for students to observe (Fig. 4).

Fig. 4. A figure of bungee jumping car in VR.

3 Research Results

This VR content tested in Chong-Lin Junior High School in New Taipei City on December 30, 2011. Total of 24 students completed the course study. The results of the pre- and post-learning tests are shown in the table below (Table 1).

Table 1. Comparison table of test questions and learning outcomes.

No	Question	pre-test correct rate	post-test correct rate
1	Under the same condition of shape and height of the orbits are, if you take a roller coaster on the earth, the moon, and Mars, which planet will you go down the fastest?	65%	100%
2	Comparing the roller coaster moving to a height of 50 m on the earth, the moon, and Mars, which planet has the greatest gravitational potential energy?	58%	92%
3	Raise a 5 kg object by 2 m from the ground, if the gravitational acceleration is 9.8 m/s2, how many joules is the potential energy of the object?	58%	81%

4 Conclusion

Comparing the test results before and after learning, the immersive of virtual reality is used to explain the physical phenomena of the law of energy conservation, which can promote students' understanding of abstract formulas.

References

1. Wang, Z.-J., Shang, H.-F., Briody, P.: Investigating the impact of using games in teaching children English. Int. J. Learn. Dev. **1**, 127–141 (2011)
2. Piaget, F.J., Kohlberg, G., et al.: Moral Development. Psychol. Perspect. Hum. Dev. **7**, 1–25 (2005). https://warwick.ac.uk/fac/cross_fac/iatl/study/ugmodules/ethicalbeings/theoretical_approach_intro_reading.pdf Accessed 1 April 2023
3. Tilhou, R., Taylor, V., Crompton, H.: 3D virtual reality in K-12 education: a thematic systematic review. In: Yu, S., Ally, M., Tsinakos, A. (eds.) Emerging Technologies and Pedagogies in the Curriculum. Bridging Human and Machine: Future Education with Intelligence, pp. 169–184. Springer, Singapore (2020). https://doi.org/10.1007/978-981-15-0618-5_10
4. Burke, B.N.: The ITEEA 6E learning ByDesign™ model: maximizing informed design and inquiry in the integrative STEM classroom. Technol. Eng. Teach. **73**(6), 14–19 (2014)

An Efficient Edge-Based Index for Processing Collective Spatial Keyword Query on Road Networks

Ye-In Chang[1], Jun-Hong Shen[2](✉), and Sheng-Yang Lin[1]

[1] Department of Computer Science and Engineering, National Sun Yat-Sen University, Kaohsiung 804, Taiwan
changyi@cse.nsysu.edu.tw
[2] Department of Information Management, National United University, Miaoli 360301, Taiwan
shenjh@nuu.edu.tw

Abstract. Spatial keyword queries find extensive applications in geographic information systems like Facebook and Instagram. The collective spatial keyword query (CSKQ) plays a crucial role among the various types of queries. This query aims to retrieve a set of Points of Interest (POIs) that collectively cover the specified keywords while being in proximity to both the query location and other objects. To evaluate the spatial cost of a set of POIs in CSKQ, we introduce the Edge-Based Collective Nine-Area Tree Index (EBCNA). By incorporating edge information and POIs into the NA-tree structure, the EBCNA offers a comprehensive solution. All edge information, including POIs, is stored in the leaf nodes, and each edge links to its adjacent edges via pointers. This design enables direct retrieval of edge information and POIs without repeatedly trailing back to the root node. Through a comparative analysis, we have demonstrated our proposed method's superior performance compared to the existing one.

Keywords: Collective spatial keyword query · road network · spatial database

1 Introduction

With the growing popularization of geo-positioning technologies and geo-location services, many spatial-textual objects are used in many applications (e.g., Twitter and Facebook). Those geo-tagging services combine the location information with textual descriptions. Through the aforementioned development, some services efficiently process spatial keywords query (SKQ) that concern textual relevance and spatial closeness between POIs (Points of Interest) and the query location.

The spatial road network belongs to the category of geographic graphs, wherein nodes are situated along road networks [1–3]. A road network can be represented as a graph comprising a collection of vertices (or nodes), edges, and weights (or network distances) assigned to these edges. In this context, each vertex signifies an endpoint or a road intersection within the network, while each edge represents a road segment. Furthermore, the weight associated with each edge corresponds to the respective road

J. C. Hung et al. (Eds.): FC 2023, LNEE 1134, pp. 152–159, 2024.
https://doi.org/10.1007/978-981-99-9342-0_17

segment's length (or network distance). Additionally, when considering a set of spatial-textual objects, namely points of interest (POIs) located on the road network, each POI possesses both a spatial location and a textual description.

The collective spatial keyword query (CSKQ) is an essential variant of spatial keyword queries. The purpose of CSKQ is to find a set of objects that collectively incorporate the query keywords, and those objects are near the query location and close to each other object [4, 5]. Namely, we must evaluate the keyword matching and the spatial proximity of query location and objects. For instance, we issue a query with keywords {*School, Park, Restaurant*}, and we have an object $o1$ with keyword *{School}* and an object $o2$ with keywords {*Park, Restaurant*}. An example of a keyword matching approach of CSKQ is $\{o1, o2\}$, which collectively covers the query keywords. Besides the keyword-matching approach, spatial proximity also needs to be concerned.

In the literature, Gao et al. introduced a renowned algorithm for addressing the collective spatial keyword query (CSKQ), employing the CCAM index structure for storing points of interest (POIs) on the road network [4]. Building upon the CCAM index structure [6], the authors proposed algorithms to tackle the CSKQ problem. The first algorithm, Network Expansion Based (NEB), identifies the nearest objects encompassing the queried keywords relative to the query location. These objects are subsequently utilized as the result set. The closeness of the result set is evaluated using a cost function. The goal of NEB is to find an upper bound of the cost used by the other algorithms to find a better answer. Their exact algorithm, called Sliding Window (SW) algorithm, is to find the optimal result set with the lowest cost calculated by the cost function. In their proposed algorithms, upon issuing a CSKQ, the process necessitates the traversal of numerous edges. In order to retrieve the requisite edge information, an iterative search of the B + -tree structure from the root node becomes imperative. However, this recurrent search operation significantly escalates the overall search time.

Therefore, to reduce the search time, this paper presents the edge-based collective nine-area tree (EBCNA) index structure to shorten the search time in the leaf node and enhance the efficiency of collective spatial keyword query processing on road networks. By incorporating edge information and POIs into the NA-tree structure [7], the EBCNA offers a comprehensive solution. All edge information, including POIs, is stored in the leaf nodes, and each edge links to its adjacent edges via pointers. This design enables direct retrieval of edge information and POIs without repeatedly trailing back to the root node.

The rest of this paper is organized as follows. Section 2 presents the proposed algorithms. Section 3 evaluates the performance efficiency. Section 4 presents the concluding remarks of the study.

2 The Proposed Algorithms

In the proposed algorithms, we use an ECBNA (edge-based collective nine-area tree) index, a revised version of the NA-tree proposed by Chang et al. [7], to build our road network model. We proposed the basic expanding algorithm and the nearest keyword first exact algorithm to process CSKQ.

2.1 Edge-Based Collective Nine-Area Tree

We define the road networks as an undirected graph $G = (V, E)$, where V is a set of vertices and E is a set of edges. The vertex v in V is denoted by $v = (vid, pos)$, where *vid* is the vertex ID number and *pos* coordinates the vertex in 2D space. Every vertex v in V corresponds to the junctions of edges and the endpoints of edges on the road network. The edge e in E is represented as $e = (eid, v_s, v_e, length, obj)$. *Eid* is the ID number of the edge. $v_s \in V$ is the start point of the edge. $v_e \in V$ is the endpoint of the edge. *Length* is the length of the edge. *Obj* is the spatial object which contains both spatial and textual information belonging to the edge.

An NA-tree [7] is a structure according to data location and is organized by spatial numbers. The spatial space is decomposed into four equal-sized regions, and an NA-tree might have nine children according to the decomposition of the spatial space, as shown in Fig. 1. Figure 2 shows a running example of the road network. The length of the edge between two vertexes is marked in red color, and the black circle indicates an object with spatial keywords and the corresponding distance to the vertex with the lower vertex number along the edge. For example, object $o3$ is associated with the keyword c, and the distance from vertex $v2$ to $o3$ is 6.

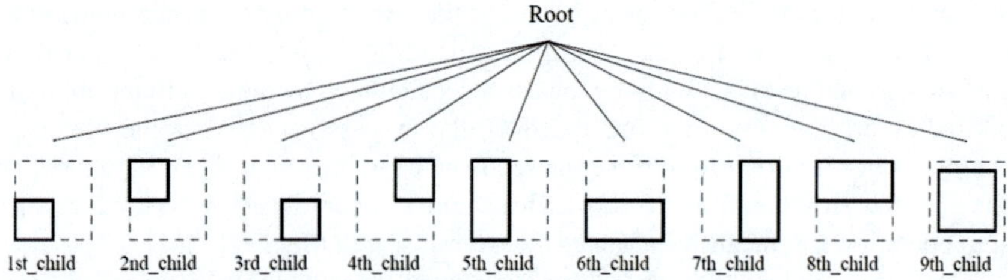

Fig. 1. NA-tree

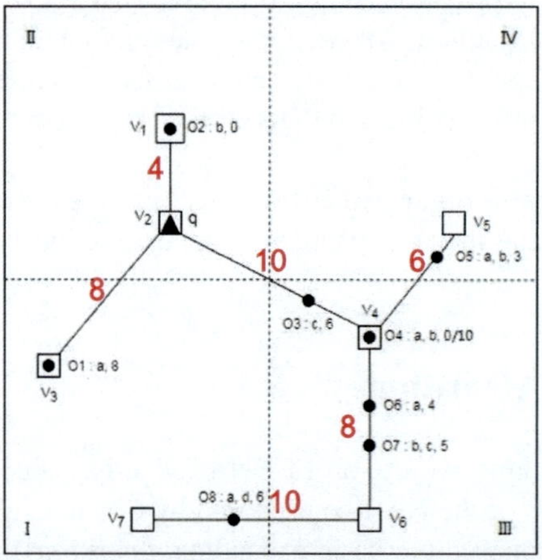

Fig. 2. A running example of the road network

Inspired by the NA-tree representation approach, we use two endpoints of a line segment to represent line segments. Figure 3 shows the corresponding EBCNA index structure. For example, the edge $v2v3$ is located at the fifth child of the NA-tree, according to Figs. 1 and 2. In Fig. 3, the fifth leaf node stores the information about the edge from vertex $v2$ to vertex $v3$ and the edge from vertex $v3$ to vertex $v2$. The second column indicates the length of the edge. The third column indicates the object lists along the edge. The last column contains a reference to the other leaf node that stores information about the adjacent edge connected to the endpoint of the current edge.

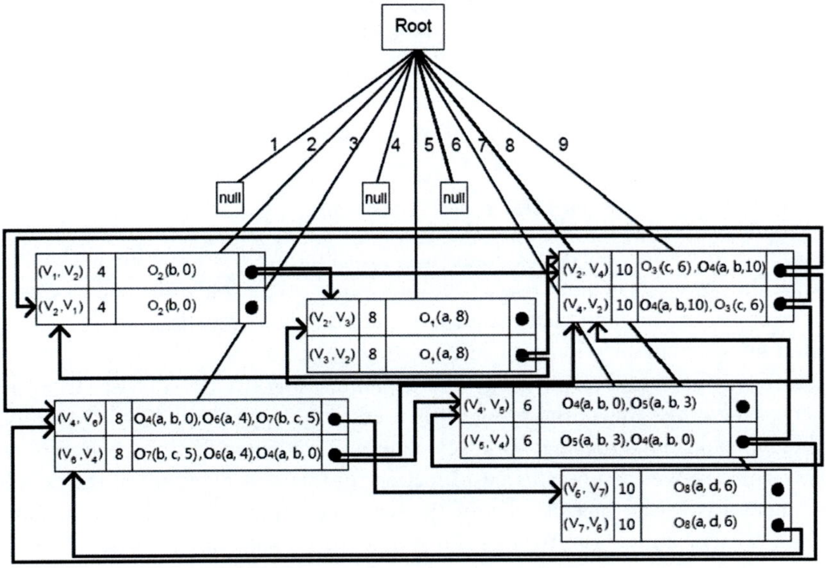

Fig. 3. The EBCNA index structure

2.2 The Basic Expanding Algorithm

We present the basic expanding algorithm to find a result set with a reasonable cost function value and to serve as the upper bound of the following exact algorithm. The cost function is as follows [4]: Cost(R) = $(1 - \alpha) \times max_{o\in V}\ d(q, o) + \alpha \times max_{oi,oj\in V}\ d(o_i, o_j)$, where R denotes the set of the result objects. $Max_{o\in V}\ d(q, o)$ represents the maximal distance between the query point and any object in R. $max_{oi,oj\in V}\ d(o_i, o_j)$ represents the maximal distance between every two of objects in R. α is a user-defined parameter between 0 and 1 to determine the importance between $max_{o\in V}\ d(q, o)$ and $max_{oi,oj\in V}\ d(o_i, o_j)$. The Cost($R$) function evaluates the spatial cost between the query point and the result set R. α is set to 0.5.

In Fig. 2, a triangle on node $v2$ denotes the CSKQ query q with keywords $\{a, b, c\}$. *Min_Q* is a minimal priority queue that keeps objects in the queue based on their distances to the query location. *Find_Key* is a set that records the query keywords.

Based on the EBCNA index structure, we expand the road network from query q and find the first shortest edge $v1v2$. The edge $v1v2$ has one object $o2$ containing its keyword b, and the distance = 0 from object $o2$ to the node with a smaller ID, *i.e.*, $v1$. Thus, we have $Find\_Key = \{a, b, c\}$ - $\{b\} = \{a, c\}$ and edge $v1v2$ is dequeued from queue *Min_Q*. We keep the same way to expand the road network and find the next shortest edge, $v2v3$, in queue *Min_Q*. On edge $v2v3$, object $o1$ with keyword a is found, and the distance = 8 from object $o1$ to the node with a smaller ID, *i.e.*, $v2$. Thus, we have $Find\_Key = \{a, c\}$ - $\{a\} = \{c\}$ and edge $v2v3$ is dequeued from queue *Min_Q*. Since there is still one keyword c in set *Find_Key*, we keep expanding the road network and find the next shortest edge $v2v4$ in queue *Min_Q*. On edge $v2v4$, object $o3$ with keyword c is found, and the distance = 6 from object $o3$ to the node with a smaller ID, *i.e.*, vertex $v2$. Thus, we have $Find\_Key = \{c\}$ - $\{c\} = \{\}$. Then, we have the result set $R = \{o1, o2, o3\}$ and the function Cost(R) = $(1–0.5) \times d(q, o1) + 0.5 \times d(o1, o3) = 0.5 \times 8 + 0.5 \times 14 = 11$, which can be the upper bound of the following exact algorithm.

2.3 The Nearest Keyword First Exact Algorithm

We propose the nearest keyword first exact algorithm, *NKF*, to find the optimal result set with the lowest function cost. In the first phase, we expand the road network to identify objects containing query keywords from the query location. These objects are inserted into a minimal priority queue. We then select the first object in the queue as the new query point and employ the basic expanding algorithm to obtain the result set. The cost function of this initial result set is used as the upper bound. Subsequently, we continue to expand the road network, selecting the next object in the queue and finding a new result set. We calculate the cost function for this new result set. If the calculated cost function is lower than the cost of the basic expanding algorithm, we update the current lowest cost function. We repeat this process of expanding the road network and finding different result sets until the distance between the discovered object and the query location exceeds the current lowest cost function. In the second phase, we record the objects with the exact query keywords in the minimal priority queue for each query keyword. Next, we find the rest of the combinations and calculate their function cost. Finally, the result set with the lowest function cost becomes optimal.

For the same example in Fig. 2, a CSKQ q is issued at vertex $v2$. Figure 4 shows the procedure of the first phase. We expand the road network starting from vertex $v2$, the query location. Initially, we trace the shortest edge, $v1v2$, and discover an object, $o2$, with the keyword b on this edge. Object $o2$ is inserted into the minimal priority queue *Min_Q*, and the remaining keywords to be found are updated as $Find\_Key = \{a, b, c\}$ - $b = \{a, c\}$. Next, we employ the basic expanding algorithm from object $o2$ to identify objects close to $o2$ and collectively contain the keywords $\{a, c\}$. Consequently, we obtain the first result set, $R = \{o1, o2, o3\}$, and compute its function cost, which is $0.5 * d(q, o1) + 0.5 * d(o1, o3) = 0.5 * 8 + 0.5 * (8 + 6) = 11$ (i.e., the upper bound), as listed in Step 1 of Table 1.

Moving forward, we expand the road network again and reset *Find_Key* to $\{a, b, c\}$. On the next shortest edge, $v2v3$, we find object $o1$ with the keyword a. Object $o1$ is added to the *Min_Q* queue, and *Find_Key* is updated as $\{a, b, c\}$ - $a = \{b, c\}$, as listed in Step 2 of Table 1. We apply the basic expanding algorithm from object $o1$ and obtain the result set {$o1$, $o2$, $o3$} with a function cost of 0.5 * $d(q, o1)$ + 0.5 * $d(o1, o3)$ = 0.5 * 8 + 0.5 * (8 + 6) = 11. Since the function cost is not lower than the current lowest function cost, we continue expanding the road network and reset *Find_Key* to {a, b, c}.

Following the same approach as above, on edge $v2v4$, we find object $o3$ with the keyword c and object $o4$ with the keywords {a, b}, as listed in Step 3 of Table 1. Initially, we insert object $o3$ into the *Min_Q* queue, and *Find_Key* is updated as {a, b, c} - c = {a, b}. Employing the basic expanding algorithm from object $o3$, we find the result set {$o3$, $o4$} with a function cost of 0.5 * $d(q, o4)$ + 0.5 * $d(o3, o4)$ = 5 + 2 = 7. Since the function cost of 7 is lower than the current lowest function cost of 11, we update the current lowest function cost to 7. Finally, we reset *Find_Key* to {a, b, c} and insert object $o4$ into the *Min_Q* queue, resulting in *Find_Key* being updated as {a, b, c} - {a, b} = {c}. In step 4, $d(q, o4)$ = 10 is also not lower than the current lowest function cost = 7. The first phase terminates once the distance between the object and the query location is equal to or larger than the current lowest function cost.

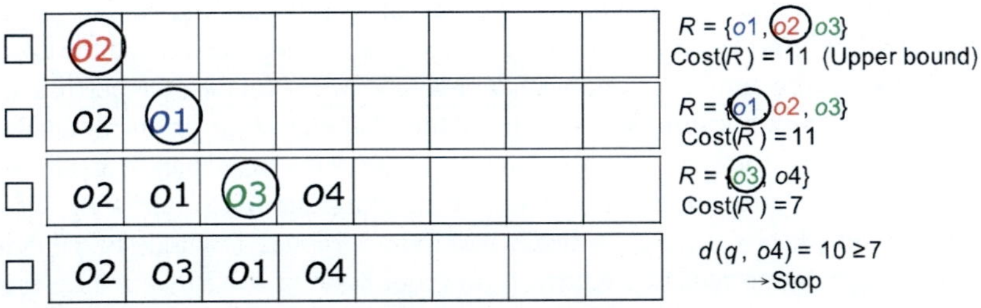

Fig. 4. The first phase

Table 1. The steps of the network expansion

Step	Edge	Object	Keywords	*Find_Key*	Result Set *R*
1	$v1v2$	$o2$	b	{a, c}	{$o1$, $o2$, $o3$} Cost(*R*) = 11
2	$v2v3$	$o1$	a	{b, c}	{$o1$, $o2$, $o3$} Cost(*R*) = 11
3	$v2v4$	$o3$	c	{a, b}	{$o3$, $o4$} Cost(*R*) = 7
4	$v2v4$	$o4$	{a, b}	c	$d(q,o4) = 10 \geq 7$ → Stop

In the second phase, we store objects with each query keyword in the *Min_Q* queue. Specifically, objects with keyword a are {$o1$, $o4$}, objects with keyword b are {$o2$, $o4$}, and the object with keyword c is {$o3$}. We then generate combinations of these objects to create a set encompassing all query keywords. Ultimately, we identify the result set {$o3$, $o4$} with the lowest function cost 7. Hence, the optimal result set is {$o3$, $o4$}.

3 Performance Evaluation

We evaluate the performance efficiency of query time for the collective spatial keyword query processing on road networks using the proposed EBCNA index structure and the CCAM index structure [4]. We perform our experiment on the Oldenburg real road network dataset, which contains 6,105 vertices and 7,035 edges at 10,000 * 10,000 (https://users.cs.utah.edu/~lifeifei/SpatialDataset.htm). Each data object contains three keywords randomly selected from the datasets with 50 keywords. The density of data objects with keywords is set to 0.3. The number of data objects is set to 2110. The threshold value of the leaf node in the $B+$ -tree for the CCAM index structure is set to 4096 bytes. The threshold value of the leaf node in the NA-tree is set to 200. α is set to 0.5 for the cost function.

First, we compare the query time performance for the proposed expanding algorithm utilizing the EBCNA index structure and the NEB algorithm employing the CCAM index structure across varying query keywords, ranging from 3 to 5. Table 2 lists this comparison of the query time. On average, the proposed basic expanding algorithm exhibits a 64.6% improvement over the NEB algorithm. It is observed that the execution time increases as the number of query keywords increases. Notably, the performance of our basic expanding algorithm surpasses that of the NEB algorithm, primarily attributable to the distinct data structures employed for real data. This discrepancy arises because our EBCNA index structure is an edge-based indexing approach, allowing for direct linkage to other edges. Conversely, the CCAM index structure is a node-based indexing structure, necessitating repeated returns to the root node of the B + -tree during road network expansion.

Table 2. The comparison of the query time (sec) between the proposed basic expanding algorithm and the NEB algorithm

Number of Keywords	3	4	5
BASIC	0.076	0.077	0.083
NEB	0.206	0.225	0.230
% Improvement	63.1%	66.8%	63.9%

Second, we compare the query time performance for the proposed NKF algorithm utilizing the EBCNA index structure and the SW algorithm employing the CCAM index structure across varying query keywords, ranging from 3 to 5. Table 3 lists this comparison of the query time. On average, the proposed NKF algorithm exhibits a 94.4% improvement over the SW algorithm. The performance of the proposed NKF algorithm is better than the SW algorithm since our NKF algorithm considers not only the distance between the query location and objects but also the distance between any two objects. The method can reduce the number of objects in the minimal priority queue in the first phase. Thus, we can reduce the calculation of combinations in the second phase.

Table 3. The comparison of the query time (sec) between the proposed NFK algorithm and the SW algorithm

Number of Keywords	3	4	5
NFK	0.904	4.590	16.489
SW	5.555	1959.956	3619.980
% Improvement	83.7%	99.8%	99.6%

4 Conclusions

This paper introduces the EBCNA index structure to process collective spatial keyword queries (CSKQ) efficiently. This edge-based approach divides the road network into nine distinct areas and employs two spatial numbers to represent each edge. Unlike the node-based approach proposed by Gao et al. [4], the EBCNA enables efficient access to spatial objects during the expansion of the road network, eliminating the need for repeated returns to the root node. We introduce the basic expanding algorithm, which aims to find a result set with an acceptable cost function value and is the upper bound for the subsequent exact algorithm. We also propose an exact algorithm to deal with the CSKQ problem with the lowest function cost. The performance evaluation confirms the superiority of the proposed algorithms over existing ones.

Acknowledgments. This research was supported by grants MOST 105–2221-E-110–084, MOST 107–2221-E-110–064, and NSTC110–2410-H-239–019 from the National Science and Technology Council, Taiwan.

References

1. Chang, Y.-I., Tsai, M.-H., Wu, X.-L.: An edge-based algorithm for spatial query processing in real-life road networks. Int. J. Model. Optim. **5**(4), 308–312 (2015)
2. Fang, H., et al.: Effective spatial keyword query processing on road networks. In: Sharaf, M., Cheema, M., Qi, J. (eds.) Databases Theory and Applications. ADC 2015. LNCS, vol. 9093, pp. 194–206. Springer, Cham (2015). https://doi.org/10.1007/978-3-319-19548-3_16
3. Kuang, X., et al.: TK-SK: textual-restricted *K* spatial keyword query on road networks. In: Sharaf, M., Cheema, M., Qi, J. (eds.) Databases Theory and Applications. ADC 2015, LNCS, vol. 9093, pp. 167–179. Springer, Cham (2015). https://doi.org/10.1007/978-3-319-19548-3_14
4. Gao, Y., Zhao, J., Zheng, B., Chen, G.: Efficient collective spatial keyword query processing on road networks. IEEE Trans. Intell. Transp. Syst. **17**(2), 469–480 (2016)
5. Xue, J., Wu, C., Zhao B., Hu, Y.: Collective spatial keyword query on time dependent road networks. In: Proceedings of Tenth International Conference on Advanced Cloud and Big Data, pp. 7–12 (2022)
6. Shekhar, S., Liu, D.-R.: CCAM: a connectivity-clustered access method for networks and network computations. IEEE Trans. Knowl. Data Eng. **9**(1), 102–119 (1997)
7. Chang, Y.-I., Liao, C.-H., Chen, H.-L.: NA-trees: a dynamic index for spatial data. J. Inf. Sci. Eng. **19**(1), 103–139 (2003)

Multi-feature Data Generation for Design Technology Co-Optimization: A Study on WAT and CP

Shih-Nung Chen[1](✉) and Shi-Hao Chen[2]

[1] Department of Information Communication, Asia University, Taichung, Taiwan, R.O.C.
nung@asia.edu.tw

[2] DigWise Technology Corporation, Ltd., Hsinchu, Taiwan, R.O.C.
hock.chen@digwise-tech.com

Abstract. This study explores the use of Generative Adversarial Networks (GANs) to generate wafer-level Wafer Acceptance Test (WAT) and Chip Probe (CP) test data in semiconductor manufacturing processes, and their application in relevant process and Design-Technology Co-Optimization (DTCO). The generated virtual silicon data includes device performance, physical-electrical characteristics, distribution of wafer process parameters, and implicit information on wafer-level features such as uniformity and defects. This approach enables interdisciplinary teams to overcome data acquisition barriers while ensuring data confidentiality, and it holds significant potential for the development of advanced Electronic Design Automation (EDA) tools in co-optimizing process and chip design flows.

Keywords: Generative Adversarial Network (GAN) · Wafer Acceptance Test (WAT) · Chip Probe (CP) · Design-Technology Co-Optimization (DTCO) · virtual silicon data · Electronic Design Automation (EDA)

1 Introduction

The Design-Technology Co-Optimization (DTCO) methodology has been widely discussed and applied in physical design processes to enhance the overall productivity and competitiveness of semiconductor chips. As shown in Fig. 1, it can be analogized to a massive neural network optimization process. Our focus is on optimizing productivity through inference, which includes chip monitoring, WAT-CP-SLT testing, feature correlation analysis, machine learning, and binning strategies with compensation techniques, among others. On the other hand, in the back-propagation optimization phase, we concentrate on optimizing design and process recipes, which encompass chip model calibration using actual measurements, process parameters tuning, timing extraction based on WAT measurements for the device library, customization and optimization of the device library, On-Chip Variation (OCV) regression for local variations, and optimization of design margins and sign-off strategies.

J. C. Hung et al. (Eds.): FC 2023, LNEE 1134, pp. 160–168, 2024.
https://doi.org/10.1007/978-981-99-9342-0_18

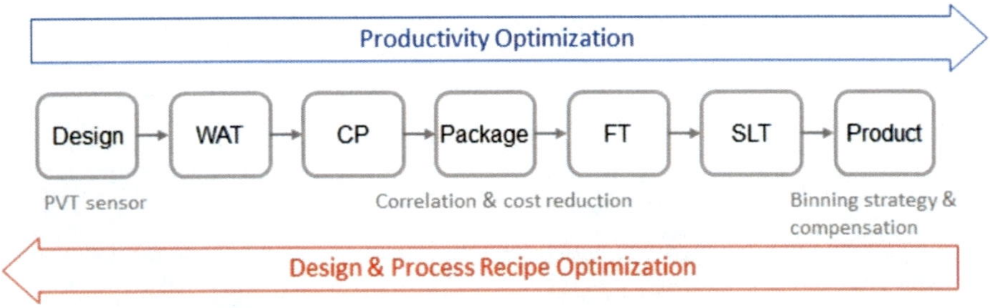

Fig. 1. Design-Technology Co-Optimization (DTCO)

However, obtaining and exchanging valuable test data is often challenging and poses barriers to the overall advancement of industry technologies. Therefore, this study proposes an innovative virtual silicon technology that leverages deep learning models to rapidly and accurately generate a large amount of chip data, while reflecting the parameter distribution, defects, and features in wafer manufacturing processes. This study introduces a GAN-based approach that trains and encapsulates multidimensional WAT and CP test data using compact GAN models to generate highly realistic chip data with multidimensional features. This technology plays a crucial role in optimizing chip design and improving the manufacturing process. It brings significant benefits such as enhancing production efficiency, reducing costs, and improving product quality.

2 Background

Traditional models often simplify events by assuming Gaussian distributions, disregarding the fact that physical quantities in real chips often exhibit skewed-normal or log-normal distributions. Additionally, these models frequently overlook the interdependencies among vectors in high-dimensional spaces. As shown in Fig. 2, even if each dimension's feature follows the distribution of the parent population when projected individually, the combined distribution in high-dimensional space may lose the interrelationship between them, somewhat akin to rolling dice.

Due to the multitude of process parameters involved in wafer manufacturing, the relationship between process parameters and wafer or chip-level test data becomes highly intricate, making it challenging for traditional methods to effectively model and analyze. As shown in Fig. 3, even with a comprehensive understanding of the distribution and interrelationship of chip-level features, the lack of wafer-level coordinate information results in the loss of characteristics related to the actual wafer fabrication uniformity. This is a prevalent issue in current simulation analysis modeling. In fact, the distribution of feature vectors in high-dimensional space lacks authenticity, leading to significant discrepancies between production data and simulated data.

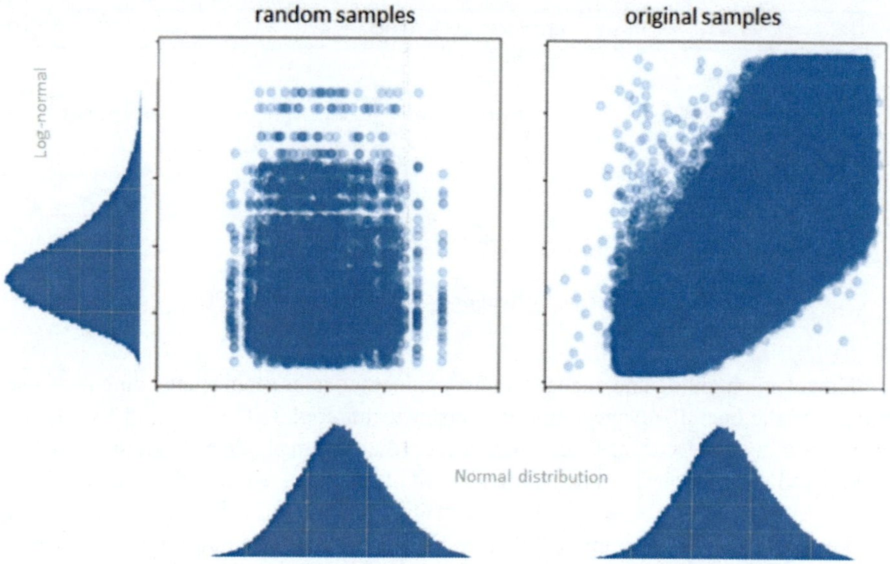

Fig. 2. Interdependence and Correlation Among Features

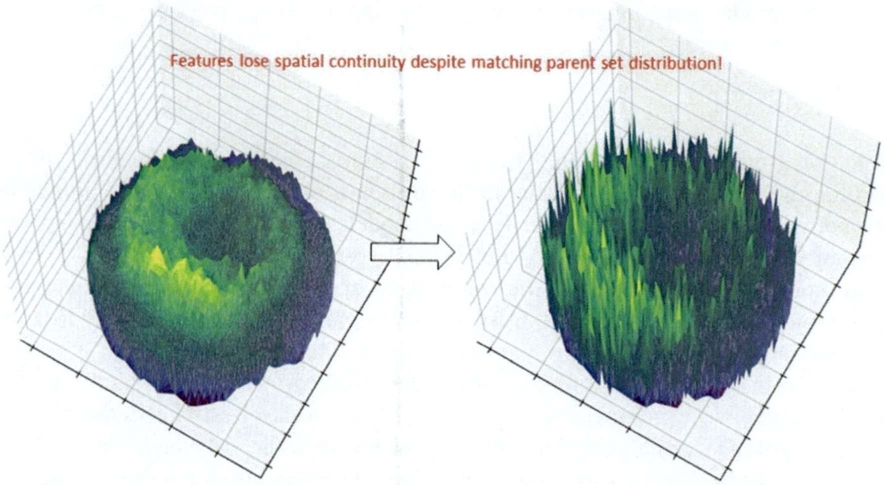

Fig. 3. Distribution of Physical Features of Chips at the Wafer-Level Losing Realism

Existing literature has explored the use of deep learning models for simulating and predicting wafer manufacturing processes. Studies [1] and [2] propose a method for wafer defect detection based on a Deep Convolutional Generative Adversarial Network (DCGAN). This approach utilizes a DCGAN to learn the distribution of defect images on wafers and employs the generated model for defect detection and classification. Furthermore, studies such as [3–5], and [6] demonstrate the potential of Generative Adversarial Networks (GANs) for various other applications in the manufacturing domain.

However, our research differs from existing literature in several aspects. We have successfully achieved the generation of highly realistic virtual silicon data and proposed a platform for chip and wafer-level data analysis and co-optimization based on GAN models. To better capture the variations in wafer-level processes, we have incorporated additional physical features into the construction of the training dataset.

3 System Architecture

This study employs a GAN model to capture the uniformity characteristics of defects and parameters in the wafer manufacturing process using a large volume of multi-dimensional data. Firstly, we transform the original multi-dimensional data into two-dimensional images, and set multiple feature dimensions (parameter C), as shown in Fig. 4. The size of parameter C is correlated with the network size, and the training time exhibits non-linear growth. Based on computations performed on a personal computer CPU, we select the feature dimensionality C to be between 10 and 18.

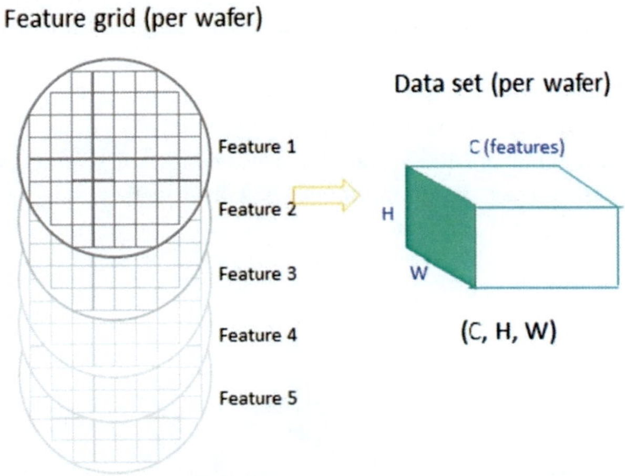

Fig. 4. Transformation and Integration of Wafer-Level Multidimensional Training Dataset

In this study, we further augment the training dataset using small angle rotation transformations to simulate the occurrence of rotational defects and process parameters distributions in the wafer manufacturing process, as shown in Fig. 5. This approach enables us to accurately capture the key features in the wafer manufacturing process, thereby improving the training effectiveness of the model.

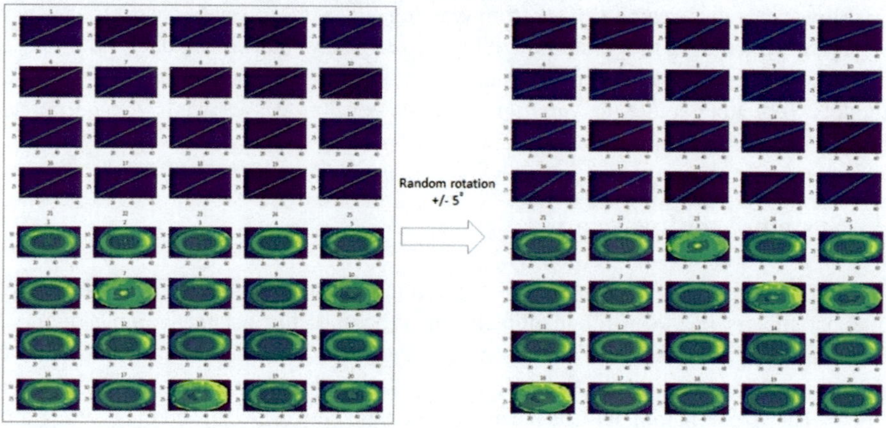

Fig. 5. Augmentation of Training Dataset through Small Angle Rotation Transformations

In our study, we utilized a Convolutional Neural Network (CNN) to construct a Generative Adversarial Network (GAN) model, as shown in Fig. 6, for generating chip data with various process features while incorporating potential defects. The generator component of the model consists of multiple convolutional layers and Tanh activation layers to generate wafer images. Simultaneously, the discriminator component also includes multiple convolutional layers and Sigmoid activation layers to distinguish between real and generated data. These design components work together to achieve the goal of generating high-quality silicon data.

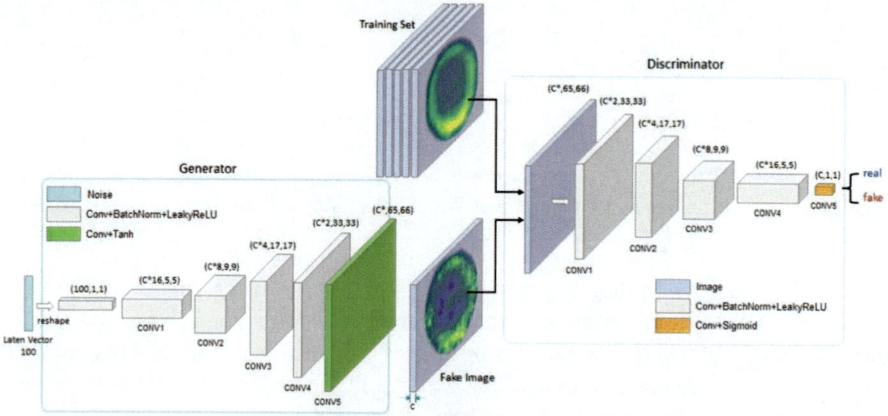

Fig. 6. GAN Model

During the model training, we utilized the gradient descent optimization algorithm to minimize the difference between the generated chips and real chips. To enhance the stability of the model, we employed techniques such as batch normalization and the LeakyReLU activation function. Through several hundred iterations, our GAN model was capable of generating highly realistic silicon data, including the chip's position on the wafer, the uniformity of physical features at the wafer level, and the defects present in the chip manufacturing process. These training outcomes provide a reliable data foundation for simulating and analyzing the chip manufacturing process.

4 Experimental Results and Collaborative Optimization Platform

This section will showcase the generated chip data using the GAN model and conduct a detailed analysis, while establishing a Design-Technology Co-Optimization (DTCO) platform. Our dataset consists of approximately 12 million chip data points, with the exclusion of 3σ outliers and missing chip data, deliberately retaining chips with uniformity defects as the training set for the GAN model. In addition to visualizing the data, we also utilize quantitative metrics to evaluate the quality of the generated silicon data. For instance, we use Jensen-Shannon Divergence to compare the similarity of probability distributions between the generated data and real chip data. Additionally, we leverage the Kernel Density Estimation (KDE) metric to quantify the numerical differences between probability distributions of different features. These evaluation methods ensure a reliable and accurate assessment of the generated silicon data quality.

To protect the confidentiality of chip technology and wafer process data, we have uniformly normalized the charts and figures of our research results, limiting the numerical range between 0 and 1.

The experimental results show that the scatter plots between the features of the generated silicon data by the GAN model and the real chip data are highly similar, capturing the process adjustment and variability in the early stages, as shown in Fig. 7. Further analysis using the Jensen-Shannon Divergence index reveals that the probability distributions of the generated silicon data closely align with the characteristics of the real data, ranging from 0.98 to 1.0 across different dimensions, as shown in Fig. 8.

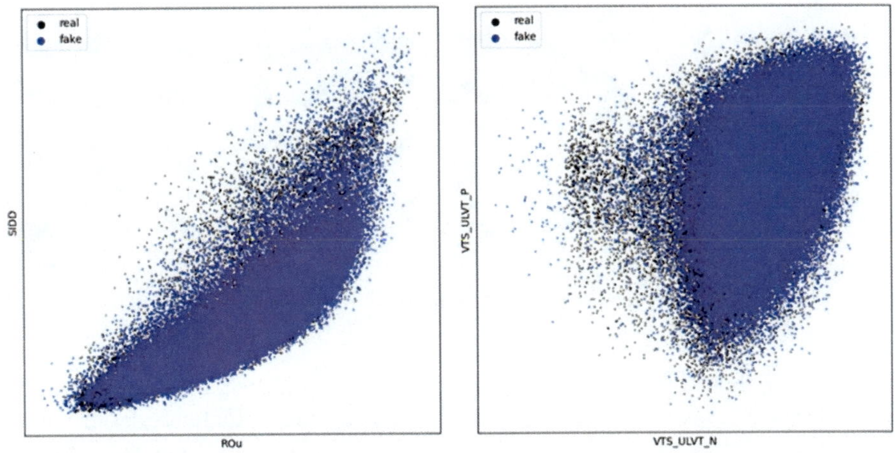

Fig. 7. Scatter Plots of Features for Generated Data and Real Data

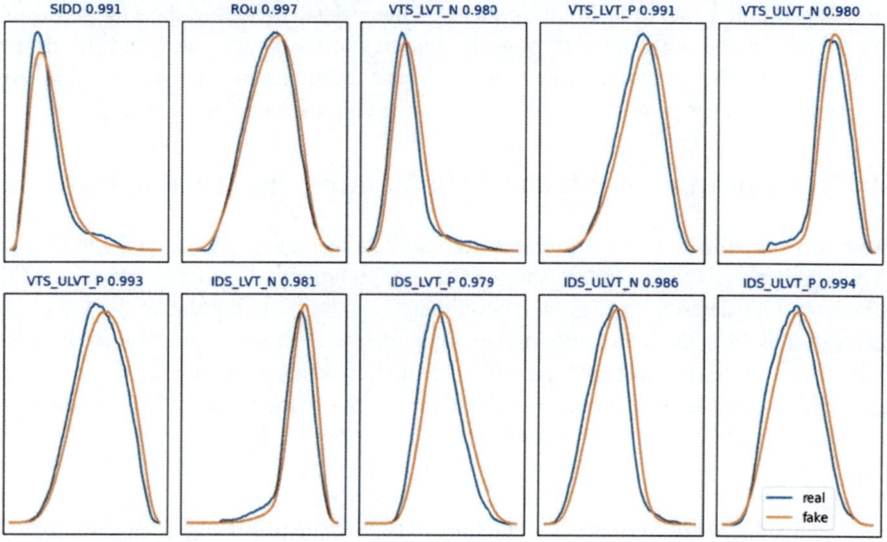

Fig. 8. Probability Distribution of Generated Data and Real Data for Each Feature

Furthermore, the combination of generated data in high-dimensional space still preserves the correlations of the original parent population, as shown in Fig. 9.

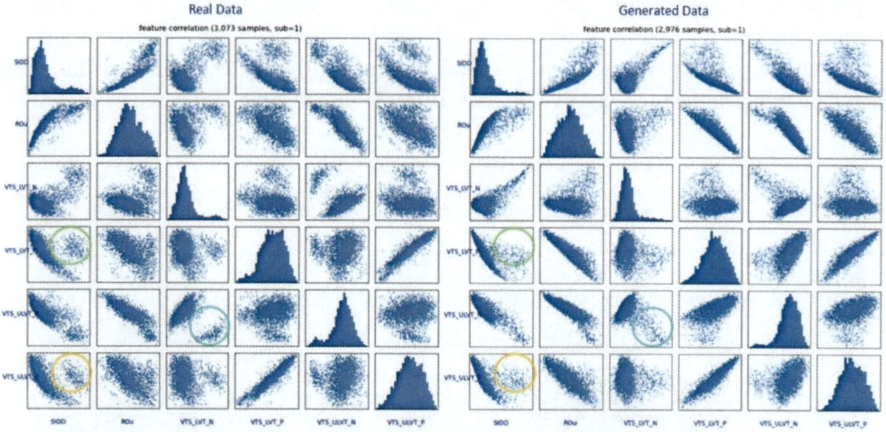

Fig. 9. Preservation of Feature Combination in High-Dimensional Space

Figure 10 shows the compromise space between yield and design margin based on the generated large dataset of chip data, providing specific guidance for future design recipes and capacity optimization. The model demonstrates good stability and generalization performance across different training and testing sets.

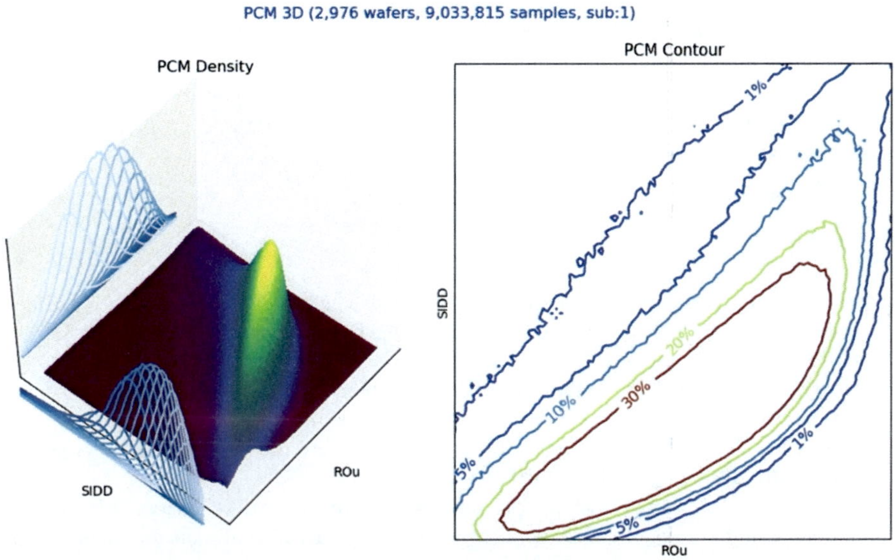

Fig. 10. Probability Density Distribution of Multidimensional Features

5 Conclusions

This study presents the application of Generative Adversarial Networks (GANs) in chip and wafer test data modeling and silicon virtualization. The study explores the utilization of virtual silicon data in Design-Technology Co-Optimization (DTCO) and showcases several related design and process co-optimization schemes. The research aims to assist process and chip design engineers in generating more realistic design examples, facilitating trade-offs between different process recipes and binning strategies for overall capacity optimization. Additionally, the study contributes to the optimization of process parameters and design margins, enabling better energy efficiency designs. However, GANs also face challenges, including selecting appropriate generator and discriminator architectures and handling higher-dimensional and complex data. Furthermore, training the generator and discriminator models requires significant time and computational resources.

To enhance the transformation of data into trainable models, we employ a method that converts multidimensional data into two-dimensional images with multiple feature channels. This approach enables us to simulate uniformity defects and variations in process parameters that may occur in wafer manufacturing during the training process. The research findings substantiate the effectiveness of this proposed method in supporting chip design, product optimization, and process improvement. It leads to enhanced production efficiency, cost reduction, and improved product quality, thereby offering valuable contributions to the industry.

In summary, GANs hold enormous potential for the development of advanced EDA tools. By harnessing GAN-generated models to capture the intricate mapping between process and design, we can achieve enhanced efficiency in process and design optimizations. Nevertheless, further research and development efforts are necessary to address current challenges and limitations. We eagerly anticipate increased attention from researchers in this domain, as they continue to explore and propose innovative solutions for the future.

References

1. Wang, J., et al.: AdaBalGAN: an improved generative adversarial network with imbalanced learning for wafer defective pattern recognition. IEEE Trans. Semicond. Manuf. **32**(3), 310–319 (2019)
2. Hu, G., et al.: Unsupervised fabric defect detection based on a deep convolutional generative adversarial network. Text. Res. J. **90**(3–4), 247–270 (2020)
3. Kusiak, A.: Convolutional and generative adversarial neural networks in manufacturing. Int. J. Prod. Res. **58**(5), 1594–1604 (2020)
4. Singh, R., et al.: Generative adversarial networks for synthetic defect generation in assembly and test manufacturing. In: 2020 31st Annual SEMI Advanced Semiconductor Manufacturing Conference (ASMC), pp. 1–5. IEEE (2020)
5. Hertlein, N., et al.: Generative adversarial network for early-stage design flexibility in topology optimization for additive manufacturing. J. Manuf. Syst. **59**, 675–685 (2021)
6. Qin, J., et al.: A novel temporal generative adversarial network for electrocardiography anomaly detection. Artif. Intell. Med. **136**, 102489 (2023)

Joint Multi-view Feature Network for Automatic Diagnosis of Pneumonia with CT Images

Hao Cui, Fujiao Ju(✉), and Jianqiang Li

Faculty of Information Technology, Beijing University of Technology, Beijing 100124, China
jfj2017@bjut.edu.cn

Abstract. Automated recognition of pneumonia from chest CT plays an important role in the subsequent clinical treatment for patients. While a few pioneering works only focus on several random slices from chest CT image, thus they have ignored the anatomical dependency information of local lesions. Considering it, this paper explores a novel automatic classification method for pneumonia detection based on fusing regional and global information, which not only improves detection performance, but also provides explainable diagnostic basis for radiologists. Firstly, identifying the interested local region by a lesion detection module, then we extracts the correlation relationship between local regions through a graph attention module. The image-level classification results can be acquired by fusing the information of global and local region. To realize the detection of full CT sequence, a person-level classifier is designed in the proposed model. In the experiment, we collected 781 chest CT sequences in total corresponding to 274 cases of viral pneumonia patients, 285 cases of bacterial pneumonia patients and 222 cases of healthy people. The experimental results show that our model achieves the accuracy of 95.5%, with 95.6% precision and 0.991 AUC. The recall and F1 score are 95.8% and 95.7% respectively, which outperformed previous works. Therefore, our method can be regarded as an efficient assisted tool in the diagnose of pneumonia.

Keywords: Pneumonia diagnosis · Chest CT sequence · Graph network · Multi-view feature fusion

1 Introduction

Pneumonia is one of the greatest threats for human health, which can spread world-wide especially for viral pneumonia with strong transmissibility. The clinical practice shows that the main factors causing pneumonia are virus, bacteria, fungi and mycoplasma [1]. Due to the distinction of etiology and clinical manifestations of different types of pneumonia [2], the treatment measures for patients should be formulated according to their image findings. At present, the diagnosis of different types of pneumonia can be preliminarily determined by Chest X-Ray(CXR). However, the resolution of CXR is too low that it is difficult to detect

J. C. Hung et al. (Eds.): FC 2023, LNEE 1134, pp. 169–180, 2024.
https://doi.org/10.1007/978-981-99-9342-0_19

the lesions, resulting in misdiagnosis [3]. As chest Computed Tomography(CT) scan provides more detailed cross-sections of the organ, it is regarded as the most effective and commonly used imaging technique for diagnosing many diseases [4,5]. Moreover, the chest CT scans of different pneumonia patients show different characteristics. For bacterial pneumonia patients, the chest CT scans mainly show pulmonary parenchymal involvement, lung consolidation and central lobular nodules or tree buds sign, while the chest CT infected by virus mainly shows ground glass opacity, grid opacity and paving stone sign [6]. Therefore, radiologists often make preliminary etiological judgments based on the patient's CT scans. In recent years, pneumonia patients have gradually increased, accompanied by a significant increase in the number of chest CT scans, which leads to a serious shortage of experienced radiologists [7]. Due to the complexity of lung structure, the lesions of different pneumonia are difficult to distinguish, thus it becomes a serious challenge for doctors to diagnose quickly and accurately, especially when respiratory infectious breakout in a large scale. Therefore, it is extremely important to establish a automatic system to assist the clinical practice in speeding up screening and improving the diagnose accuracy.

This study is based on the collected chest CT scans containing viral pneumonia, bacterial pneumonia patients and healthy person in the Beijing area. In this paper, we explore a joint multi-view feature network (JMFNet) for automatic diagnosis of pneumonia with CT images. Firstly, extracting the global features of CT images by a backbone, then we find the interested local regions by adding a object detection branch on backbone. Next, a graph attention module is used to extract the correlation information between local regions. By fusing the information of global slice and local region, we can obtain the classification results of each slice. Finally, for a full chest CT sequence, we design a person-level prediction mechanism to simulate the diagnostic process in clinical operations.

The main contributions of this paper are summarized as follows:

- To simulate the radiologist's diagnosis process, we construct a automatic pneumonia diagnosis system based on full CT sequence prediction, so as to avoid missed diagnosis or misdiagnosis;
- To focus on the key regions, we extract the interested lesions by a object detection branch and mine the relative position dependencies between lesion regions;
- To improve the accuracy of the automatic diagnosis system, we fuse the multi-view information from global and local region to construct recognition model and then provide interpretable analysis to assist doctor diagnosis.

2 Related Work

In clinical practice, radiologists usually observe CXR or CT scans to realize the diagnose of pneumonia [4]. To establish automatic diagnose system, the researchers try to explore machine learning algorithms, which can be mainly divided into two categories, traditional pattern recognition algorithms and deep learning algorithms. In general, the traditional algorithms manually extract features and then construct classifier [8]. In [9], Jin C et al. use the public CXR dataset for feature selection, then compare the prediction results of decision tree, random forest and support vector machine models. Cheng Jin et al. [10] extract 12 characterizing features of CT slices and test that there is a significant difference between community acquired pneumonia (CAP) and COVID-19 by constructing the lasso regression model. However, the feature extractions based on pattern recognition algorithms mainly depend on the experience of experts and feature transformation, which lack abstract hidden information and result in limitations in performance improvement. As Convolutional Neural Networks (CNN) [11] have been successfully applied to the medical imaging, some researchers have attempted to combine traditional pattern recognition algorithms with deep learning to detect lung disease through CT scan or CXR. In [12], the authors propose CheXNet to extract CXR features by deep pretraining model and manual techniques. The most important features can be selected by combining principal component analysis (PCA) and recursive feature elimination (RFE). The final comparison shows that XGBoost classifier has the best performance.

Due to the excellent feature extraction ability of deep learning, researchers have intend to utilize deep learning algorithms to build lung disease diagnose models. Many works regard CT images classification as a coarse-grained classification task, and randomly select CT slices as input. For example, Shouliang Qi et al. [13] pick ten CT slices from a patient's CT sequence for feature extraction. Each slice is processed by ResNet50 [14] for feature extraction so as to obtain the prediction of each slice. The final prediction of a patient is generated by aggregating slice predictions. In general, the slices with lesions are unknown, thus randomly selecting slices may result in misdiagnosis, which is also inconsistent with the diagnose habits of radiologists. In addition, the most of the existing methods realize image-level prediction only using global features. The lesions in CT slices are easily confused, which lead to the low variance between different types of pneumonia. Therefore, subtle lesions containing the correlation between lesion areas are also important and critical. In [15], Ying Song et al. extract the main region of lung and design a details relation extraction neural network to obtain the image predictions. The person diagnosis are achieved by aggregating the image predictions.

Different pneumonia have different lesion distribution in the lung window. For viral pneumonia, lesions are distributed in the subpleura and lesions of bacterial pneumonia are distributed in the lung parenchyma. As graph neural networks (GNN) [16] performs well in acquiring the correlation between target nodes. Several studies have begun to utilize GNN to extract correlation information in medical image analysis. In [17], the authors attempt to construct the graph by calculating the Euclidean distance between the extracted features vectors to predict the viral pneumonia via CT. To utilize the similarity and co-occurrence between lesions, in this study, we segment the interested local region and construct the graph attention network [18] based on lesions. By analyzing the joint multi-view features, we design a automatic diagnosis of pneumonia with CT images. The specific model is introduced in the next section.

3 Method

3.1 Data Collection and Processing

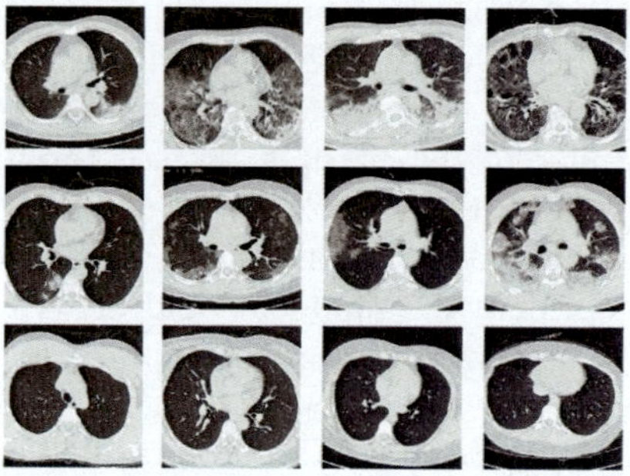

Fig. 1. The images in chest CT sequence of different pneumonia patients. The three rows show the lung images of bacterial, viral patient and healthy person from top to bottom, respectively.

This study is based on 781 plain chest CT sequences provided by Beijing Ditan Hospital and all patients are confirmed by clinical examination. The images of CT sequence are scanned at 5mm interval, and performed with a spiral scan after inspiratory breath hold. Each image contains detection time, detection specification and other parameters. In order to pay more attention to the information in lung window, we crop the original CT images to ensure that the contents of the lung window are preserved and the noise is removed. As a 3D CT sequence of one patient contains more then 300 images, we delete the images without lung parenchyma. Considering adjacent images are highly similar, we select 64 representative images with lesions to speed up the calculation [19]. Finally, the collected dataset includes 274 viral pneumonia patients with 16387 images, 285 bacterial pneumonia patients with 15848 images and 222 healthy person with 14208 images. Figure 1 shows several CT images of different types of pneumonia patients.

3.2 Model Architecture

This paper proposes an automatic pneumonia diagnosis system based on graph attention mechanism utilizing chest CT sequence. We firstly introduce the process of image-level pneumonia detection, which is based on fusing multi-view information of the interested regions and whole image. After that, the designed

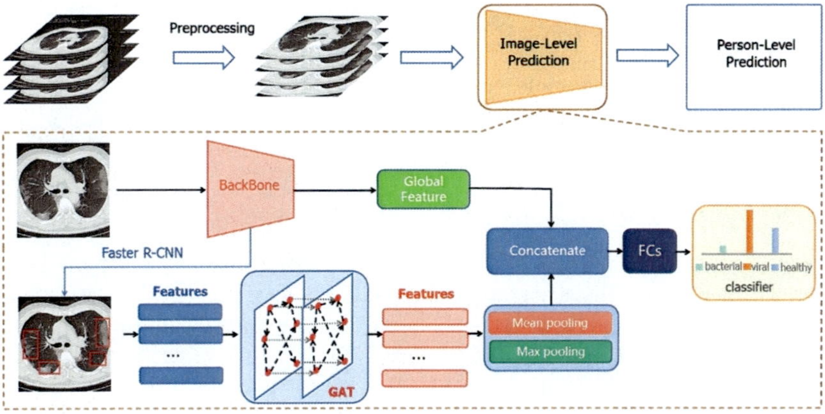

Fig. 2. The architecture of JMFNet. Extracting the global features of CT images and then a fast R-CNN branch is used to detect the top-k local lesion regions. Then the obtained local region features are fed to a graph attention neural network. Two pooling operations are performed on the aggregated k features, and they are combined with the global features and sent into the dense network for the final image-level classification. Finally, the person-level prediction is obtained based on all image classification results.

person-level classifier is introduced based on the image-level classification results. The model architecture is shown in Fig. 2.

In the proposed model, the collected CT images are preprocessed by converting dicom format to png images and removing boundary regions around the lungs. Then all images are input to the image-level prediction module. In this module, we extract the feature map and global features by a backbone network. The feature maps are then input to Faster R-CNN [20] to detect lesion regions. To explore the relationship between the anatomical regions in the CT image, we use a graph attention network (GAT) [16] to learn their dependencies. Finally, we combined the localized region features and global features to realize image-level prediction. Based on the image-level prediction results, we design a person-level classifier.

Defining a chest CT dataset as $X = \{x_1, x_2, ..., x_N\}$, x_i represents a patient CT slice. All the images are input to a backbone network to extract global features. To focus on local lesion information, a Faster R-CNN branch [20] is added on the backbone and we embed Feature Pyramid module (FPN) [21] in Faster R-CNN to detect lesions with different size. In this way, we obtain top k detection regions with the highest score as local information. As the detected interested regions have no uniform size, thus we need to normalize the size of regions to a unified value for the subsequent operations. In the experiments, the detected regions are resized to $112 * 112$. Finally, the features of the top k anatomical regions in image x_i as H_i can be denoted as,

$$H_i = f(x_i) = \{h_1, h_2, ..., h_{k-1}, h_k\} \in \mathbb{R}^{k \times d}.$$

We take those k region features as the nodes of the graph to explores the correlation relationship. The new embedding expression h_i' can be updated as follows,

$$h_i' = \sigma\left(\sum_{j \in N_i} a_{ij} W h_j\right) \tag{1}$$

where $W \in \mathbb{R}^{d \times d}$ is the learned weight matrix, N_i represents the notes in the neighbor of node i and $\sigma(\cdot)$ denotes the nonlinear activation function. In addition, a_{ij} is the normalized importance weight coefficient of node j for node i, which can be calculated as follows:

$$a_{ij} = softmax_j(e_{ij}) = \frac{\exp(e_{ij})}{\sum_{k \in N_i} \exp(e_{ik})}$$

where

$$e_{ij} = LeakyReLU\left(a^T W h_i \parallel W h_j\right),$$

which represents the importance of the node j to node i and a is a weight vector.

By introducing the attention mechanism into the graph network, the correlation between different focal areas can be better integrated into the model. The neighbor nodes in (1) can be determined by similarity matrix S_{ij},

$$S_{ij} = \frac{h_i \cdot h_j}{\|h_i\| \, \|h_j\|} \tag{2}$$

We set a similarity threshold θ to judge whether the nodes are the first-order neighbors. If $S_{ij} > \theta$, node j is a neighbor of node i, otherwise node j does not belong to the neighbor of node i.

The obtained k region embeddings h' have been transferred to mean pooling and max pooling operation. Finally, we concatenated the global features and graph embeddings of local region to a 1-D vector and then sent them into a full connection layer and Softmax classifier. The loss function in the image-level prediction is computed by cross-entropy function,

$$\ell = -\sum_{c=1}^{C} \sum_{i=1}^{N} y_{ic} \log y_{ic}' \tag{3}$$

where C is the class number, y_{ic} and y_{ic}' are ground truth and predicted label, respectively.

3.3 The Person-Level Classifier

In the image-level prediction, we not only obtain the prediction label of each training image, but also get the probability belonging to the corresponding label. By utilizing the image-level prediction results, we design the person-level prediction method, which contains two steps. The first step is the judgment for healthy person. If all the image-level prediction results of a full CT sequence are healthy, we can conclude that the person-level prediction of the test sample is health person. Otherwise, the test sample is from bacterial or viral pneumonia, which can be determined by the second step. In this step, deleting the slices with healthy prediction results, then we perform averaging operations on the remaining slices. Comparing the average distribution probability of bacterial and viral pneumonia, and finding out the largest value, then the test sample is divided into the label corresponding to the largest average value.

4 Experiment and Results

4.1 Implementation and Evaluation

For the requirement of hospital, we have designed two classification tasks: discriminating between viral and bacterial infected patients, and separating viral patients from bacterial patients and healthy person. In the experiments, we extract 4 or 6 interested lesion regions in each CT image and the five-fold validation is used to assess robustness and the performance of the model. In each fold, we randomly split the dataset according to the ratio of 6:2:2 for training, validation and testing.

For the initialization of the model, the learning rate is set to 0.0001 and Adam optimizer is selected to update the parameters. All the codes are implemented in Pytorch and performed on a server with Nvidia GeForce GTX 3090 GPUs. To assess the performance of the proposed model, the accuracy, precision, F1 score and recall are used as metrics to measure the evaluation of the classification results.

4.2 Experiment Results and Analysis

In this section, we have introduced the detailed description of binary and three-classification tasks.

Binary Classification Task. For the classification of viral and bacterial pneumonia, we divide the dataset into three groups in the each fold: one group contains 164 viral and 171 bacterial pneumonia patients for training, the second group is for validation with 56 viral and 57 bacterial pneumonia patients, the last group contains 54 viral and 57 bacterial pneumonia patients for testing. We compared the results of different backbones consist of VGG-16 [22], GoogLeNet [23], DenseNet-121 [24], ResNet-50 [14] and ResNet-101 [14]. All the architectures are pre-trained on ImageNet for transfer learning.

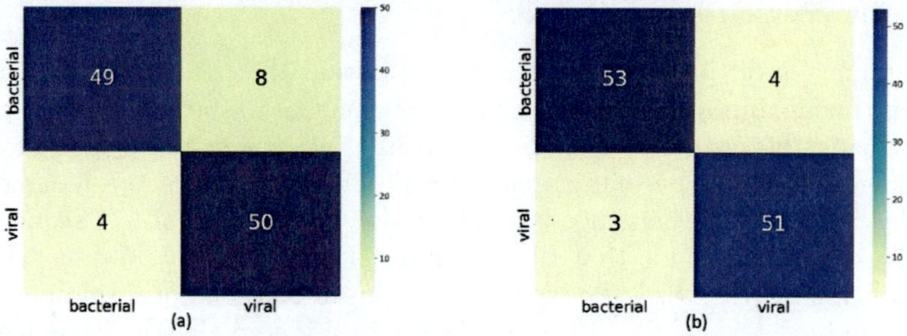

Fig. 3. Confusion matrix for binary classification tasks

Table 1 shows the results of different architectures. From the table we can see that the architecture ResNet-50 can achieve the highest classification accuracy of 86.5% and 88.4% for precision. For the recall and F1 score, the highest results are 87.3% and 86.5% acquired by DenseNet-121 and GoogLeNet, respectively. ResNet-50 have obtained 86.8% recall and 86.4% F1 score, which are just lower slightly than that of DenseNet-121. Thus, we choose ResNet-50 as the backbone of the proposed model for global feature extraction. In the proposed model, we intend to extract the anatomical dependency information of local lesions by adding a Faster R-CNN branch on the backbone. To verify its superiority, we design ablation experiments and the results are listed in Table 2. 'JMFNet w/o Local' means the proposed model without fusing local dependency information, which only extracts global features of slices for image-level classification. 'JMFNet w/o GAT' represents the model containing Faster R-CNN branch with global feature extraction module. 'JMFNet' lists the results of our proposed model, which has achieved obvious improvement. The ablation experiments have valid the effectiveness and superiority of fusing multi-view features. Except for ablation experiment, we also compare the classification results of other three methods, including DeCoVNet [25], AD3D-MIL [26] and DRENet [15]. Finally our model also obtain the higher results for all indicators. Figure 3 (a) shows the confusion matrix of the proposed model. A total of 12 test samples are classified incorrectly. By analysis, we deduce that the reason for low accuracy is that the training dataset only contains the slices with viral and bacterial lesions in image-level prediction. However, the full CT sequence of a test sample consists

Table 1. Comparison of different architectures

Method	Accuracy	Precision	Recall	F1 score
VGG-16	0.842	0.801	0.843	0.843
GoogLeNet	0.863	0.815	0.853	**0.865**
DenseNet-121	0.826	0.767	**0.873**	0.835
ResNet-50	**0.865**	**0.884**	0.868	0.864

Table 2. Comparison of different methods for binary classification task

Method	Accuracy	Precision	Recall	F1 score
DeCoVNet	0.813	0.739	0.825	0.780
AD3D-MIL	0.848	0.762	**0.906**	0.828
DRENet	0.883	0.885	0.884	0.883
JMFNet w/o Local	0.865	0.884	0.868	0.864
JMFNet w/o GAT	0.875	0.890	0.882	0.884
JMFNet (our method)	**0.892**	**0.893**	0.893	**0.892**
JMFNet w/o healthy slices	**0.9369**	**0.9371**	**0.9367**	**0.9367**

of 200–230 slices, most of which may not contain lesions. Therefore, the slices without lesions would be identified as viral or bacterial infections, resulting in wrong results. Considering this, the healthy slices should be added to the training dataset in the image-level prediction even for the binary classification. The subsequent experiments have confirmed this point.

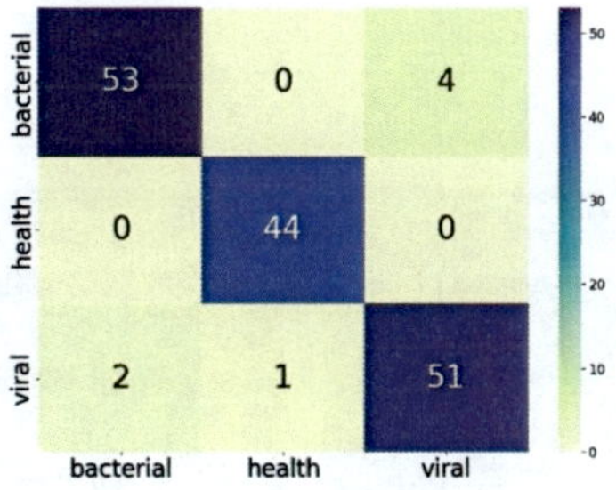

Fig. 4. Confusion matrix for three-classification task

Three-Classification Task. According to the above analysis, we need to add healthy slice samples to the training set and train a three-classification model in the image-level prediction, even for the binary classification of virus and bacteria pneumonia. The training set contains 164 viral, 171 bacterial pneumonia patients and 134 healthy person. Table 2 lists the results for the classification of viral and bacterial pneumonia after adding healthy samples in image-level prediction. Comparing with the above results, the accuracy has increased from 89.2% to 93.69% and all other indicators have improved significantly. Figure 3 (b) shows the improved confusion matrix for binary classification task. Comparing the two sub-figures we can see that the number of samples classified incorrectly has decreased to 7 samples. For the three-classification of viral, bacterial patients and healthy persons, keeping the above training process, we only need to add

Table 3. Comparison of Different Method for Three-classification task

Method	Accuracy	Precision	Recall	F1 score
DeCoVNet	0.897	0.882	0.850	0.861
AD3D-MIL	0.906	0.937	0.841	0.861
DRENet	0.942	0.945	0.946	0.945
JMFNet w/o Local	0.890	0.903	0.900	0.895
JMFNet w/o GAT	0.931	0.935	0.931	0.932
JMFNet(our method)	**0.955**	**0.956**	**0.958**	**0.957**

healthy person samples to the test set. The number of healthy CT sequences are 44 in the test set. Table 3 shows the comparison results of different methods and our model has still achieved the best results. Figure 4 shows the confusion matrix for three-classification task. For bacterial patients, there are four misdiagnosed cases, which are regarded as viral infection. Two viral patients were diagnosed with bacterial infection and one was identified as a healthy person.

4.3 Interpretability

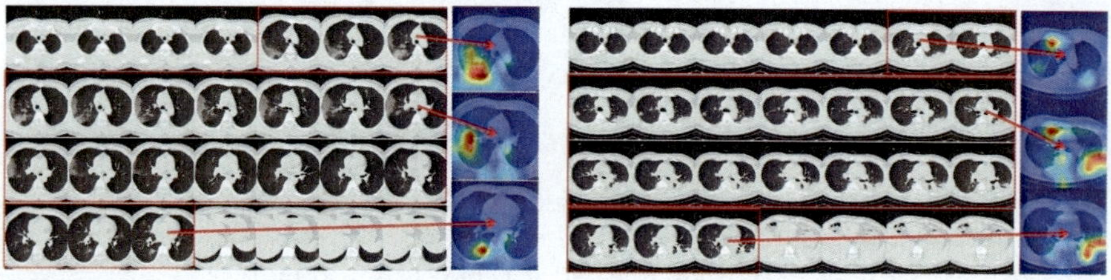

Fig. 5. Visualization of patients with pneumonia by different etiologies on CT images. The difference in CT appearance between viral and bacterial pneumonia is shown from left to right.

To provide explainable diagnostic basis for radiologists, we show the several important slices and their visual maps by Gradient-weighted Class Activation Mapping (Grad-CAM) [27] in Fig. 5. For viral and bacterial pneumonia, the location of the lesions are different. The lesions of viral pneumonia mainly appears in the subpleura, while those of bacterial pneumonia are mainly concentrated in the lung parenchyma [28]. From the figure we can see that the proposed model has payed more attention to the lesion regions, which can clearly reflect the location information of the lesions. The visualization results are consistent with the clinical diagnosis basis, which further verifies the reliability of our method.

5 Conclusion

With the development of computer vision technology, artificial intelligence-assisted disease diagnosis performs excellent functions in clinical preliminary screening. In this study, we propose a joint multi-view feature network for automatic diagnosis of pneumonia based on chest CT sequences. We implemented the detection of focus regions and features aggregation by fusing CNN and graph attention module. Aims to explore the location relationship and dependency relations of local lesions, we add a Faster-RCNN branch on the backbone and construct graph attention network to extract the embeddings of lesions. The accuracy has improved by 0.65% comparing with the that without considering local features. The experiments demonstrate the feasibility and superiority of the proposed method.

References

1. Dueck, N.P., et al.: Atypical pneumonia: definition, causes, and imaging features. Radiographics **41**(3), 200131 (2021)
2. Lang, M., et al.: Pulmonary vascular manifestations of COVID-19 pneumonia. Radiol. Cardio. Imag. **2**(3), e200277 (2020)
3. Shi, F., et al.: Review of artificial intelligence techniques in imaging data acquisition, segmentation and diagnosis for COVID-19. IEEE Rev. Biomed. Eng. PP.99, 1 (2021)
4. Traub, M., et al.: The use of chest computed tomography versus chest X-ray in patients with major blunt trauma. Inj. Int. J. Care Inj. **38**(1), 43–47 (2007)
5. Lei, J., et al.: CT Imaging of the 2019 Novel Coronavirus (2019-nCoV) Pneumonia. Radiology **295**(1), 18 (2020)
6. Xi, Z., et al.: Dandelion and focal crazy paving signs: the lung CT based predictors for evaluation of the severity of coronavirus disease. Curr. Med. Res. Opin. **37**(2), 219–224 (2021)
7. Parag, P., Hardcastle, T.C.: Interpretation of Emergency CT Scans of the Head in Trauma: Neurosurgeon vs Radiologist". In: World Journal of Surgery: Official Journal of the Societe Internationale de Chirurgie, Collegium Internationale Chirurgiae Digestivae, and of the International Association of Endocrine Surgeons 46–6 (2022)
8. Mahadevkar, S.V., et al.: A review on machine learning styles in computer vision-techniques and future directions. IEEE Access **10**, 107293–107329 (2022). https://doi.org/10.1109/ACCESS.2022.3209825
9. Gupta, V.K., et al.: Prediction of COVID-19 confirmed, death, and cured cases in India using random forest model. Big Data Min. Anal. **4**(2), 116–123 (2021). https://doi.org/10.26599/BDMA.2020.9020016
10. Jin, C., et al.: Development and evaluation of an AI system for COVID- 19 (2020)
11. Krizhevsky, A., Sutskever, I., Hinton, G.: ImageNet classification with deep convolutional neural networks. In: Advances in Neural Information Processing Systems 25.2 (2012)
12. Abdel-Fattah Sayed, S., Mohamed Elkorany, A., Sayed Mohammad, S.: Applying different machine learning techniques for prediction of COVID-19 severity. IEEE Access **9**, 135697–135707 (2021). https://doi.org/10.1109/ACCESS.2021.3116067

13. Shouliang Qi, A.B., et al.: DR-MIL: deep represented multiple instance learning distinguishes COVID-19 from community-acquired pneumonia in CT images. Comput. Methods Programs Biomed. **211**, 106406 (2021)
14. He, K., et al.: Deep residual learning for image recognition. In: IEEE Conference on Computer Vision and Pattern Recognition (2016)
15. Song, Y., et al.: Deep learning enables accurate diagnosis of novel coronavirus (COVID-19) With CT Images. IEEE/ACM Trans. Comput. Biol. Bioinform. **18**(6), pp. 2775–2780 (2021). https://doi.org/10.1109/TCBB.2021.3065361
16. Kipf, T.N., Welling, M.: Semi-supervised classification with graph convolutional networks (2016)
17. Yu, X., et al.: ResGNet-C: a graph convolutional neural network for detection of COVID-19. Neurocomputing **452**, 592–605 (2020)
18. Velikovi, P., et al.: Graph attention networks (2017)
19. Takebe, H.: Development of similar CT image retrieval technology based on lesion natures and their three-dimensional distribution (2017)
20. Ren, S., et al.: Faster R-CNN: towards real-time object detection with region proposal networks. IEEE Trans. Pattern Anal. Mach. Intell. **39**(6), 1137–1149 (2017)
21. Lin, T.Y., et al.: Feature pyramid networks for object detection. In: 2017 IEEE Conference on Computer Vision and Pattern Recognition (CVPR) (2017)
22. Simonyan, K., Zisserman, A.: Very deep convolutional networks for large-scale image recognition. In: Computer Science (2014)
23. Szegedy, C., et al.: Going deeper with convolutions. In: IEEE Computer Society (2014)
24. Huang, G., et al.: Densely connected convolutional networks. In: IEEE Computer Society (2016)
25. Wang, X., et al.: A weakly-supervised framework for COVID-19 classification and lesion localization from chest CT. IEEE Trans. Med. Imag. **39**(8), 2615–2625 (2020). https://doi.org/10.1109/TMI.2020.2995965
26. Han, Z., et al.: Accurate screening of COVID-19 using attention based deep 3D multiple instance learning. IEEE Trans. Med. Imag. **39**(8), 2584–2594 (2020). https://doi.org/10.1109/TMI.2020.2996256
27. Selvaraju, R.R., et al.: Grad-CAM: visual explanations from deep networks via gradient-based localization. In: 2017 IEEE International Conference on Computer Vision (ICCV), pp. 618–626 (2017). https://doi.org/10.1109/ICCV.2017.74
28. Johansson, N., Kalin, M., Hedlund, J.: Clinical impact of combined viral and bacterial infection in patients with community-acquired pneumonia. Scandinavian J. Infect. Dis. **43**(8), 609–615 (2011)

Ensemble Deep Learning Techniques for Advancing Breast Cancer Detection and Diagnosis

Adam M. Ibrahim[1], Ayia A. Hassan[1], Jianqiang Li[1], and Yan Pei[2](✉)

[1] Faculty of Information Technology, Beijing University of Technology, Beijing 100124, China
adammohamed@email.bjut.edu.cn, ayaadamhassan@emails.bjut.edu.cn, lijianqiang@bjut.edu.cn

[2] Computer Science Division, University of Aizu, Aizuwakamatsu, Fukushima 965-8580, Japan
peiyan@u-aizu.ac.jp

Abstract. The integration of deep learning (DL) and digital breast tomosynthesis (DBT) presents a unique opportunity to improve the reliability of breast cancer (BC) detection and diagnosis while accommodating novel imaging techniques. This study utilizes the publicly available Mammographic Image Analysis Society (MIAS) database v1.21 to evaluate DL algorithms in identifying and categorizing cancerous tissue. The dataset has undergone preprocessing and has been confirmed to be of exceptional quality. Transfer learning techniques are employed with three pre-trained models - MobileNet, Xception, DenseNet, and MobileNet LSTM - to improve performance on the target task. Stacking ensemble learning techniques will be utilized to combine the predictions of the best-performing models to make the final prediction for the presence of BC. The evaluation will measure the performance of each model using standard evaluation metrics, including accuracy (ACC), precision (PREC), recall (REC), and F1-score (F1-S). This study highlights the potential of DL in enhancing diagnostic imaging and advancing healthcare.

Keywords: Breast Cancer · Deep Learning · Ensemble Learning · Detection · Artificial Intelligence

1 Introduction

BC is a frequent and lethal illness, making risk prediction difficult. Mammography is the most expensive early detection technology, and a standardized and community-based screening approach has been proposed to address this [1,2]. Cancer risk prediction methods use multiple risk factors, such as molecular genetics,

This study is supported by the National Key R&D Program of China with the project no. 2020YFB2104402.

J. C. Hung et al. (Eds.): FC 2023, LNEE 1134, pp. 181–192, 2024.
https://doi.org/10.1007/978-981-99-9342-0_20

imaging, and public health data, to accurately predict the likelihood of BC based on individual diagnostic imaging screenings [3]. Breast density is not a reliable predictor of BC risk, as it is used to determine the frequency of screening [4]. Mammography screening is essential to reduce death rates from breast cancer, but age is the main factor used to select people for screening. Interest is growing in customized screening methods [5]. Risk stratification using disease prediction models can identify women at risk of developing BC, allowing tailored surveillance to maximize benefit [6]. A technique used in histopathology photos to find cancer is the BC detection factor [7]. Cancer risk models are used to assess cancer risk and project outcomes, based on the elevated risk of BC linked to various characteristics, without any connection to the type of mammography used [8].

Cancer is a major global public health issue, with increasing prevalence in both industrialized and developing countries [9]. Breast disease is the unchecked, potentially cancerous proliferation of breast cells, and microscopic histopathological examinations are dependent on visual interpretation by experts, which is subjective and dependent on the observer's knowledge [10]. The process of multi-classification cancer diagnosis utilizing histology images is difficult and time consuming due to the lack of qualified pathologists in many low-income nations. This can lead to incorrect findings due to the intricacy of the pictures and the pathologist's limited ability to comprehend a large quantity of data [11,12]. Misinterpretation of screening mammography can lead to overdiagnosis, which can cost people money. To increase the efficacy of screening mammography, a new methodology was developed combining picture characteristics and a forecasting technique. Bidirectional screening mammography density imbalance was used as a signal to assess the likelihood of BC in computed tomography images [13]. The experiment tested if a DL-based method could outperform more established frameworks for identifying cancer risk, as patients often have repeated mammography examinations during BC monitoring [14]. Predicting the results of a single abnormal mammogram is the screening task, but we did not use a large number of priors as inputs to the models [15]. Research opportunities for intelligent forecasting of biological subtypes have become available due to the rapid development of BC detection technologies. However, forecasting for biological subgroups remains a challenging problem.

1.1 Related Works

According to the ratings of the remaining data, writers in [16] used a deep feed-forward network to train a RankDeepSurv model for predicting relapse in patients with nasopharyngeal cancer. The RankDeepSurv model outperformed more conventional survival analysis techniques, achieving a C-index of 0.681. In [17], the authors utilized images of tumor cell extracts and presented a DL network that combines convolutional neural networks (CNNs) with recurrent models to predict the prognosis of colorectal cancer.

The study analyzed 420 colorectal cancer patient tumor samples and demonstrated that DL algorithms can extract more predictive information from tissue shape compared to conventional human observation techniques. In [18], DeepSurv is introduced as a deep neural network (DNN) for survival analysis based on Cox regression hazard models. It predicts the correlation between a person's variables and clinical outcomes, utilizing link weights to determine the impact of patient variables on their risk level. DeepSurv outperforms existing advanced survival models and captures more intricate relationships between participant characteristics and failure risk. According to research published in [19], MesoNet is a deep convolutional neural network method that estimates the likelihood of survival for mesothelioma patients without the assistance of a toxicologist to pinpoint specific areas. MesoNet can identify regions in the stroma and histology that are linked to patient outcomes. The study suggests that DL algorithms can potentially discover previously unknown features as predictive biomarkers of clinical outcomes. In [20], researchers outline three techniques, including the use of a solitary training batch to assess CNN training effectiveness, applying a dispersed stochastic neighbor modeling approach to reveal class separations in deep layer activations, and employing DeepDream with specific settings to visualize deep neuron activations in the VGG19 DL model's 46 layers. Researchers in [21] present three residual DNN models as options for estimating methylation conditions without the need for a separate tumor segmentation step. The study shows that ResNet50 outperformed ResNet18 and ResNet34, achieving a statistically significant accuracy (ACC) of 94.90%.

2 Methodology

This section will outline our approach, which will be broken down into a series of steps illustrated in Fig. 1.

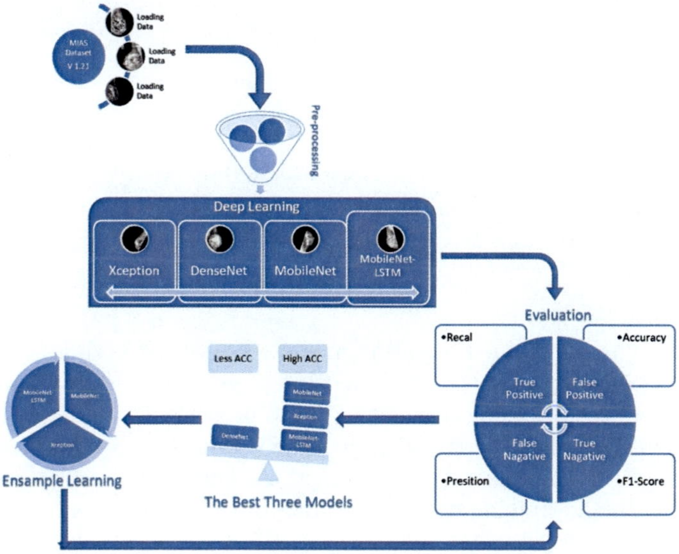

Fig. 1. The proposed framework for breast cancer classification involves collecting mammography images from the MIAS dataset, comparing the performance of each model, identifying the best performing models for feature extraction, and training multiple models using ensemble learning. The predictions are then pooled to create a final forecast.

2.1 Dataset

The Mammographic Image Analysis Society (MIAS) database v1.21 has been utilized to identify and diagnose breast cancer. It consists of 322 digitized mammograms, with 208 labeled benign and 114 labeled malignant. The ground truth data includes information about the size and shape of any masses or microcalcifications, the degree of speculation, and the presence of architectural distortions. These data play a crucial role in comprehending the normal structure of breast tissue.

2.2 Preprocessing

Preprocessing steps are crucial for medical image analysis tasks as they greatly influence the model's performance. Inaccurate preprocessing can introduce image artifacts or anomalies, potentially compromising the model's accuracy (ACC). It is vital to ensure that the selected preprocessing methods are suitable for the specific task and that the data is of exceptional quality. Although the dataset has already undergone preprocessing and is available online, it is still important to verify that the applied preprocessing techniques are appropriate for the task and that the preprocessing process has not introduced any artifacts or errors.

2.3 Deep Learning

In this section, we will elaborate on the DL techniques that were utilized in our approach.

MobileNet. The MobileNet architecture was implemented, which includes convolutional and pooling layers, followed by multiple fully connected layers. The ReLU activation function was used, and the model was trained for 100 epochs with a batch size of 16. The Adam optimizer was utilized, along with the definite cross-entropy loss function and the accuracy (ACC) metric for evaluation. To fine-tune the pre-trained model, only the weights of the fully connected layers were trained while keeping the convolutional layers frozen.

Xception. The pre-trained Xception model was utilized with a predetermined input shape and size, and the "include top" parameter was set to False. Fresh layers were added, including dropout layers to prevent overfitting and fully connected layers with ReLU activation functions. The output layer consisted of two nodes with a softmax activation function. The model was trained for 100 iterations, with the training data being randomly shuffled before each iteration.

MobileNet-LSTM. This approach utilized the MobileNet architecture as a feature extractor and incorporated an LSTM layer for sequence processing. The model was constructed using the Adam optimizer with a batch size 16 and trained for 100 iterations, with the training data being randomly shuffled before each iteration. To incorporate sequence processing, the output of the fully connected layers was reshaped into a 3D tensor with a shape of (batch size, time steps, and input dimension). The LSTM layer had 256 units and a dropout rate of 0.3. Additionally, a softmax output layer with two nodes was added. The model was trained for 100 iterations, with the training data being randomly shuffled before each iteration.

DenseNet. The DenseNet201 model is pre-trained on the ImageNet dataset, and for our specific classification task, the last few layers are replaced with new layers. The code starts by loading the pretrained DenseNet201 model, specifying the input shape of the images, excluding the top layers, and setting the pooling method to average. Next, new layers are defined on top of the pretrained model, taking inputs from the pretrained model and producing outputs for the classification task. The model is then compiled using the Adam optimizer, categorical cross-entropy loss, and the accuracy (ACC) metric. It is fitted on the training data for 100 epochs, with a batch size of 16. The data is shuffled after each epoch.

2.4 Ensemble Learning

After comparing the performance of the DL models, the best three models will be selected based on their evaluation metrics. An ensemble learning technique, such as voting or stacking, will be employed to combine the predictions of these three models and generate a final prediction for the presence of breast cancer (BC) in mammography images. This ensemble learning approach aims to improve the model's overall performance by leveraging each individual model's strengths and mitigating their weaknesses.

2.5 Evaluation

During the assessment phase, common evaluation metrics such as accuracy (ACC), precision (PREC), recall (REC), and F1-score (F1-S) will be utilized to evaluate the performance of each model. These metrics are computed using the confusion matrix, which summarizes the model's performance across four categories: true positives (TP), false positives (FP), true negatives (TN), and false negatives (FN). Accuracy (ACC) measures the proportion of correctly classified samples. Precision (PREC) calculates the ratio of correctly classified positive samples to the total number of positive predictions. Recall (REC) computes the proportion of correctly classified positive samples to the total number of positive samples in the dataset. F1-score (F1-S) is a harmonic mean of precision and recall, providing a balanced measure of the model's performance. These metrics are useful for datasets with imbalanced classes and can be calculated using specific formulas, as depicted in Eqs. 1–4.

$$ACC = \frac{TP + TN}{TP + TN + FP + FN} \quad (1)$$

$$PREC = \frac{TP}{TP + FP} \quad (2)$$

$$REC = \frac{TP}{TP + FN} \tag{3}$$

$$F1 - S = 2 \cdot \frac{PREC \cdot REC}{PREC + REC} \tag{4}$$

3 Results and Comparison

3.1 Results

The results indicate that MobileNet achieved the highest accuracy (ACC) of 93.34%, followed by MobileNet-LSTM with an ACC of 92.64%. Xception achieved an ACC of 88.62%, and Densnet achieved an ACC of 87.23%. Analyzing the confusion matrix Fig. 2 of each model, we can observe that MobileNet had the fewest misclassifications, with only 18 false negatives and 30 false positives. Xception had a higher number of false positives with 52 misclassifications, while Densnet had the same number of false positives but more false negatives. It is important to note that these results were obtained using the same dataset and training procedure, indicating that the variations in performance can be attributed to the architectural differences of the models. MobileNet and MobileNet-LSTM employ lightweight architectures optimized for mobile devices, which may contribute to their superior performance. On the other hand, Xception is a deeper and more complex model, which may have made it more challenging to train effectively with the limited dataset. Densnet, despite having a similar number of layers to Xception, has a different architecture that might have affected its lower performance. Additionally, although transfer learning was employed to initialize the models' weights using pretrained weights from ImageNet, the specific pretrained model used could have influenced the overall performance. The results suggest that the MobileNet architecture is well-suited for this classification task and outperforms other architectures such as Xception and Densnet. However, further experimentation with different architectures and datasets may yield different outcomes.

Multiple models were trained using the ensemble learning approach, and their predictions were pooled to obtain a final forecast. The three best-performing models, MobileNet, MobileNet-LSTM, and Xception, were chosen to create the ensemble model. The ensemble model combined the predictions of the three models using a simple voting scheme, where the class with the highest number of votes was considered the final prediction. The ensemble model achieved an ACC of 94.45%, which is a significant improvement over the individual models' performances. Comparing the confusion matrices of the individual models and the ensemble model, it can be observed that the ensemble model had fewer misclassifications. This is because the ensemble model took into account the strengths and weaknesses of each individual model and made a final prediction based on their combined expertise. In conclusion, the ensemble model achieved the highest ACC, indicating that combining the predictions of multiple models resulted in a more accurate and robust model. The classification

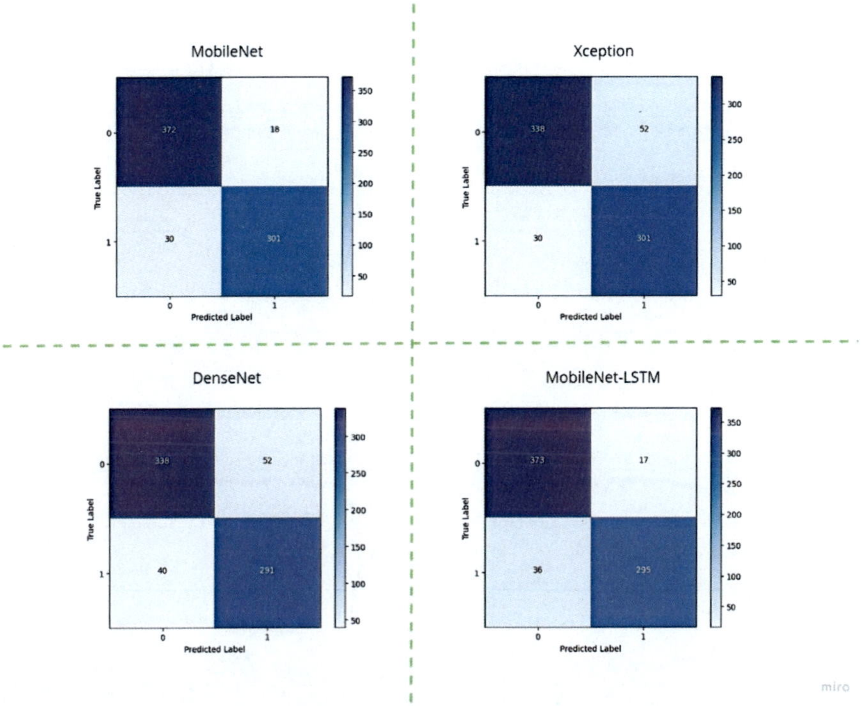

Fig. 2. MobileNet had the fewest misclassifications, with 18 false negatives and 30 false positives. Xception had more false positives, while Densnet had the same number but more false negatives. MobileNet-LSTM had 17 false negatives and 36 false positives.

report demonstrates that all models have achieved high ACC, PREC, REC, and F1-S scores in detecting malignant and benign tumors. The MobileNet and MobileNet-LSTM models obtained the highest PREC scores of 0.93 and 0.91 for benign tumors, respectively, while the ensemble model achieved the highest PREC score of 0.95 for malignant tumors. The MobileNet-LSTM model also achieved the highest REC score of 0.96 for benign tumors, whereas the Xception model achieved the highest REC score of 0.91 for malignant tumors. On the other hand, the Densenet model exhibited the lowest ACC and F1-S scores among the individual models. Based on Fig. 3, the MobileNet-LSTM model was trained for 100 epochs with an unspecified batch size. The training ACC started at 60.12% in the first epoch and steadily increased to 98.92% by the end of the training. The loss function decreased from 0.7559 to 0.0270 throughout the training process. As for the Xception model, the ACC steadily improved from 55.81% to 81.50%, and then exhibited more fluctuations but continued to show overall improvement, reaching a maximum ACC of 94.59% by the 30th epoch. After that, the model's ACC seemed to plateau and remained relatively stable, fluctuating around 93–94%. Overall, the training process appears to have produced well-performing models. The results presented in Figure Fig. 4 demon-

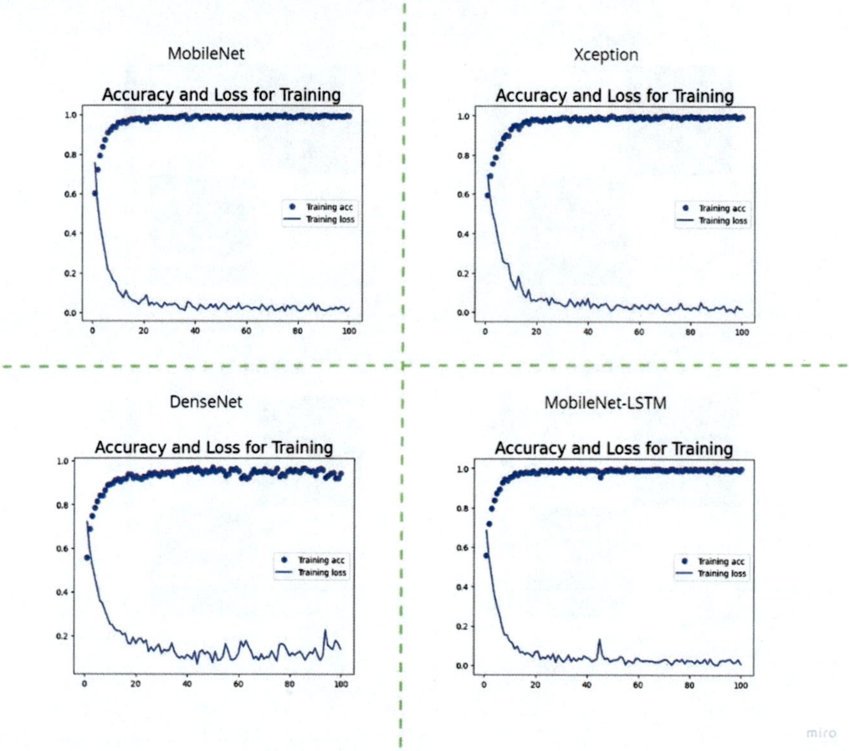

Fig. 3. The proposed algorithms improved their training and testing accuracy with each epoch. After 100 epochs, the hybrid model MobileNet-LSTM achieved an accuracy of 0.99. The training cross-entropy loss function consistently decreased with each epoch, with the final epoch showing a minimal decrease.

strate the performance of four different models: MobileNet, Xception, DenseNet, and MobileNet-LSTM, as evaluated using the receiver operating characteristic (ROC) curve analysis. The ROC curve provides a graphical representation of the trade-off between sensitivity (true positive rate) and specificity (true negative rate) of a binary classifier as the decision threshold is varied. The ROC curve for MobileNet exhibits an AUC of 0.99, indicating excellent performance in distinguishing between positive and negative classes. Xception achieves an AUC of 0.96, DenseNet achieves an AUC of 0.93, and the MobileNet-LSTM model achieves an AUC of 0.98, which is similar to Xception. The results suggest that both MobileNet and MobileNet-LSTM models outperform the other two models in terms of AUC, while all models demonstrate good performance in distinguishing between the two classes. DenseNet exhibits the lowest performance with an AUC of 0.93. These findings are crucial for selecting an appropriate model for a given task, as the trade-off between performance and computational complexity needs to be carefully considered.

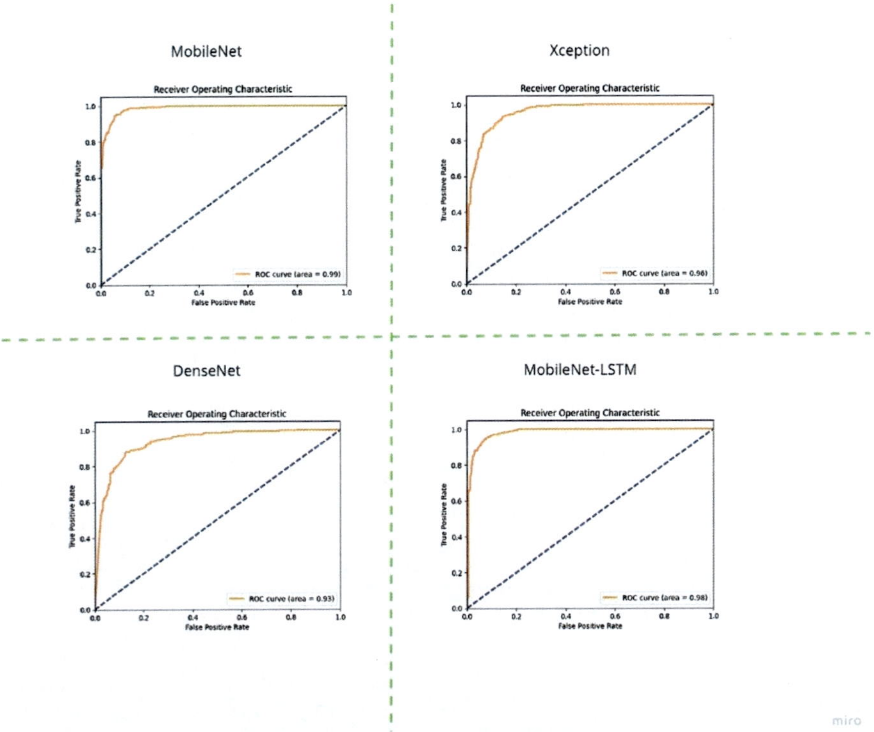

Fig. 4. The hybrid MobileNet-LSTM model achieves an AUC of 0.98, indicating excellent classification performance and a lower number of false-negative results compared to standard CNN-based models.

In Table 1, this study evaluated five models using various performance metrics such as ACC, PREC, REC, F1-S, and ROC AUC. The ensemble model achieved the highest ACC score of 94.45%, followed by MobileNet with a score of 93.34%, MobileNet-LSTM with a score of 92.64%, Xception with a score of 88.62%, and DenseNet with a score of 87.23%. Additionally, the ensemble model exhibited the highest PREC, REC, and F1-S scores, as explained in Table 2. Since the ensemble model is not a binary classifier, the ROC AUC was not available for evaluation. Overall, the ensemble model outperformed the other models in terms of ACC and other performance metrics.

Table 1. Summary of results for MobileNet, Xception, DenseNet, MobileNet-LSTM, and Ensemble models.

Method	ACC	PREC	REC	F1-S	Roc Auc
MobileNet	0.9334	0.9300	0.9500	0.9400	0.9900
Xception	0.8862	0.9200	0.8700	0.8900	0.9600
DenseNet	0.8723	0.8900	0.8700	0.8800	0.9300
MobileNet-LSTM	0.9264	0.9100	0.9600	0.9300	0.9800
Ensemble	0.9445	0.9400	0.9600	0.9500	–

Table 2. Summary of cancer prediction studies using DL models

Article	Cancer Type	Methodology	Result/Performance
[16]	BC	Deep neural network	Cancer prediction, C-index 0.704
[17]	Colorectal cancer	VGG16	1Cancer diagnosis, HR 2.3, CI 95 percent 1.79–3.03, AUC 0.69
[18]	BC	Neural Network	Cancer prediction, CI 0.67
[19]	BC	CNNs	Cancer prediction, ACC 87%
[20]	Colorectal cancer	VGG19, GoogLeNet, Resnet50, AlexNet, SqueezeNet	CI 95 classification of 9 tissues
[21]	Glioblastoma multiforme	Deep neural network	Cancer prediction, ResNet50: 94.90% (±3.92%); ResNet34 (34 layers): 80.72% (±13.61%)
Our Study	Breast cancer	MobileNet, Xception, DenseNet, MobileNet-LSTM and Ensemble Learning	Ensemble Learning: 94.54%

4 Conclusion

This study demonstrates the potential of digital breast tomosynthesis (DBT) in enhancing breast cancer (BC) detection and diagnosis. Four pre-trained models, namely MobileNet, Xception, DenseNet, and MobileNet-LSTM, have shown promising results in identifying and categorizing cancerous tissue. The ensemble model, which combines the predictions of the best-performing models, achieved the highest accuracy (ACC) and outperformed all individual models. These findings highlight the effectiveness of transfer learning and ensemble learning techniques in improving the reliability of BC detection and diagnosis, especially in the context of novel imaging techniques.

5 Future Work

Future research in medical imaging and AI for BC detection and diagnosis could explore several directions. Firstly, investigating larger datasets could provide more diverse and representative samples for training and testing models. Secondly, incorporating clinical and patient-specific data, such as patient demographics, medical history, and genetic information, could improve the accuracy and personalized nature of the models. Additionally, evaluating the proposed models in clinical settings and comparing their performance with human experts could provide valuable insights for their practical implementation. Furthermore, exploring the potential of deep learning models in predicting treatment response and assessing the risk of recurrence could assist in developing personalized treatment plans for BC patients. Lastly, it is crucial to establish ethical and regulatory frameworks to ensure the responsible and safe integration of deep learning models into clinical practice, addressing issues related to data privacy, interpretability, and patient well-being.

References

1. Shah, S.H., Iqbal, M.J., Ahmad, I., Khan, S., Rodrigues, J.J.P.C.: Optimized gene selection and classification of cancer from microarray gene expression data using deep learning. Neural Comput. Appl. **32**(22), 17457–17468 (2020)
2. Gouda, W., Almurafeh, M., Humayun, M., Jhanjhi, N.Z.: Detection of COVID-19 based on chest X-rays using deep learning. Healthcare **10**(4), 343 (2022)
3. Ismael, S.A., Mohammed, A., Hefny, H.: An enhanced deep learning approach for brain cancer MRI images classification using residual networks. Artif. Intell. Med. **102**, 101779 (2020)
4. Dif, N., Elberrichi, Z.: A new deep learning model selection method for colorectal cancer classification. Int. J. Swarm Intell. Res. **11**(2), 72–88 (2020)
5. Brohi, S.N., Pillai, T.R., Brohi, N.N., Jhanjhi, N.Z.: A multilayer perceptron model for the classification of breast cancer cells. Int. J. Comput. Digit. Syst. **10**(2), 104–115 (2021)
6. Khamparia, A., Singh, P.K., Rani, P., Samanta, D., Khanna, A., Bhushan, B.: An internet of health things-driven deep learning framework for detection and classification of skin cancer using transfer learning. Trans. Emerg. Telecommun. Technol. **32**, e3963 (2021)
7. Welikala, R.A., et al.: Automated detection and classification of oral lesions using deep learning for early detection of oral cancer. IEEE Access **8**, 132677–132693 (2020)
8. Humayun, M., Alsayat, A.: Prediction model for coronavirus pandemic using deep learning. Comput. Syst. Sci. Eng. **40**, 947–961 (2022)
9. Pacal, I., Karaboga, D., Basturk, A., Akay, B., Nalbantoglu, U.: A comprehensive review of deep learning in colon cancer. Comput. Biol. Med. **126**, 104003 (2020)
10. Murtaza, G., et al.: Deep learning-based breast cancer classification through medical imaging modalities: state of the art and research challenges. Artif. Intell. Rev. **53**, 1655–1720 (2020)
11. Chi, W., Ma, L., Wu, J., Chen, M., Lu, W., Gu, X.: Deep learning-based medical image segmentation with limited labels. Phys. Med. Biol. **65**, 235001 (2020)
12. Qin, R., et al.: Fine-grained lung cancer classification from PET and CT images based on multidimensional attention mechanism. Complexity **2020**, 6153657 (2020)
13. Manne, R., Kantheti, S., Kantheti, S.: Classification of skin cancer using deep learning, convolutional neural networks-opportunities and vulnerabilities-a systematic review. Int. J. Modern Trends Sci. Technol. **01**(12), 2455–3778 (2020)
14. Shon, H.-S., Batbaatar, E., Kim, K.-O., Cha, E.-J., Kim, K.-A.: Classification of kidney cancer data using cost-sensitive hybrid deep learning approach. Symmetry **12**(1), 154 (2020)

15. Shon, H.-S., Batbaatar, E., Kim, K.-O., Cha, E.-J., Kim, K.-A.: Automated detection and classification of oral lesions using deep learning to detect oral potentially malignant disorders. Cancers **13**(11), 2766 (2021)
16. Jing, B., et al.: A deep survival analysis method based on ranking. Artif. Intell. Med. **98**, 1–9 (2019)
17. Bychkov, D., et al.: Deep learning based tissue analysis predicts outcome in colorectal cancer. Sci. Rep. **8**, 3395 (2018)
18. Katzman, J.L., Shaham, U., Cloninger, A., Bates, J., Jiang, T., Kluger, Y.: DeepSurv: personalized treatment recommender system using a cox proportional hazards deep neural network. BMC Med. Res. Methodol. **18**(1), 24 (2018)
19. Alanazi, S.A., et al.: Boosting breast cancer detection using convolutional neural network. J. Healthc. Eng. **2021**, 5528622 (2021)
20. Kather, J.N., et al.: Predicting survival from colorectal cancer histology slides using deep learning: a retrospective multicenter study. PLoS Med. **16**(1), e1002730 (2019)
21. Korfiatis, P., Kline, T.L., Lachance, D.H., Parney, I.F., Buckner, J.C., Erickson, B.J.: Residual deep convolutional neural network predicts MGMT methylation status. J. Digit. Imaging **30**(5), 622–628 (2017)

Enhanced Multipath QUIC Protocol with Lower Path Delay and Packet Loss Rate

Chih-Lin Hu[1(✉)], Fang-Yi Lin[1], Wu-Min Sung[1], Nien-Tzu Hsieh[1], Yung-Hui Chen[2], and Lin Hui[3]

[1] Department of Communication Engineering, National Central University, Taoyuan City 320317, Taiwan
clhu@ce.ncu.edu.tw, {fangyi.lin,wumin.sung,neintzu.hsieh}@g.ncu.edu.tw
[2] Department of Computer Information and Network Engineering, Lunghwa University of Science and Technology, Taoyuan City 333326, Taiwan
cyh@mail.lhu.edu.tw
[3] Department of Computer Science and Information Engineering, Tamkang University, New Taipei City 25137, Taiwan
121678@mail.tku.edu.tw

Abstract. Consider the high dynamics of traffic loading and resource provision on network hosts that forward data flows along a particular path between two endpoints. The Quick UDP Internet Connect (QUIC) protocol performs better than TCP for its effects in shortening the time of connection establishment and data transmission between two endpoints. Recent studies attempted to exploit the notion of multipath QUIC that forwards the data over multiple paths. Using the multipath QUIC can not only augment the total bandwidth capacity but also avoid traffic congestion on some paths. In this paper, our study proposes a novel multipath QUIC scheme which is able to minimize the flow completion time of multipath QUIC by jointly utilizing two measures of path delay and packet loss rate on a path. Experimental results show that the proposed algorithm is superior to other scheduling schemes, including naive QUIC and Lowest-RTT-First QUIC.

Keywords: Quick UDP Internet Connect (QUIC) · Multipath Transport · HTTP · Content Distribution · Internet Protocol · Internet Services

1 Introduction

In 2013, the IETF organization proposed the RFC 9000, i.e., Quick UDP Internet Connect (QUIC) – a UDP-based multiplexed and secure transport protocol. QUIC is often known as the transport layer for HTTP/3. It is recommended to develop HTTP/3 with QUIC and UDP in place of conventional HTTP/1.1 and 2 with TCP or UDP for internet services and applications in wireless and mobile environments. QUIC provides applications with flow-controlled streams for encrypted, multiplexed and reliable communication, low-latency connection

J. C. Hung et al. (Eds.): FC 2023, LNEE 1134, pp. 193–203, 2024.
https://doi.org/10.1007/978-981-99-9342-0_21

establishment, and network path migration. Compared with TCP, QUIC need not the 3-way handshake mechanism, so it can greatly reduce the time of network connection establishment and transmission latency. With multiplexing and path migration, it can strengthen the control of congested networks, making it more suitable for emerging mobile services in Wi-Fi and 4G/5G environments.

However, the performance of QUIC can be affected in the case of delivering large-size data between two endpoints [1]. This is because the packet pacing policy is basically used to vary the transmission speed of each stream when numerous packets enter that stream. The overall completion time of a data flow in a stream will vary as well. Thus, data throughput of each flow through a link may not reach to the full bandwidth capacity. Recent studies used Multipath QUIC (MPQUIC) to deal with the above concerns subject to the restriction of a single path. MPQUIC sends data through different paths and uses the aggregate bandwidth of different paths. It also likely modifies the *path scheduler* policy for increasing the transmission speed and thence decreasing the path delay that definitely corresponds to the end-to-end transmission delay of a QUIC stream between two endpoints in a network.

In this paper, our study proposes a novel MPQUIC scheme which aims to reduce both path delay and packet loss rate simultaneously. The proposed MPQUIC scheme is able to obtain a shorter flow completion time for each stream in the network. We investigate the proposed MPQUIC scheme in comparison with naive QUIC and Lowest-RTT-First (LRF) scheduling schemes. Experiments are driven using the Mininet emulator and the Abilene topology. Performance results show that the stable and efficient effects of our proposed scheme as demonstrated by the cumulative distribution function (CDF) of flow completion time. In addition, our proposed scheme obtains lower path delay and packet loss rate than the other schemes.

The rest of this paper is organized as follows. Section 2 describes related work. Section 3 details the problem formulation and the path selection algorithm. Section 4 describes the relative performance. Finally, the conclusion is given in Sect. 5.

2 Related Work

The concept of MPQUIC which arranges QUIC connections to go on different paths according to network characteristics. Recent studies proposed several MPQUIC scheduling methods. In [2], an environment-aware MPQUIC packet scheduling method was proposed to perform collaborative scheduling for optimize the overall system transmission time through the Round-Robin (RR) manner in sequential cycles. [3] emphasized on the fair allocation of aggregate bandwidth based on stream priority, thereby avoiding the delay of any individual stream due to heterogeneous paths. [4] developed a Priority Bucket method, which divides streams into different buckets according to stream priority. The priority and size factors can be extracted from HTTP/2 expression. When streams with the same priority exist in the same bucket, they are served in first-come-first-served order. In [5], the Peekaboo method based on reinforcement learning was proposed. It decided on a scheduling sequence by referring to the properties of temporal certainty and randomness of current path characteristics.

The above literature review shows that previous studies on MPQUIC mainly used the QUIC-default Round-Trip Time (RTT) to determine the path selection. Our study considers two network-oriented factors, i.e., delay and packet loss rate of a path. Accordingly, we formulate a weighting normalization method to calculate the weights of paths, which can be used to facilitate path selection and thus minimize the flow completion time over MPQUIC streams.

3 Design of Path Selection Scheme

Give a network topology $G(V, L)$. For every link $l_{i,j} \in L$ from v_i to v_j, the available bandwidth, the delay of the link, and the packet loss rate w.r.t $l_{i,j}$ are denoted as $b_{i,j}$, $t_{i,j}$ and $o_{i,j}$, respectively. Then, $b_{i,j}^{max}$ denotes the maximum amount of bandwidth that $l_{i,j}$ can use.

Let F contain a set of all streams in $G(V, L)$, $\mathcal{P}_f^*$ represent a multipath set in use for a stream $f \in F$, $P_f^*[m]$ be the set of links in the mth path, and likewise $\mathcal{P}_f^*[m][n]$ be the nth link of the mth path. Thus, for the stream and path selection, we take $x_{l_{i,j}}^f$ to be a binary indicator whose value equals to 1 if a stream f passes through a link $l_{i,j}$, or 0 for other conditions.

We define several expressions regarding the relationship between links and paths, as follows:

$$b_f^P = min\Big(b_{i,j} \times x_{i,j}^f\Big), \quad \forall l \in l_{i,j},\ x_{i,j}^f \neq 0,\ f \in F \tag{1}$$

$$b_{i,j}^{max} \geq \sum_{f \in F} b_{i,j} \times x_{i,j}^f, \quad \forall l \in l_{i,j} \tag{2}$$

$$t_f^P = \sum_{l_{i,j} \in L} t_{i,j} \times x_{i,j}^f, \quad \forall\ f \in F, l \in l_{i,j} \tag{3}$$

$$o_f^P = 1 - \prod_{l_{i,j} \in L} \Big(1 - o_{i,j} \times x_{i,j}^f\Big), \quad \forall l \in l_{i,j},\ x_{i,j}^f \neq 0,\ f \in F \tag{4}$$

$$y(\mathcal{P}_f^*) = \begin{cases} 1, & \bigcup P_f^* \neq \emptyset, \\ 0, & \bigcup P_f^* = \emptyset. \end{cases} \tag{5}$$

Formula (1) indicates the available bandwidth of a stream f in the set of paths P, and then takes the minimum value. (2) indicates that the bandwidth passed by a link cannot be greater than the maximum bandwidth available of the link. (3) means the sum of transmission delays on a link w.r.t. a stream f. (4) is to multiply the successful rate of each link to get the overall successful rate on a path, so as to obtain the packet loss rate of this path.

To transform a single-path stream into a multipath stream by (5), $y(\mathcal{P}_f^*)$ indicates whether any link and path in the set of paths $\mathcal{P}_f^*$ can be reused or not. Here, we further discuss two cases, as follows.

Case 1 When the links and paths in $\mathcal{P}_f^*$ are not reused.

The sum of the available bandwidth of each path can be calculated by (6). For $y(\mathcal{P}_f^*) = 0$ and $\forall v_j \in V$, we formulate (7) to check the link condition of v_i and v_j: (i) the total number of positive multipaths, (ii) the total number of negative multipaths, and (iii) a balanced state if both v_i and v_j are intermediate relays.

$$b_f^* = \sum_{P \in \mathcal{P}_f^*} b_f^P, \quad \forall f \in F,\ y(\mathcal{P}_f^*) = 0 \tag{6}$$

$$\sum_{l_{i,j} \in L} x_{i,j}^f - \sum_{l_{j,i} \in L} x_{j,i}^f = \begin{cases} |\mathcal{P}_f^*|, & \text{if } v_i \text{ is a start point of } f, \\ -|\mathcal{P}_f^*|, & \text{if } v_i \text{ is a target point of } f, \\ 0, & \text{if } v_i \text{ is a relay point of } f. \end{cases} \tag{7}$$

□

Case 2 When the links and paths in $\mathcal{P}_f^*$ can be reused

Let $z_{l_{i,j}}^{\mathcal{P}_f^*}$ indicate whether $l_{i,j}$ is reused in $\mathcal{P}_f^*$, so $z_{l_{i,j}}^{\mathcal{P}_f^*}$ equals to 1 as $l_{i,j} \subseteq \bigcup \mathcal{P}_f^*$, or 0 as $l_{i,j} \nsubseteq \bigcup \mathcal{P}_f^*$.

Then, let $n(l_{i,j}, P_f^*)$ indicate the number of times that $l_{i,j}$ is reused by some paths in $\mathcal{P}_f^*$:

$$n(l_{i,j}, P_f^*) = \begin{cases} \sum_{m \in |\mathcal{P}_f^*|} \sum_{n \in |\mathcal{P}_f^*[m]|} l_{i,j} \wedge P_f^*[m][n] - 1, & \forall f \in F,\ l_{i,j} \in L,\ z_{l_{i,j}}^{P_i^*} = 1, \\ 0, & \forall f \in F,\ l_{i,j} \in L,\ z_{l_{i,j}}^{P_i^*} = 0. \end{cases} \tag{8}$$

Thus, the bandwidth of a link is divided into two parts: the link bandwidth that has been reused $\bar{b_f^*}$, and the link that has not been reused $\hat{b_f^*}$, as follows.

$$b_f^* = \bar{b_f^*} + \hat{b_f^*}, \quad \forall f \in F, \quad \text{subject to} \tag{9a}$$

$$\bar{b_f^*} = min(b_f^P), \quad \forall f \in F,\ P \in \mathcal{P}_f^*,\ y(\mathcal{P}_f^*) = 1,\ z_{l_{i,j}}^{P_i^*} = 1. \tag{9b}$$

$$\hat{b_f^*} = \sum_{P \in \mathcal{P}_f^*} b_f^P, \quad \forall f \in F,\ y(\mathcal{P}_f^*) = 1,\ z_{l_{i,j}}^{P_i^*} = 0, \tag{9c}$$

Formula (9a) adds the two parts together, which yields the total amount of bandwidth that a path set can provide.

Formula (10) clarifies the link relation in three conditions. (i) If v_i is a start point of a stream f, the total of paths that a steam can still use is given by $|\mathcal{P}_f^*|$ minus the number of times $l_{i,j}$ that is currently used by some paths in $\mathcal{P}_f^*$, i.e., $n(l_{i,j}, P_f^*)$. (ii) If v_i is a target point, the calculation is in opposition to (i). (iii) Finally, if v_i is a relay w.r.t. $\forall\ y(\mathcal{P}_f^*) = 1$ and $v_j \in V$, there are three sub-cases (a)(b)(c). Explicitly, (a) multiple paths converge at this relay point, then

$n(l_{j,i}, P_f^*) - n(l_{i,j}, P_f^*)$ is negative. (b) multiple paths to divert from this point, this outcome is positive. (c) in a balanced state, the outcome equals to 0.

$$\sum_{l_{i,j}\in L} x^f_{l_{i,j}} - \sum_{l_{j,i}\in L} x^f_{l_{j,i}} = \begin{cases} |\mathcal{P}_f^*| - n(l_{i,j}, P_f^*), & \text{if } v_i \text{ is a start point of } f, \\ -|\mathcal{P}_f^*| + n(l_{j,i}, P_f^*), & \text{if } v_i \text{ is a target point of } f, \\ n(l_{j,i}, P_f^*) - n(l_{i,j}, P_f^*), & \text{if } v_i \text{ is a relay point of } f. \end{cases} \tag{10}$$

□

Note that under the multipath scenario, the delay time and packet loss rate of a path are not affected by whether a path is reused subject to (1). The delay time and packet loss rate w.r.t. any $P \in \mathcal{P}_f^*$, denoted as t_f^* and o_f^*, are given as:

$$t_f^* = \max(t_f^P)\ , \quad \forall f \in F,\ P \in \mathcal{P}_f^*,\ y(\mathcal{P}_f^*) = 0 \tag{11}$$

$$o_f^* = \sum_{P\in\mathcal{P}_f^n} \frac{o_f^P}{|\mathcal{P}_f^*|}, \quad \forall\ f \in F, y(\mathcal{P}_f^*) = 0 \tag{12}$$

By (11), given a set of final selected multipaths, the delay time is represented by the maximum delay time on the path for $\forall P \in \mathcal{P}_f^*$. The outcome of (12) indicates the average of packet loss rate for those selected paths in $\mathcal{P}_f^*$.

We now figure out the comparison between user requirements and actually available provision, as explained below.:

$$b_f \leq b_f^*, \quad \forall f \in F \tag{13}$$

$$t_f \geq t_f^*, \quad \forall f \in F \tag{14}$$

$$o_f \geq o_f^*, \quad \forall f \in F \tag{15}$$

(13) ensures that the multipath bandwidth is available for streaming f, while (14) and (15) enforce that both transmission delay and packet loss rate in the selected path need to be smaller than the tolerable bounds as requested by f.

In accordance with the above formulae and constraints of the multipath provision, we develop an optimal multipath selection problem of minimizing the flow completion time subject to user requirements, as expressed below:

$$\begin{aligned} &\arg\ \min \sum_{f\in F} t_f^*, \\ &\text{s.t.} \\ &\quad x^f_{i,j} = 1, && \forall l_{i,j} \in L, \\ &\quad z^{\mathcal{P}_f^*}_{l_{i,j}} \in (0,1), && \forall \mathcal{P}_f^*,\ l_{i,j} \in L, \\ &\quad y(\mathcal{P}_f^*) \in (0,1), && \forall \mathcal{P}_f^* \in \mathcal{P}, \\ &\quad \text{Eqs. (13), (14), (15).} \end{aligned} \tag{16}$$

Our study learns that such a multipath selection problem for QoS-based data streaming is known as NP-Complete [6,7]. We attempt to develop an optimal-approximate solution to figure out a set of appropriate multipaths using heuristic

Algorithm 1. Path Set Selection with Joint Path Delay and Packet Loss Rate

```
input : G(V, L): network topology,
        k: the number of paths in the multipath,
        α: a coefficient of path delay,
        β: a coefficicient of packet loss.
output: P*_f: the set of multipath.
while Flow f comes into the system do
    P_f = {∅};
    A[ ][ ] = null;
    while P_f = {∅} do
        P_f ← getDefaultPathSet(P, f) ;
        foreach p ∈ P_f do
            A[p][0] ← getPathBW(P[p]) ;                    ▷ (1)
            A[p][1] ← getPathDelay(P[p]) ;                 ▷ (3)
            A[p][2] ← getPathPL(P[p]) ;                    ▷ (4)
        end foreach
    end while
    if (P_f = {∅} or |P_f| < k) then
        Reject f ;
    else
        P*_f ← getkPath(P_f, α, β, f, k, A) ;              ▷ Go to Alg. 2
        if P*_f = ∅ then
            P*_f ← getShorestkPath ∈ P_f ;
        end if
    end if
end while
```

strategies with two design factors, i.e., path delay and packet loss rates. Particularly, we describe a weighting normalization method in (17) with two tuning parameters α and β to change the relative influence of path delay and packet loss rate over MPQUIC streams.

$$p_w = \alpha \times \frac{t_f}{t_f^*} + \beta \times \frac{o_f}{o_f^*}. \tag{17}$$

In what follows, we specify the algorithmic procedures for finding the paths for MPQUIC streams.

With Algorithm 1, the system initializes the set of available paths $\mathcal{P}_f$ for a data stream f, as well as prepares an empty two-dimensional matrix $A[\,][\,]$. At first, when $\mathcal{P}_f$ is empty, the system refers to (1), (3) and (4) to determine the values of data stream bandwidth, delay, and packet loss rate, which are stored in $A[\,][\,]$. Then, the system checks a condition of whether the set of available paths for f contains equal to or more than k paths. As this condition is valid, the system proceeds to Algorithm 2 with a set of candidate paths for f.

Algorithm 2 is the path selection procedure for finding the k-shortest paths based on QoS requirements. This procedure refers to Yen's k-shortest path algorithm [8] with QoS-specific conditions. To find the k-shortest paths, the procedure runs several routes sequentially: (a) define variables p_w, b_f^* and $\mathcal{P}_f^*[\,][\,]$,

Algorithm 2. Finding k Shortest Paths over MPQUIC Streams

```
Function getkPath(P_f, α, β, f, k, A) is
    p_w[ ] = null;
    b*_f = 0;
    P*_f[ ][ ] = null;
    foreach p ∈ P_f do
        p_w[p] ← getPathWeight(P[p], α, β, A);                ▷ (17)
    end foreach
    p_w ← sortByDescendingOrder(p_w) ;
    P*_f ← selectPathTopk(p_w, k) ;
    b*_f ← getMultiPathBW(P*_f);                              ▷ (6) and (9a)
    while b*_f ≤ b_f do
        if minBWPath(P*_f) ≥ maxBWPath(P_f − P*_f) then
            P*_f = ∅;
            break;
        end if
        P*_f ← P*_f − minBWPath(P*_f);
        P*_f ← P*_f + maxBWPath(P_f − P*_f);
        b*_f ← getMultiPathBW(P*_f);                          ▷ (6) and (9a)
    end while
    return P*_f;
end
```

(b) calculate the weight value p_w of a stream by (17), (c) sort the weights of streams in descending order, and (d) update the available bandwidth of each link according to (6) and (9a). Then, the procedural routine goes into a while-loop with a condition as b_f^* is smaller than the bandwidth b_f asked by a stream f. If the minimum bandwidth of $\mathcal{P}_f^*$ exceeds the currently available path $\mathcal{P}_f$, $\mathcal{P}_f^*$ is still to be null. Then, the routine updates the set of available paths $\mathcal{P}_f^*$ and the bandwidth b_f^*, removes the path of the smaller bandwidth from $\mathcal{P}_f^*$, adds a path with the larger bandwidth, updates b_f^*, and then pushes the value of $\mathcal{P}_f^*$ back to Algorithm 1 to allocate available paths. Eventually, the data flow is passed through those suitable and multiple paths in the current network.

4 Performance Results

This section shows the performance of our proposed method in comparison with QUIC and multipath QUIC LRF [9].

4.1 Experimental Setting

Experiments were divided into three sorts with different sizes per data flow: 100 and 200 MB, and produced three measure results of the overall flow completion time, path delay, and packet loss rate. We employed the Mininet to adjust simulation parameters. Explicitly, we set $k = 3$, delay coefficient $\alpha = 0.5$ and packet loss coefficient $\beta = 0.5$ as calculating the weighted value p_w. We adopted the

Abilene topology [10]: there are 11 nodes and 14 links, the size of each packet is between 960 and 1200 bytes, the path bandwidth is set to 100 Mbps, the delay is from 0 to 100 ms by the binomial distribution, and packet loss rate is set to 0.001%. All experimental cases were run in 20 times to have the results on average.

4.2 Flow Completion Time

Figures 1 and 2 exhibit the flow completion time in terms of the cumulative distribution function (CDF). As observed, the performance by naive QUIC is the worst, because QUIC only transmits data through a single path, as compared with the other schemes that take multiple paths. It is visible that our scheme outperforms LRF. LRF is based on finding the path with the minimum RTT for transmitting the top-priority data first. Thus, LRF behaves like a greedy way and only focuses on the RTT condition without referring to other network characteristics. Relatively, our proposed scheme considers both path delay and packet loss rate of path candidates. By using a weighting normalization method, it is able to calculate P_w. The higher P_w, the higher priority the data needs to be scheduled for transmission first. Our proposed scheme with weighting effects can minimize the flow completion time, resulting in a remarkable comparison with LRF.

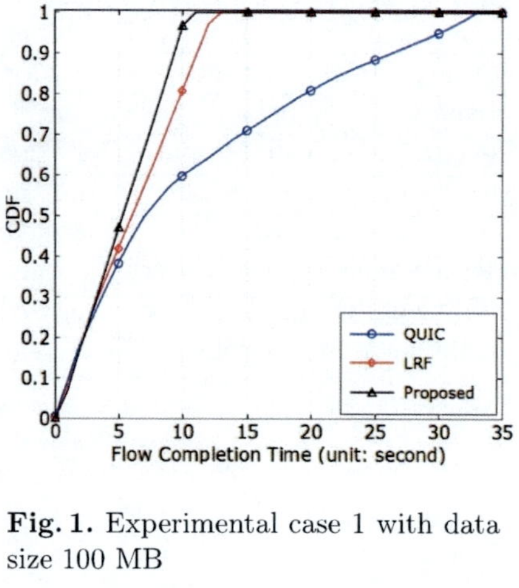

Fig. 1. Experimental case 1 with data size 100 MB

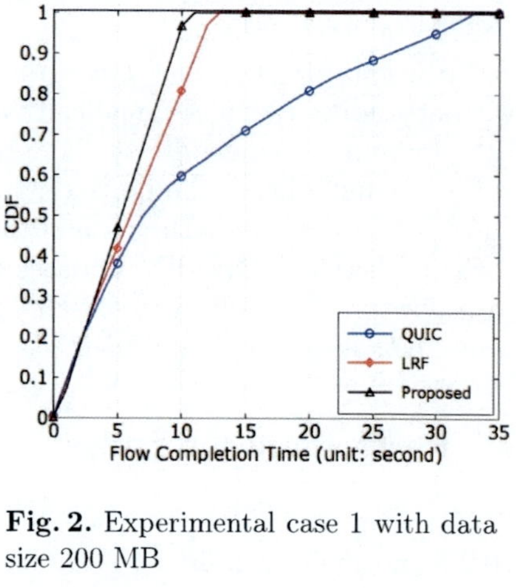

Fig. 2. Experimental case 1 with data size 200 MB

4.3 Packet Loss Rate

Figures 3 and 4 present the packet loss rate of the overall system performance. As observed, the packet loss rate of QUIC is higher than the other multipath schemes, for the major reason that only the resource allocation of a single path is used. In the case of data size 100 MB per stream, the packet loss rates of QUIC and LRF are similar, but become different when the data size per stream increases to 200 MB. LRF searches for the path of the minimum RTT, which may cause the problem of packet loss in the rear tail of data stream. By contrast, our scheme can distribute the data to multiple paths efficiently, thereby being less susceptible to the increase of data size per stream. As seen, our scheme is able to cope with the packet loss rate to be lower than 1 % regardless the increasing data size from 100 MB to 200 MB.

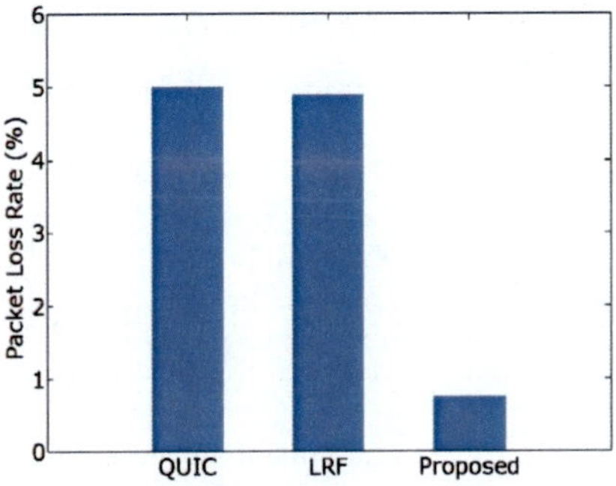

Fig. 3. Experimental case 2 with data size 100 MB

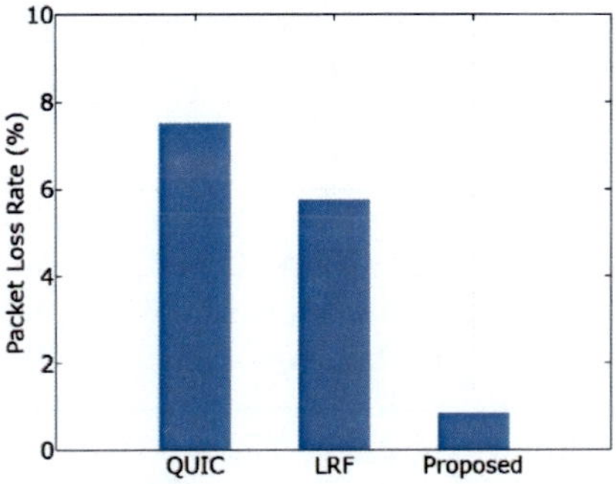

Fig. 4. Experimental case 2 with data size 200 MB

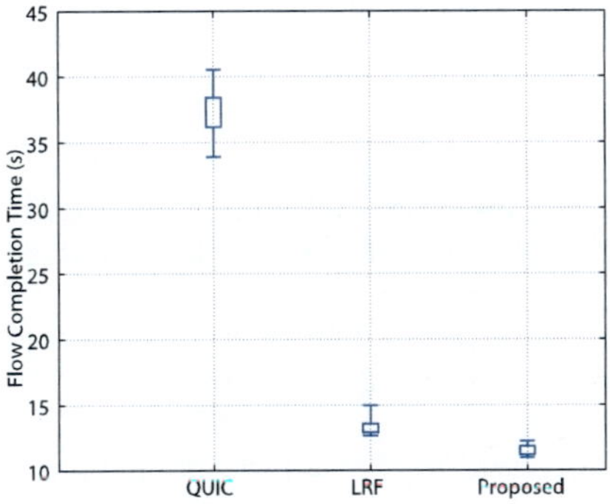

Fig. 5. Experimental case 3 with data size 100 MB

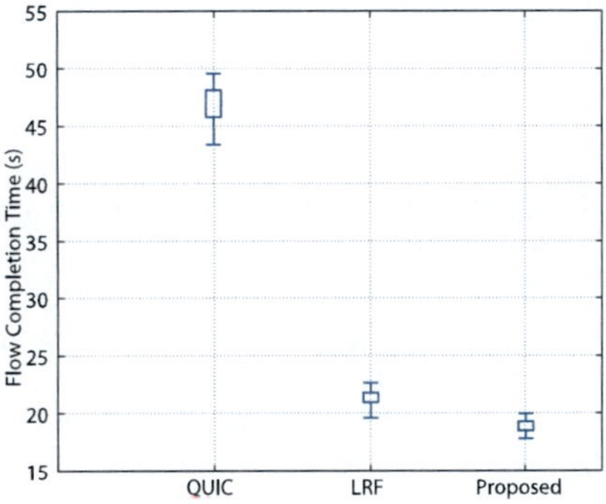

Fig. 6. Experimental case 3 with data size 200 MB

4.4 Overall System Stability

Figures 5 and 6 depict the quartile distribution of flow completion time when the experiment launched 20 data flows one by one repeatedly. Obviously, QUIC needs to take much more time to accomplish the transmission of per data flow. The time gap between QUIC and two multipath QUIC scheme is apparent. LRF has not only a larger completion time but also a wider quartile distribution than our scheme. Our scheme still keeps a minor quartile distribution with the lowest flow completion time, which shows the stable transport performance.

5 Conclusion

In this paper, we have proposed a novel data transport scheme based on MPQUIC. Our MPQUIC scheme is able to jointly sustain transmission delay and packet loss rate with respect to data flows. Performance study is conducted by comparing the proposed scheme with two prior schemes, i.e., QUIC and LRF. It is remarkable that our proposed scheme performs efficiently and stably in terms of the flow completion time in the system. Our future research will try to implement MPQUIC and measure the transport performance in more complicated network scenarios with emerging AR/VR applications, particularly in mobile environments.

Acknowledgment. This work was supported in part by the National Science and Technology Council, Taiwan (R.O.C.), under Contracts MOST-109-2221-E-008-051, NSTC-111-2221-E-008-064 and NSTC-111-2410-H-262-001.

References

1. Megyesi, P., Krämer, Z., Molnár, S.: How quick is QUIC?. In: Proceedings of 2016 IEEE International Conference on Communications (ICC 2016), pp. 1–6 (2016)
2. Jing Wang, Y.G., Xu, C.: A stream-aware multipath QUIC scheduler for heterogeneous paths. In: Proceedings of 2019 ACM 3rd Asia-Pacific Workshop on Networking, pp. 43–49 (2019)
3. Rabitsch, A., Hurtig, P., Brunstrom, A.: A stream-aware multipath QUIC scheduler for heterogeneous paths. In: Proceedings of of the Workshop on the Evolution, Performance, and Interoperability of QUIC, pp. 29–35 (2018)
4. Shi, X., Zhang, F., Liu, Z.: PriorityBucket: a multipath-QUIC scheduler on accelerating first rendering time in page loading. In: Proceedings of the 11th ACM International Conference on Future Energy Systems, pp. 572–577 (2020)
5. Wu, H., Alay, Ã., Brunstrom, A., Ferlin, S., Caso, G.: Peekaboo: learning-based multipath scheduling for dynamic heterogeneous environments. IEEE J. Sel. Areas Commun. **38**(10), 2295–2310 (2020)
6. Hu, C.-L., Hsu, C.-Y., Sung, W.-M.: FitPath: QoS-based path selection with fittingness measure in integrated edge computing and software-defined networks. IEEE Access **10**, 45 576–45 593 (2022)
7. Karp, R.M.: Reducibility among combinatorial problems. Miller, R.E., Thatcher, J.W. (eds.) Complexity of Computer Computations, pp. 85–103. Boston, MA, USA: Plenum Press (1972) ISBN 0-306-30707-3
8. Yen, J.Y.: Finding the K shortest loopless paths in a network. Manage. Sci. **17**(11), 712–716 (1971)

9. Viernickel, T., Froemmgen, A., Rizk, A., Koldehofe, B., Steinmetz, R.: Multipath QUIC: a deployable multipath transport protocol. In: Proceedings of 2018 IEEE International Conference on Communications (ICC), pp. 1–7 (2018)
10. Knight, S., Nguyen, H.X., Falkner, N., Bowden, R., Roughan, M.: The internet topology zoo. IEEE J. Sel. Areas Commun. **29**(9), 1765–1775 (2011)

Implementation of a Deep Learning-Based Application for Work-Related Musculoskeletal Disorders' Classification in Occupational Medicine

Yu-Wei Chan[1(✉)], Yi-Cyuan Tseng[2], Yu-An Chen[2], Yu-Tse Tsan[3(✉)], Chen-Yen Liu[2], Shang-Zhe Lu[2], Li-Fan Xu[2], and Chao-Tung Yang[4]

[1] Department of Information Management, Providence University, Taichung, Taiwan
ywchan@gm.pu.edu.tw

[2] Department of Computer Science and Information Engineering, Providence University, Taichung, Taiwan
{s1090372,s1092823,s1090356,s1091886}@gm.pu.edu.tw

[3] Institute of Occupational Medicine, Department of Emergency Medicine, Taichung Veterans General Hospital, Taichung, Taiwan
janyuhjer@gmail.com

[4] Department of Computer Science, Tunghai University, Taichung, Taiwan
ctyang@thu.edu.tw

Abstract. This research aims to develop an AI-based Ergonomics risk hazard posture recognition system to help reduce the risk of injury to workers and improve work safety in factories and warehouses. The background shows that ergonomic risk hazards are one of the most important risk factors in the workplace, among which the risk of posture hazards is higher when the human body is carrying objects. Otherwise, KIM-LHC (Key Indicator Methods - Lifting/Holding/Carrying) was used as the basis for posture determination, and the human posture information was converted into data by Movenet, and then build the neural network classification model used to recognize and analysis human pose, finally integrated into the app built by flutter. The app built by Flutter is finally integrated. In order to verify the performance of the system, it conducted experiments by actual video recording, and the results showed that the verification accuracy of the app could reach over 97%, and successfully identified the dangerous postures that might cause injury risks to workers, and the app was easy to understand and practical. In summary, this research developed an AI-based Ergonomics risk-hazard posture recognition system, which is important for improving workplace safety.

Keywords: Ergonomics · work-related musculoskeletal disorders · KIM-LHC · occupational medicine

1 Introduction

Ergonomics [1] is the research of the relationship between people and the interactions between tools, machines, equipment and the environment in human daily

J. C. Hung et al. (Eds.): FC 2023, LNEE 1134, pp. 204–210, 2024.
https://doi.org/10.1007/978-981-99-9342-0_22

life at work. Aim to enhance these interactions by adapting them to the individual's abilities, constraints, and requirements. Ergonomics is a specialized field that aims to improve the congruence between human interactions and tools, machines, equipment, and the environment through thoughtful design. If not well implemented, poorly designed Ergonomics can lead to various direct and indirect effects on workers. These include contributing to human errors, accidents, musculoskeletal injuries, illnesses, reducing the quality of work life, poor production performance, and inducing worker fatigue, all of which can seriously affect the health, safety, and overall well-being of workers. Additionally, occupational accident survey statistics from various regions including the United States, Japan, Europe, and Korea highlight the impact of musculoskeletal injuries and illnesses. These conditions, accumulated over time, have led to a significant number of lost workdays. On average, these cases account for 38% in the European Union, 32% in the United States, 41.2% in Japan, and 40% in the United Kingdom. In recent years, the overall loss caused by repetitive musculoskeletal injuries and illnesses is about US$216 billion in the EU, accounting for 1.6% [3] of the overall GDP of the EU; and about US$168 billion in the US, accounting for 1.53% [4] of the US GDP. According to the survey report [5], the number of people suffering from work-related musculoskeletal injuries in Korea increased from 124 in 1998 to 6,234 in 2009; in the 1990s, musculoskeletal injuries accounted for only 10% of all occupational injuries in Korea; But increased to 70% in 2009. Although this is related to the inclusion of musculoskeletal injuries as an occupational disease, it also shows the prevalence of musculoskeletal injury problems. Therefore, in order to eliminate or reduce work-related musculoskeletal injuries, advanced industrial countries have been making efforts to promote the prevention and control of repetitive musculoskeletal injuries in recent years.

Heavy lifting and repetitive lifting operations are the most common causes of occupational diseases in the workplace. These operations often result in work-related musculoskeletal disorders (WMSD) [6] and cumulative musculoskeletal disorders (CTD) [7,9]. The proportion of injuries varies by site, with the main causes of these disorders being prolonged repetitive operations and poor posture, leading to fatigue and inflammation of the associated musculoskeletal tissues, which can result in injury. Due to the high prevalence and long duration of musculoskeletal injuries, they have a significant impact on workers, businesses, and the nation.

Thus, society shoulders a heavier burden related to labor insurance and social relief, alongside the substantial depletion of medical and social resources. In the past, the mainly dependence on physicians' clinical expertise to discern whether repetitive tasks resulted in musculoskeletal injuries and to quantify the harm, substantially strained on-site who were tasked with precisely determining the scope of musculoskeletal injuries and ailments within a limited time frame., posed a formidable challenge. Summary, this methodology frequently culminated in inexact evaluations, exorbitant costs, and laborious procedures.

This research goal is to develop a tool leveraging the KIM-LHC [8] as a reference, focusing specifically on body parts most susceptible to injuries (see

Fig. 1). This tool is designed to facilitate prompt and accurate determinations by physicians, thereby enhancing the precision and efficiency of their assessments regarding the scope of musculoskeletal injuries and disease hazards. As a result, it anticipate to drastically curtail the cost and time associated with injury and disease evaluation, poised to facilitate the implementation of early intervention measures. Also minimize further harm, expedite recovery, and lessen medical costs. Finally, the intent is to alleviate the impact on workers, businesses, and society, while promoting a healthier and safer work environment.

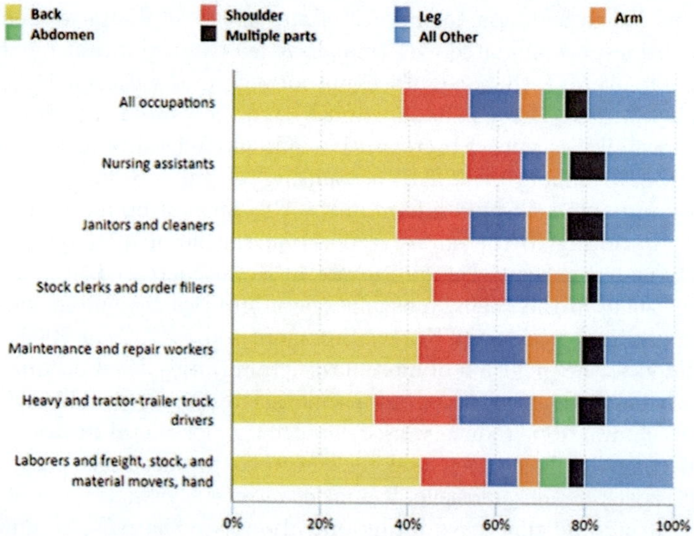

Fig. 1. The Back injuries prominent in work-related musculoskeletal disorder cases in 2016 [2].

The rest of this paper is organized as follows. Section 2 introduces the KIM-LHC method. In Sect. 3, the system design and implementation are presented. Finally, concluding remarks are given in Sect. 4.

2 The KIM-LHC Method

Musculoskeletal injuries are among the most common occupational diseases suffered by workers, and these occupational diseases often result in significant medical burdens and sequelae. The Key Indicator Method (KIM) was released as a draft in 2007 and contains three assessment scales (KIM-LHC, KIM-MHO, and KIM-PP), among KIM-LHC can use Body posture rating judgment table (see Fig. 2) cooperate with other Rating Judgment calculate risk score to quickly conduct a rapid assessment of Lifting/Holding/Carrying work situation is in a safe state or not, effectively Reduce the chance of WMSD in workers.

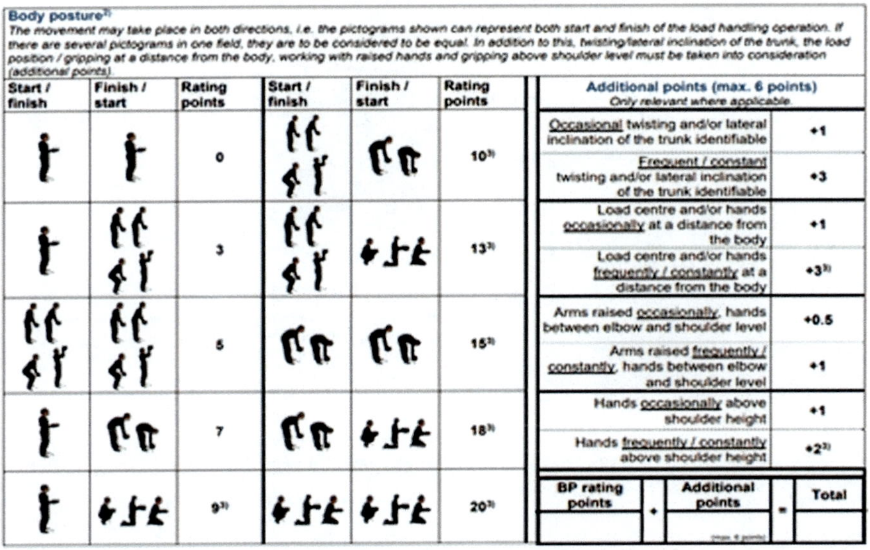

Fig. 2. The Body Posture Rating Table.

3 System Design and Implementation

3.1 System Design

In this part, we design a system to quickly assess the risk. The system is divided into four modules, which are presented in the following: The system flowchart is shown in Fig. 3.

- The Video Image Collection and Processing Module: This module is mainly responsible for video image collection, image cropping and image scaling operations.
- The Pose Detection Module: This module is based on MoveNet [10,11] to extract human 17-keypoints skeleton and obtain 2D coordinates of human keypoints.
- The Pose Classification and Risk Assessment Module: This module uses the neural network to determine the key point coordinates for classification, and it is also responsible for integrating the output of the neural network with the input job description data (e.g., time rating, load rating, and work condition rating) to calculate the posture risk value according to the KIM-LHC scale and obtain the corresponding risk assessment level.
- The Evaluation and Feedback Module: This module is responsible for generating WMSD real-time feedback analysis reports and recommendations, including KIM scores, risk level results and feedback recommendations. Users can quickly query the results and complete the whole evaluation process.

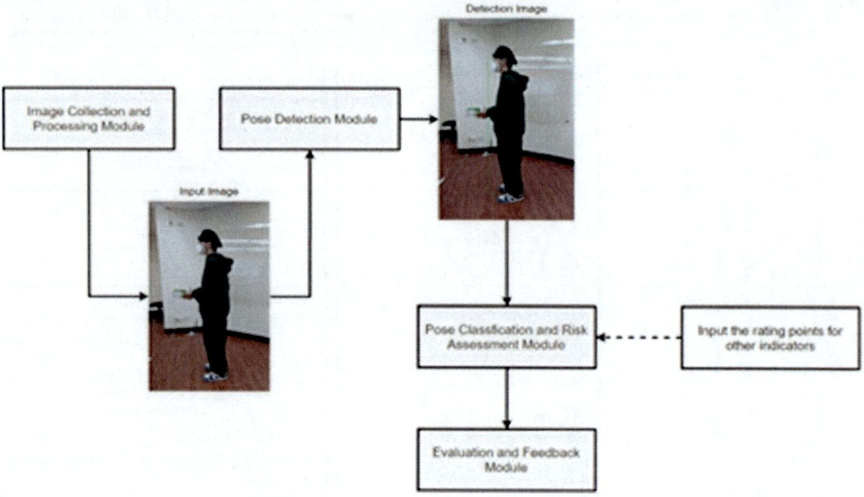

Fig. 3. The System Flowchart.

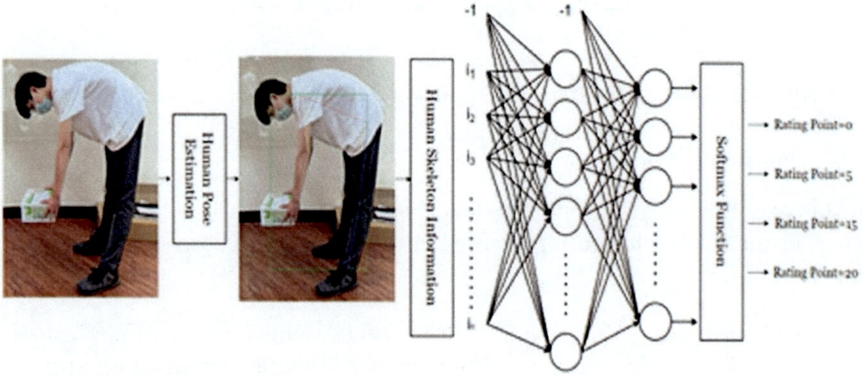

Fig. 4. The training flow of the neural classification model of this work.

3.2 System Implementation

The implementation steps are given below:

- Step 1. The Dataset acquisition: According to KIM-LHC, we designed different posture hazard level to shoot human key point dataset.
- Step 2. Building classification model: We use MoveNet to construct the human skeleton and key points, and then use the acquired human key point coordinates are used to train the classification model. The training flow of the neural classification model of this work is shown in Fig. 4
- Step 3. Constructing APP:
  - The Photography: Make each frame into MoveNet through camera plugin.
  - The User status: Change status and save information via buttons.

  - The Report download: calculate risk rating, give user advice and help doctor treatment.
- Step 4. The Calculation of Musculoskeletal Injury Risk: The musculoskeletal injury risk calculation (See Fig. 5.) will be performed according to the approach designed in [8] and feedback will be given according to different levels of risk.
- Step 5. Model deployment: The trained classification model is lightly processed, converted to tflite and ported to the APP.
- Step 6. Model optimization: Optimize the computational speed of neural networks on cell phones.
- Step 7. System testing: Conduct contextual tests and debug step by step.

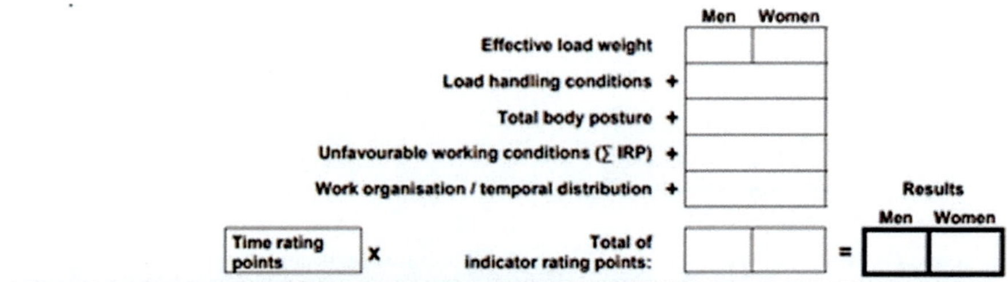

The risk score calculated and the table below can be used as the basis for a rough evaluation:					
Risk	Risk range		Intensity of load[1]	a) Probability of physical overload b) Possible health consequences	Measures
	1	< 20 points	low	a) Physical overload is unlikely. b) No health risk is to be expected.	None
	2	20 - < 50 points	slightly increased	a) Physical overload is possible for less resilient persons. b) Fatigue, low-grade adaptation problems which can be compensated for during leisure time	For less resilient persons, workplace redesign and other prevention measures may be helpful.
	3	50 - < 100 points	substantially increased	a) Physical overload is also possible for normally resilient persons. b) Disorders (pain), possibly including dysfunctions, reversible in most cases, without morphological manifestation	Workplace redesign and other prevention measures should be considered.
	4	≥ 100 points	high	a) Physical overload is likely. b) More pronounced disorders and/or dysfunctions, structural damage with pathological significance	Workplace redesign measures are necessary. Other prevention measures should be considered.

Fig. 5. The Risk Calculation Table.

4 Conclusion

In this work, we have developed an ergonomic risk hazard posture recognition system based on the human pose estimation technology, in which the KIM-LHC was used as the basis for posture judgement and the MoveNet pose estimation model was also used for recognition and classification. In addition, we have built a cross-platform App with Flutter framework. Through actual video recording, the system can successfully identify dangerous postures that may cause injury risks to workers. This system is important to improve workplace safety.

In the future, we plan to combine the system with KIM-PP (Key Indicator Methods-Pushing/Pulling) and KIM-MHO (Key Indicator Methods-Manual

Handling Operating Tasks) of the KIM scale, hope this system can be further applied to various industries and work situations. Through actual verification and system improvement, we can realize the rapid detection of work safety to prevent workers from suffering from musculoskeletal injuries such as WMSD due to wrong work posture.

References

1. Schneider, S., Susi, P.: Ergonomics and construction: a review of potential hazards in new construction. Am. Ind. Hyg. Assoc. J. **55**(7), 635–649 (1994)
2. U.S. Bureau of Labor Statistics (2018). https://www.bls.gov/opub/ted/2018/back-injuries-prominent-in-work-related-musculoskeletal-disorder-cases-in-2016.htm
3. Crawford, J.O.: The Nordic musculoskeletal questionnaire. Occup. Med. **57**(4), 300–301 (2007)
4. US Department of Health and Human Services: Musculoskeletal Disorders and Work-place Factors: A Critical Review of Epidemiologic Evidence for Work-Related Musculoskeletal Disorders of the Neck, Upper Extremity and Low Back. In: Bernard, B.P. (ed.) Public Health Service Centers for Disease Control and Prevention. National (1997)
5. Kim, K.H., Kim, K.S., Kim, D.S., et al.: Characteristics of work-related musculoskeletal disorders in Korea and their relatedness evaluation. J. Korean Med. Sci. **25**, 77–86 (2010)
6. Vieira, E.R., et al.: Work-related musculoskeletal disorders among physical therapists: a systematic review. J. Back Musculoskelet. Rehabil. **29**(3), 417–428 (2016)
7. Wang, J., Chen, D., Zhu, M., Sun, Y.: Risk assessment for musculoskeletal disorders based on the characteristics of work posture. Autom. Constr. **131**, 103921 (2021)
8. Bundesanstalt für Arbeitsschutz und Arbeitsmedizin (BAuA). https://www.baua.de/EN/Topics/Work-design/Physical-workload/Key-indicator-method/pdf/KIM-LHC-Lifting-Holding-Carrying.html. Accessed 10 Oct 2019
9. Lancashire County Council. https://www.lancashire.gov.uk/media/929659/
10. Bajpai, R., Joshi, D.: MoveNet: a deep neural network for joint profile prediction across variable walking speeds and slopes. IEEE Trans. Instrum. Measur. **70**, 1–11 (2021)
11. Li, Z., Zhang, R., Lee, C.H., Lee, Y.C.: An evaluation of posture recognition based on intelligent rapid entire body assessment system for determining musculoskeletal disorders. Sensors **20**(16), 4414 (2020)

Single-to-Multi Music Track Composition Using Interactive Chaotic Evolution

Ying Kai Hung[1], Yan Pei[1(✉)], and Jianqiang Li[2]

[1] Graduate School of Computer Science and Engineering, University of Aizu, Aizuwakamatsu, Fukushima 965-8580, Japan
{m5251111,peiyan}@u-aizu.ac.jp

[2] Faculty of Information Technology, Beijing University of Technology, Beijing 100124, China
lijianqiang@bjut.edu.cn

Abstract. This research presents a new music generation model and a novel MIDI data format for MIDI music generation. This innovative data format allows us to process MIDI music in a manner analogous to video analysis. Initially, the model employs Convolutional Neural Networks (CNN) as an encoder to effectively capture local and global features within the musical data. Subsequently, we utilize a Transformer as a decoder, leveraging its self-attention mechanism to handle the long-term dependencies present in music data. In the training process, an interactive chaotic algorithm is introduced to update the model's weights, assisting the model in avoiding entrapment in local optima. This enhances the learning efficiency of the model and improves the quality of the generated output, enabling the model to generate music, including accompaniment, that aligns with human aesthetics from any given melody.

Keywords: Music Composition · MIDI · Convolutional Neural Networks · Transformer · Interactive Evolutionary Computation · Interactive Chaotic Evolution

1 Introduction

In recent years, deep learning has been extensively applied in music-related research, leading to advancements in tasks such as genre recognition, song index recommendation, and music style transformation. These tasks, which previously required significant time and music theory experience, can now be expedited through the application of deep learning, reducing the associated time costs. When it comes to creating music on a computer, the most convenient data format for creation and recording is the Musical Instrument Digital Interface (MIDI) music information. MIDI records pitch, music intensity, volume, and instrument timing as a digital signal, and deep learning models are commonly used to process MIDI data for various tasks that involve handling long sequences. However, a challenge arises due to the extensive length of the data, making it difficult to

J. C. Hung et al. (Eds.): FC 2023, LNEE 1134, pp. 211–221, 2024.
https://doi.org/10.1007/978-981-99-9342-0_23

effectively correlate all units and resulting in outcomes that do not align with human aesthetics.

In other digital signal domains, techniques such as MFCC or Mel spectrograms have been used to transform signal data into continuous images, effectively capturing audio features for tasks like voice conversion and voice recognition. Building on this knowledge, we propose a new music generation model and a novel method to modify the MIDI data format for MIDI music generation in this study. We convert each piece of MIDI format music into a piano roll, segment it according to the music beat, and stack each segment as a frame in a 3D data format. This allows us to process the data in a manner analogous to video analysis. The model employs a Convolutional Neural Network (CNN) as an encoder to effectively capture the global features of each frame. Subsequently, we utilize a Transformer as a decoder, leveraging its self-attention mechanism to handle the long-term dependencies in the data.

This design provides our model with high flexibility and adaptability, enabling it to efficiently generate complete music with complex structures, including accompaniment and melody. In addition to the model and data design, an interactive chaotic algorithm is introduced during the training process to update the model's weights. This algorithm simulates chaotic phenomena in nature, allowing the model to self-organize and self-adjust during the learning process, thereby generating more creative and aesthetically pleasing music. Moreover, the interactive chaotic algorithm helps the model avoid local optima, enhancing its learning efficiency and the quality of the generated music, resulting in compositions that better align with human aesthetics.

2 Related Works

Reference [1] is a study that explores the impact of different music input representations on the performance of Convolutional Neural Network (CNN) music classification models. In this paper, the researchers compared three common music input representations: Mel spectrograms, spectrograms, and constant-Q transforms. They found that all input representations could be effectively used by the CNN model. My research converted MIDI data into piano roll images and used a CNN as an encoder to compress features based on the data nature. Our approach shares some similarities with the method proposed in this paper, as we both aim to find an effective way of visualizing music as images and improve the model's ability to extract musical features.

Reference [2]: This research paper, written by Jean-Baptiste Alayrac and others, mainly discusses the method of using Transformer networks to handle video information. In this paper, the researchers proposed a new video understanding model based on the Transformer network that can directly handle raw video frames and capture long-term dependencies in the video. Their model uses a self-attention mechanism to comprehend the contextual information in the video and can automatically learn the dynamic and static features in the video. Inspired by this paper, I conceived a new data representation method that converts MIDI music into piano rolls and frames them into a series of frames, forming a three-dimensional data format similar to video. Next, we designed a model that uses a convolutional neural network (CNN) as an encoder to capture global features in each timestep, and a Transformer as a decoder to utilize its self-attention mechanism to handle long-term dependencies in the data and achieve the task of generating music.

Reference [3] is a research paper published in 2017 by Cheng-Zhi Anna Huang and others. The paper explores the use of Convolutional Neural Networks (CNN) in music generation, specifically in generating counterpoint-style music. The researchers employed deep learning techniques, particularly Convolutional Neural Networks (CNNs), to construct a model capable of generating music in the style of counterpoint. Their model can learn and imitate the rules of counterpoint-style music and generate new counterpoint-style melodies. Their research demonstrates that deep learning techniques can be effectively applied to the field of music composition, resulting in artistic and innovative music. A part of my research model also utilizes a Convolutional Neural Network (CNN) as the encoder, although there are differences in the way the input captures music features and the specific model architecture. However, both approaches aim to effectively capture global features in musical data and generate complete music with accompaniment from a single melody input.

Reference [4] is a research paper published by the Google Magenta team in 2019. This groundbreaking work introduced the Transformer architecture to the field of music generation, successfully addressing the long-term structural issues in MIDI music generation. The primary objective of Music Transformer is to handle the long-term dependency problem in music generation. To accomplish this, the researchers utilized the Transformer, a deep learning model with a powerful self-attention mechanism. Building upon this, they developed a new MIDI event representation called "Relative Global Encoding." This encoding method not only captures the rhythmic structure in music but also considers the relative timing of notes, enabling the model to generate works with longer musical structures. My research shares many similarities with the work of "Music Transformer." Firstly, it also employs a Transformer-based model and utilizes a self-attention mechanism to address the long-term dependency problem in music data. However, our research introduces an innovative approach to process MIDI data, converting each MIDI music piece into a piano roll-like format and then framing and stacking it into a 3D data structure. This method effectively captures global music features and extracts longer temporal features.

Reference [5] is a technique proposed by Yan Pei that combines chaotic dynamics and evolutionary algorithms. The main idea is to guide evolutionary algorithms in a global search within the solution space of optimization problems by leveraging the randomness generated through chaotic mapping. This approach effectively prevents the optimization process from converging to local optima and enhances the quality and diversity of optimization results. In our research, we introduce this chaotic algorithm to update the model's weights and incorporate human evaluation into the training process to enable the model to self-organize and self-adjust based on human perception, resulting in the generation of more creative and aesthetically appealing music. Our model design combines the strengths of chaos theory, the capabilities of deep learning, and human aesthetic evaluations, thereby enhancing learning effectiveness and generation quality. The outcome is music that better aligns with human aesthetics.

3 Method

This study employed several methods to generate MIDI music that aligns with human aesthetics. These methods include converting MIDI music into a 3D piano roll image, utilizing CNN as an encoder, employing Transformer as a decoder, and training the model using an interactive chaotic algorithm.

To begin, MIDI music was transformed into a 3D matrix data structure in the piano roll image format, which mimics the format used by humans when creating music scores. This approach is not limited to any specific music genre or style, making it applicable to various types of MIDI music, such as classical or pop.

The use of CNN as an encoder offers the advantage of effectively capturing both local and global features of the piano roll, thereby transforming them into an input representation suitable for the Transformer decoder. This enhances the model's learning capability and improves the quality of the generated music.

By using the Transformer as a decoder and leveraging its self-attention mechanism, it is possible to address the challenge of long-term dependencies in music and generate music that is more musically coherent.

Lastly, the utilization of an interactive chaotic algorithm during training enhances the efficiency of the model's learning process and prevents it from getting trapped in local optima. Additionally, human perception can be leveraged to optimize the model and generate music of higher quality.

Overall, these methodologies, involving the conversion of MIDI music to a 3D piano roll image, the use of CNN and Transformer, and the incorporation of an interactive chaotic algorithm, contribute to the generation of music that is aesthetically pleasing and aligns with human preferences.

Here are the detailed methods:

3.1 Data Preparation

In this research, the data preparation process plays a crucial role in efficiently handling MIDI music data for subsequent learning and generation tasks. The first step involves converting the music data into a graphical representation using the piano roll format. The resulting image has a size of (128, n), where 128 represents a fixed pitch value in MIDI music, and n corresponds to the length of the MIDI reading time, determined by the chosen sampling rate. To ensure computational efficiency, a sampling rate of 0.1 s is used to read the MIDI data.

Next, the rhythmic duration is extracted from the MIDI music. The music sequences are then sliced into frames, with each frame comprising four beats, aligning with a single measure to maintain the musical structure. To standardize the varying lengths of the music, each frame is padded with zeros to a duration of four seconds, resulting in a uniform duration of five minutes for all music data. These transformed frames are stacked into a matrix of size 128 × 40 × 150. Additionally, the target data is adjusted to a consistent size of 128 × 3000 to ensure seamless usage by the model (Fig. 1).

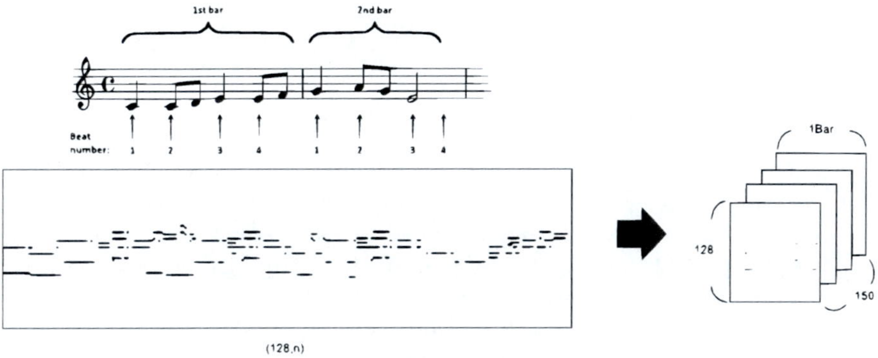

Fig. 1. Example of MIDI process

3.2 Model Architecture

The architecture of our model consists of an encoder and a decoder. The encoder utilizes a Convolutional Neural Network (CNN), while the decoder is based on the Transformer network. In the following sections, we provide a comprehensive description of each layer's structure and functionality within the model. Please refer to Fig. 2 for a visual representation of the model architecture.

Encoder (Convolutional Neural Network). The encoder component of the model processes the input data through several layers, each performing specific transformations. The description of each layer is as follows:

- **Input Layer:** The model accepts an input of size (128, 40, 150), representing the transformed music data in a piano roll-like format.
- **Conv2D Layer:** The first convolutional layer applies 32 different filters to the input data, resulting in a feature map of size (128, 40, 32). Each filter focuses on detecting specific features in the input.
- **MaxPooling2D Layer:** The max-pooling layer reduces the spatial dimensions of the feature map to (64, 20, 32) while preserving important features. This step enhances computational efficiency and helps prevent overfitting.

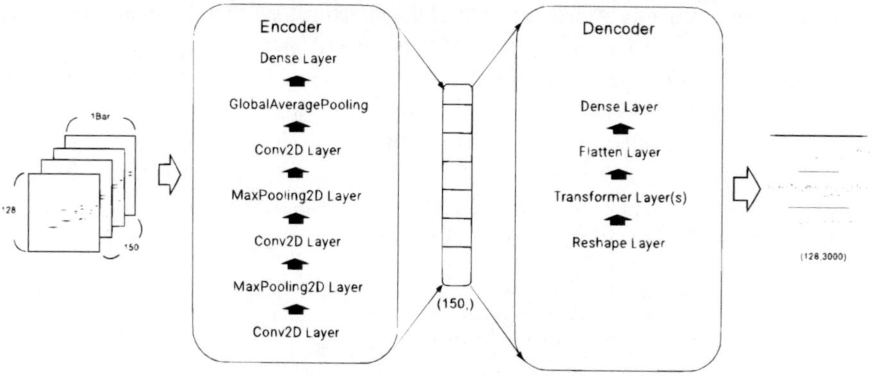

Fig. 2. Example of model architecture

- **Conv2D Layer:** The second convolutional layer applies 64 filters to the pooled feature map, producing a new feature map of size (62, 18, 64).
- **MaxPooling2D Layer:** Another max-pooling layer further reduces the spatial dimensions of the feature map to (31, 9, 64).
- **Conv2D Layer:** The final convolutional layer applies 64 filters to the pooled feature map, generating a feature map of size (29, 7, 64).
- **GlobalAveragePooling2D Layer:** This layer computes the average value of each feature map, reducing the dimensions to (64,).
- **Dense Layer:** The dense layer (also known as a fully connected layer) takes the output from the previous layer and transforms it into a (150,) vector.

Decoder (Transformer). The decoder receives the output from the encoder and processes it through several layers:

- **Input Layer:** The decoder takes an input of size (150,).
- **Reshape Layer:** The input is reshaped into a 2D matrix of size (150, 1) to be compatible with the following Transformer layers.
- **Transformer Layer(s):** The Transformer layers take the reshaped input and transform it through a series of self-attention and feed-forward neural network layers. The output is a matrix of size (150, 64). The operations in the Transformer can be represented as follows:

$$\text{Self-Attention: Attention}(Q, K, V) = \text{softmax}\left(\frac{QK^T}{\sqrt{d_k}}\right)V \tag{1}$$

$$\text{Feed-forward: FFN}(x) = \max(0, xW_1 + b_1)W_2 + b_2 \tag{2}$$

Here, Q, K, and V are the query, key, and value in the attention mechanism, respectively, and d_k is the dimension of the key. In the feed-forward network, W_1, b_1, W_2, and b_2 are the weights and biases of the two layers.
- **Flatten Layer:** The Flatten layer reshapes the 2D output matrix from the previous layer into a 1D vector of size (9600,).

- **Dense Layer:** The final dense layer transforms the flattened vector into the desired output shape of (128, 3000), which represents the generated music data in the piano roll format.

By combining Convolutional Neural Networks (used for feature extraction) and Transformers (used for sequence modeling), the architecture of this model enables it to effectively learn the features and structure of input music data, generate new music data while maintaining the input features.

3.3 Interactive Chaotic Evolution

In the process of model training, the optimizer plays a crucial role in determining the convergence rate and final performance of the model. Two key parameters of the optimizer are the learning rate and momentum, which have a significant impact on the training process. In this section, we propose an interactive chaotic evolution approach to optimize the learning rate and momentum of the ADAM optimizer using the logistic map.

The logistic map is a nonlinear dynamic system that exhibits chaotic behavior under certain conditions. It can be used to generate a sequence of pseudo-random numbers, which can then be mapped to a specific range to serve as the learning rate and momentum parameters for the ADAM optimizer. The overall algorithm for our interactive chaotic evolution approach is as follows:

1. Initialize the learning rate α and momentum β_1 and β_2 of the ADAM optimizer.
2. Train the model using the ADAM optimizer for a fixed number of iterations.
3. Check the loss function after each iteration. If the loss has not decreased significantly over the last k iterations, go to step 4. Otherwise, continue training using the current ADAM optimizer parameters.
4. Generate a sequence of pseudo-random numbers using the logistic map.
5. Map the pseudo-random numbers to a specific range to obtain the new values of the learning rate and momentum parameters for the ADAM optimizer.
6. Train the model for a fixed number of iterations using the new ADAM optimizer parameters.
7. Compare the performance of the model trained using the new ADAM optimizer parameters with that of the model trained using the previous ADAM optimizer parameters.
8. If the new model outperforms the previous one, update the ADAM optimizer parameters to the new values and continue training using the new parameters. Otherwise, continue training using the previous ADAM optimizer parameters.
9. Go back to step 3 and repeat until the model converges.

The logistic map used to generate the pseudo-random numbers is defined as follows:

$$x_{n+1} = rx_n(1 - x_n), \tag{3}$$

where r is the control parameter, x_n is the current value of the logistic map, and x_{n+1} is the next value of the logistic map. The value of r is typically set to a value between 3.6 and 4.0 to ensure that the map exhibits chaotic behavior.

The learning rate and momentum parameters for the ADAM optimizer are updated using the following equations:

$$\alpha_n = \frac{1}{1 + e^{-rx_n}}, \tag{4}$$

$$\beta_{1,n} = \frac{1}{1 + e^{-rx_{n+1}}}, \tag{5}$$

$$\beta_{2,n} = \frac{1}{1 + e^{-rx_{n+2}}}, \tag{6}$$

where α_n, $\beta_{1,n}$, and $\beta_{2,n}$ are the updated learning rate, momentum for the first moment estimate, and momentum for the second moment estimate at iteration n, respectively.

After adding chaotic algorithms, the update formulas for learning rate and momentum are as follows:

$$\alpha_{t+1} = \alpha_{\min} + (\alpha_{\max} - \alpha_{\min}) \times x_t \tag{7}$$

$$\beta_{t+1} = \beta_{\min} + (\beta_{\max} - \beta_{\min}) \times x_{t+1} \tag{8}$$

By using the interactive chaotic evolution approach, we can optimize the learning rate and momentum parameters of the ADAM optimizer in a more efficient manner, leading to better model performance and faster convergence.

4 Experiment and Evaluation

This experiment used three datasets for model training, including FreeMidi, Midi World, and the POP909 dataset curated by other researchers. There were two evaluation methods used: the first involved calculating the average distance between the model-generated audio and the original audio using the Euclidean distance method, while the second involved human evaluation to determine whether the generated audio was similar to the original audio. These evaluation methods are used to assess the performance of a model before and after the addition of interactive chaotic algorithms, to measure the expected improvements.

4.1 Dataset

This study used two self-collected MIDI music datasets, namely FreeMidi and Midi World, as well as the POP909 dataset compiled by others.

FreeMidi is a collection of 2,386 MIDI files of various genres and styles, including classical, jazz, pop, and rock. Each song is approximately 2–4 min long and consists of 64–80 measures. The total duration of the FreeMidi collection is approximately 112 h.

Midi World is another collection of MIDI files, consisting of 2,857 songs in various genres such as classical, rock, jazz, and pop. Each song is also approximately 2–4 min long and contains 64–80 measures. The total duration of the Midi World collection is approximately 140 h.

POP909 [6] is a collection of 909 MIDI files in the pop genre. Each song is approximately 2–5 min long and consists of 64–80 measures. The total duration of the POP909 collection is approximately 48 h. Below are the detailed descriptions of the datasets used (Table 1):

Table 1. Datset information

Dataset	Files	Time (min)
FreeMIDI	2186	6218
MIDIworld	3658	8431
POP909	909	2982

4.2 Evaluation

In this evaluation, 20 music pieces generated by the model were assessed for their similarity using Euclidean distance, and their originality was evaluated by human judgment (Tables 2 and 3).

Table 2. The table shows the evaluation results

Dataset	Euclidean distance	Human Evaluations
FreeMIDI	0.332	0.60
MIDIworld	0.401	0.65
POP909	0.284	0.50

Previous studies have shown that the Euclidean distance between cover songs and original songs is usually between 0.1 and 0.2. Based on our results, there is still a noticeable gap between the music generated by the model and human-created music, but these differences are not significant. It should be noted that music is subjective, and more than half of the human evaluators cannot distinguish whether the music is generated by the model or created by humans. This suggests that human creativity and evaluation of music still involve some degree of subjectivity. Therefore, the difference in the results of the Euclidean distance may be due to the fact that the generated songs have a different style from the original songs.

Table 3. The table shows using interactive chaotic algorithms expected outcomes

Dataset	Euclidean distance	Human Evaluations
FreeMIDI	0.30	0.65
MIDIworld	0.35	0.70
POP909	0.20	0.60

The expected outcome is a conservative estimate of a 10% improvement. This is based on the fact that interactive chaotic algorithms have a chaotic randomness and incorporate the subjective perception of humans, which allows the model to gain stronger randomness during training and optimize towards human aesthetic direction, thereby generating music that is closer to human-created works.

5 Future Work

For future research, we have outlined several directions to further enhance our MIDI music generation model's performance. Firstly, we plan to explore alternative optimization algorithms and generation models to achieve even better results. Specifically, reinforcement learning algorithms such as Deep Q-Network (DQN) and Actor-Critic (AC) will be investigated to enhance the model's ability to produce music with increased diversity and creativity.

Additionally, we aim to explore the potential of Generative Adversarial Networks (GANs) in generating more realistic and human-like music. GANs have demonstrated success in various image and audio generation tasks, and we believe they hold promise for music generation as well.

Moreover, we intend to incorporate user feedback into the model's training process to further refine the quality of the generated music. This will involve developing an interactive system that enables users to provide feedback on the generated music, which can be used to dynamically update the model's weights in real-time.

Finally, to showcase the generalizability and scalability of our proposed model, we plan to evaluate its performance on a larger and more diverse dataset. By doing so, we aim to demonstrate its applicability in various music-related domains such as music composition, sound design, and game development.

References

1. Costa, Y.M., Oliveira, L.S., Silla, C.N., Jr.: An evaluation of convolutional neural networks for music classification using spectrograms. Appl. Soft Comput. **52**, 28–38 (2017)
2. Neimark, D., Bar, O., Zohar, M., Asselmann, D.: Video transformer network. In: Proceedings of the IEEE/CVF International Conference on Computer Vision, pp. 3163–3172 (2021)

3. Huang, C.-Z.A., Cooijmans, T., Roberts, A., Courville, A., Eck, D.: Counterpoint by convolution. arXiv preprint arXiv:1903.07227 (2019)
4. Huang, C.-Z.A., et al.: Music transformer. arXiv preprint arXiv:1809.04281 (2018)
5. Pei, Y.: Chaotic evolution: fusion of chaotic ergodicity and evolutionary iteration for optimization. Nat. Comput. **13**, 79–96 (2014)
6. Wang, Z., et al.: Pop909: a pop-song dataset for music arrangement generation. In: Proceedings of 21st International Conference on Music Information Retrieval, ISMIR (2020)

A Fairness-Aware Load Balancing Strategy in Multi-tenant Clouds

Yu-Teng Chen and Kuan-Chou Lai(✉)

National Taichung University of Education, Taichung, Taiwan R.O.C.
kclai@mail.ntcu.edu.tw

Abstract. Load balancing is an important issue in multi-tenant clouds to ensure the load balanced on computing resources belonged to different tenants. A containerized multi-tenant environment could be managed by Kubernetes using different scheduling policies. For improving scheduling performance of Kubernetes, Apache Yunikorn project provides the fine-grain control by hierarchical resource queues to enhance the resource utilization. However, the fairness-aware load balance among tenants is missed in Yunikorn, which may result in poor resource utilization in some tenants. This paper proposes a fairness-aware load balance policy for multi-tenant environments to keep the balance of resources allocated in different tenants, and also the balance of the utilizations of different resources in a computing node. Experimental results show the superiority of the proposed policy.

Keywords: Fairness · Load Balance · Multi-tenants · Yunikorn · Resource Allocation

1 Introduction

Adopting container technique to provide services is popular in cloud computing environments. In such environments, tenants could easily deploy containerized applications on these clouds. In general, Kubernetes [2, 5] is a high-availability distributed orchestration platform to maintain containers. Although Kubernetes allocates resources to tenants, the fairness issue of the resource allocation policy in Kubernetes still could be improved in the multi-tenant environment.

Apache Yunikorn [1] provides a fine-grained control over resources among tenants, which is missed in the Kubernetes. Apache Yunikorn adopts the hierarchical resource queues and the access control list (ACL) to manage resources among different tenants. The scheduling decision in Apache Yunikorn considers the specific order of applications and nodes; therefore, Apache Yunikorn has better scheduling performance because its scheduling cycle is shorter than that of Kubernetes.

However, neither Yunikorn nor Kubernetes considers both the load balance among nodes and the fairness among tenants. The fairness among tenants [4, 7] is an important issue to avoid resource conflict in the scheduling policy. In the meantime, neither Yunikorn nor Kubernetes considers the load balance among clusters. For example, the

J. C. Hung et al. (Eds.): FC 2023, LNEE 1134, pp. 222–233, 2024.
https://doi.org/10.1007/978-981-99-9342-0_24

NodeResourceFit procedure in the Kubernetes scheduler has three policies: most, least and balance. The most and least policies calculate the score based on average resource utilization and make decision. However, average resource utilization couldn't indicate the utilization gap among heterogeneous resources resulting in the resource waste. The utilization gap occurs when certain resources are depleted when other resources are plentiful in one computing node. Although the balance policy may reduce the utilization gap among heterogeneous resources in a node, but it doesn't consider the load balance among nodes even among clusters. The similar problem is also found in Apache Yunikorn. So, the schedulers in Apache Yunikorn and Kubernetes couldn't provide an efficient load balance approach in containerized clusters.

In order to improve the fairness and load-balance in multi-tenant environments, this work proposes a fairness-aware load balancing (FALB) strategy to minimize the difference of quantities of heterogenous resources among tenants, the utilization gap [3] of heterogeneous resources in a node and the deviation between heterogeneous resource utilizations among nodes.

In the rest of this paper, the fairness problem and the load balance issue are described first; and the pseudo code of the FALB is introduced. Experimental results are shown and analyzed finally.

2 Related Works

Apache Yunikorn [1] adopts hierarchical queues to the fine-grained control and its scheduling performance is better than the Kubernetes one. However, current Yunikorn doesn't provide fairness among tenants. For evaluating fairness, Wang et al. [7] proposed the global dominant resource. But the indicator doesn't consider the elapsed time corresponding to this resource. Another previous work [6] keeps the fairness shared among tenants by computing resource quota. However, this previous work doesn't propose an indicator to solve the load balance. Pfreundschuh et al. [8] considers the profiling execution time of applications via neural network. With the profiling with neural network, this work indicates that the execution time of applications is predictable. The proposed DDRF approach in this work includes the execution time of the applications. To enhance the scheduling performance, the Apache Yunikorn project is proposed, and its scheduling strategies are simple. Carrión et al. [2] and Hilman et al. [4] describe the different scheduling objectives and they list indicators of each works such as CPU, memory, GPU. Chung et al. [3] proposed a method to minimize the resource waste. But it doesn't improve the load balance among nodes. Menouer et al. [5] adopted the Technique for Order of Preference by Similarity to Ideal Solution (TOPSIS) to achieve objectives including CPU utilization, memory utilization and the number of running container. However, it didn't reduce resource waste and the load-balance among nodes. Ramasamy et al. [6] proposed a priority queue scheduling algorithm and fair share strategies. Tenant's application is in high priority when the fair share is lower than the fair share quota. The algorithm focused on maintaining fair share among tenants but the resource utilizations did not be considered.

3 Fairness-Aware Load Balancing Strategy

This section introduces the proposed fairness-aware load balancing strategy mechanism. This work adopts a profiling system [8] to capture the application execution time for supporting the scheduling decision making. In this work, containerized applications have different execution time by using different parameters to estimate the expected execution time.

Apache Yunikorn is a cloud-native, efficient, and cost-saving resource scheduling system. It could handle the resource scheduling for running big data and machine learning applications on the Kubernetes platform. Apache Yunikorn adopts the hierarchical resource queue structure and the access control list (ACL) to provide the fine-grained control, so administrators could build customized hierarchical queues to filter tenants and manage resources. Apache Yunikorn manages the resource accessed by tenants according to ACLs. When a new application is submitted, Apache Yunikorn accepts the application when the ACL is matched the permission.

The system components of Yunikorn are as follows:

- Scheduler interface

  The scheduler interface defines the *grpc* protocol between the scheduler core and the Kubernetes shim. Common constraints for Yunikorn are also defined here.
- Scheduler core

  The scheduler core encapsulates the whole scheduling algorithms, such as application sorting, node sorting and queue sorting. The scheduler core is responsible for making the scheduling decision according to the container allocation requests. There are three sorting policies for applications: FIFO, fair and stateAware. The policies of node-sorting include fair and bin packing approaches.
- Scheduler shim

  Kubernetes shim is responsible for communicating with Kubernetes. Kubernetes shim watches events in Kubernetes clusters and translates Kubernetes events into corresponding information. Requests for resource allocation and the status of Kubernetes objects are translated based on the definitions of *grpc* protocol in the scheduler interface; and then the information is transmitted to the scheduler core (Fig. 1).

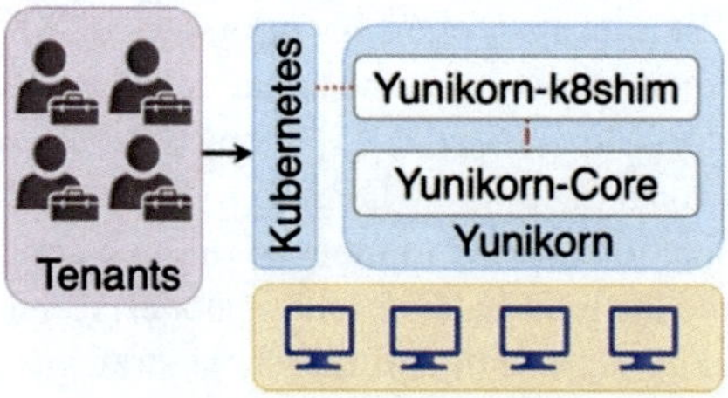

Fig. 1. System Architecture

For evaluating fairness among tenants, this work adopts the resource requirement dominant factor (RRDF) to present the dominant factor of the resource requirement.

Every tenant's RRDF is the sum of the product of resource requirements in each application request within the same tenant. The proposed approach keeps the balance of every tenant's RRDF in order to avoid that some tenants occupy excessive resources in a long time. Assume that there is a tenant set U = {$\text{tenant}_i \in$ the cluster} and $app(CPU, Mem, Exetime)$ represents the request of resource requirement of an application, where r_{CPU} is CPU request, r_{mem} is the memory request, and r_{exe} is the execution time. The variable, $scheduled_{app}$ is 1 when the application is scheduled. Equation (1) sums up the production of each handled request to compute certain tenant's RRDF. The FALB uses Eq. (2) to find out the tenant with the least RRDF and try to schedule the tenant's application to increase the tenant's RRDF.

$$RRDF_{tenant} = \sum\nolimits_{app \in tenant} (scheduled * \prod\nolimits_{r \in app} r), \tag{1}$$

where scheduled = 1 if *app is scheduled*; otherwise, it is 0.

$$tenant_i = ARGMin_{tenant \in U} (RRDF_{tenant})) \tag{2}$$

For example, a tenant submits three applications and they are scheduled. If each application request 1 CPU and 1 KB memory, the RRDF of the tenant is 3 * 10^6 (3 * 1000(vcore) * 1000(bytes)). Load-balance minimizes the utilization gap in a node by minimizing the largest difference between any two resource requirements, as shown in Eq. (3) and minimizes the resource deviation among nodes, as shown in Eq. (4) Resource waste happens when there is a large utilization gaps and some resources are exhausted. For measuring resource waste in a node, Eq. (3) calculates the max utilization gap in a node. In the meanwhile, the FALB utilizes the deviation to measure inter-node load balance when calculating the utilization gap in a node. For improving the load balance of nodes within a cluster, Eq. (4) presents the max deviation of resource utilizations of heterogeneous resources in a cluster. FALB considers the utilization gap and the deviation to reduce the resource waste within a node, and improves load-balance among nodes. For example, there is a cluster that CPU deviation is higher than memory one. The proposed FALB mechanism would try to reduce the CPU deviation and to avoid increasing the utilization gap in a node. Equation (5) finds the mean utilization of the node. This work finds the least mean utilization of a node, the minimal utilization gap in a node and the deviation of resource utilizations when there is a new request from applications, as shown in Eqs. (6), (7) and (8). However, there would be a trade-off between Eqs. (7) and (8) in some situations. This work adopts the TOPSIS (Technique for Order of Preference by Similarity to Ideal Solution) which is a multi-criteria decision analysis method to choose the best node after calculating the distance from current choice to best choice, as shown in Eq. (9).

$$UG_i = |Max_{r \in node_i}(u_{r1}) - Min_{r \in node_i}(u_{r2})|, \tag{3}$$

where take two resource types as example, but could be extended to multiple resource types.

$$\sigma_r = Max_{r \in nodes}(\sigma_r), \tag{4}$$

where σ is standard deviation of utilization of nodes.

$$MU_i = Average(\sum\nolimits_{r\epsilon node_i} \frac{utilizated\ quantity_r}{Capcity_r}) \tag{5}$$

$$MU_{min} = Min_{i\in nodes}(MU_i) \tag{6}$$

$$UG_{min} = Min_{i\epsilon nodes}(UG_i), \tag{7}$$

where UG is the utilization gap in *node i.*

$$\sigma_{min} = Min_{i\in nodes}(\sigma_{r,i}) \tag{8}$$

$$node_{best} = ARG\,Max_{i\in nodes}(TOPSIS(MU, UG, \sigma)) \tag{9}$$

The following is an example to illustrate above equations. For example, there are two nodes (one has 1 CPU, 1 GB memory, one has 2 CPU, 1 GB memory) in the cluster. The both initial mean resource utilizations are 0. When a new application request (0.5 CPU, 0.2 GB memory) is submitted, assigning the application to the specific node makes utilization gaps and standard deviations in the FALB. Utilization gaps of nodes would be 30% (i.e., (0.5/1 – 0.2/1) * 100%) and 10% (i.e., (0.5/2–0.2/1) * 100%) separately. Their standard deviations are 25 and 10. In the FALB approach, Algorithm 1 finds the application from the tenant with the minimal *RRDF*. For assigning the application to a node, Algorithm 2 chooses a best node based on the trade-off between utilization gaps, standard deviation, and the mean node's utilization. Finally, FALB assigns the tenant's application according to the starting time and node ID by TOPSIS.

Algorithm 1: Fairness in FALB

Inputs:

- *apps, the applications in the cluster.*
- *tenants, who submit applications in cluster*

Outputs:

- *Application ID, which is from tenant with the minimal RRDF*

1:Initialize every $RRDF_{tenant}$ with 0
2:For tenant in U:
3: For app in apps:
4: if app belongs to tenant and app is scheduled:
5: $RRDF_{tenant} += \prod_{r\epsilon app} r$
6:tenantID := ARG $MIN_{tenant\epsilon U}(RRDF_{tenant})$
7:Initialize heaps H, which order is increasing order of submitted timestamp of applications.
8:Put tenant's unscheduled apps into H separately.
9:return the top of H

Algorithm 2: Load-balance in FALB

Inputs:
- *nodes, nodes in cluster allow to run the application*
- *req, request of the submitted application*

Outputs:
- *node ID, the node to run the submitted application*

1:nodes:= find nodes allowing request to run
2:for node ϵ *nodes:*
3: UG_{node} *:= Eq. (3) computes the utilization gap if the app runs on the node*
4: MU_{node} *:= Eq. (5) calculates mean utilizations of the node*
5: σ_{node} *:= Eq. (4) finds the standard deviation if the app runs on the node*
6: Put $UG_{node}, MU_{node}, \sigma_{node}$ *to* $UG_{nodes}, MU_{nodes}, \sigma_{nodes}$
7: $M\grave{U}_{nodes}, U\grave{G}_{nodes}, \grave{\sigma}_{nodes}$ *:=normalize* $MU_{nodes}, UG_{nodes}, \sigma_{nodes}$ *and then divide them with 3 separately.*
10: # Based on $M\grave{U}_{nodes}, U\grave{G}_{nodes}, \grave{\sigma}_{nodes}$
11:$A^+_{MU}, A^+_{UG}, A^+_{\sigma}$*:=finding the minimal values Eq. (6)(7)(8) # best point*
12:$A^-_{MU}, A^-_{UG}, A^-_{\sigma}$*:=finding the max values # worst point*
13:for n in nodes:
14: the point ($wa\grave{i}t_n, U\grave{G}_n, \grave{\sigma}_n$*) when choosing node n*
15: append Euclidean distance between the point and best point to SM^+
16: append Euclidean distance between the point and worst point to SM^-
17: append $SM^-_n/(SM^-_n + SM^+_n)$ *to RC*
18:return the node with the minimal RC value

4 Experiment Results

This section introduces the experiment to show the performance improvement of FALB. Table 1 shows the machine specification in this experiment. The whole system consists of ten workstations and the node10 is the master node for Apache Yunikorn. Yunikorn on the master node assigns tenants' containerized applications to the slave workstations.

Table 2 shows the software information. Apache Yunikorn is responsible to schedule the pod which is basic scheduling unit in Kubernetes. Pods sharing the same application ID belong to the same application. A pod contains multiple containers and Docker is a well-known container runtime to provide operations of containers. Kubernetes is a

Table 1. Machine specification

Node number	Machine Specification		
	Product Name	CPU	Memory
node1	IBM System x	16	36G
node2	IBM System x	16	42G
node3	IBM System x	16	32G
node4	BladeCenter HS23	8	32G
node5	BladeCenter HS23	8	32G
node6	ProLiant DL360 G6	16	36G
node7	ProLiant DL360 G6	16	30G
node8	ProLiant DL360 G6	16	36G
node9	ProLiant DL360p Gen8	24	32G
node10	Pro E500 G6_WS720T	16	40G

container management platform, and a pod is a basic scheduling unit. Helm chart is the tool to manage configuration and developers could deploy applications to Kubernetes by helm.

Table 2. Software version

Software	Version
Apache Yunikorn	1.1.0
Docker	20.10.17
Kubernetes	1.21.0–00
Ubuntu	18.04
Helm	3.9.0

Table 3 indicates the resource requirement of four tenants who submits 50 applications separately. Each application is encapsulated in a pod. User1 and user2 prefer CPU. The others prefer memory.

There are two application scenarios: the stream scenario submits applications of tenants sequentially after deploying Yunikorn; In the batch scenario, Yunikorn deploys all application when all of them are submitted.

Table 4 indicates what objective strategies adopt. The mean utilization is the main scheduling objective in the original Yunikorn. The Fairness-aware Yunikorn (FA-YK), FALB-2 and FALB-3 implement the Algorithm 1 to maintain the *RRDF*. Comparing to the original Yunikorn and FA-YK, FALB series use TOPSIS to find the best node. The mean utilization and the utilization gap are the objectives in FALB-2 and FALB-3. Additionally, FALB-3 objectives include the resource deviation.

Table 3. Tenants' application information

tenants	CPU	memory(G)	execution(s)
user1	2	8	50
user2	1	4	200
user3	8	2	50
user4	4	1	200

Table 4. Active objectives among strategies

Strategies	DDRF	Mean utilization	Utilization gap	Deviation
Original Yunikorn		x		
FA-YK	x	x		
FALB-2	x	x	x	
FALB-3	x	x	x	x

Figure 2 indicates the total execution time of FALB series is better than the execution times of Yunikorn and the FA-YK. In the both scenario, two phenomena describe the benefits from *DDRF* and the effect of objectives. Maintaining tenants' *DDRF* reduces the specific resource exhaustion when a lot of same kind applications are submitted. Objectives in FALB series make the total execution time shorter than the execution time of Yunikorn and FA-YK.

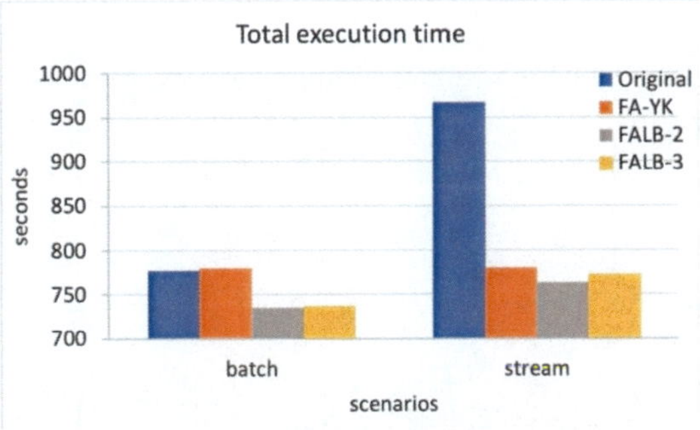

Fig. 2. Total execution time with different strategies

In the stream scenario, the FALB provides better fairness among tenants. Comparing to the FALB in Fig. 4, original Yunikorn in the Fig. 3 did not try to keep every tenant's *RRDF* close.

Figure 5 indicates the max resource deviations among original Yunikorn and FALB are between 5 and 28. Although the max resource deviation of the FALB is higher than

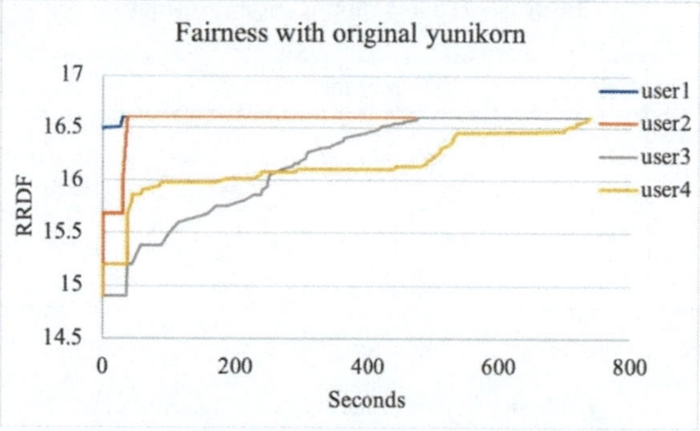

Fig. 3. Fairness in stream scenario (original)

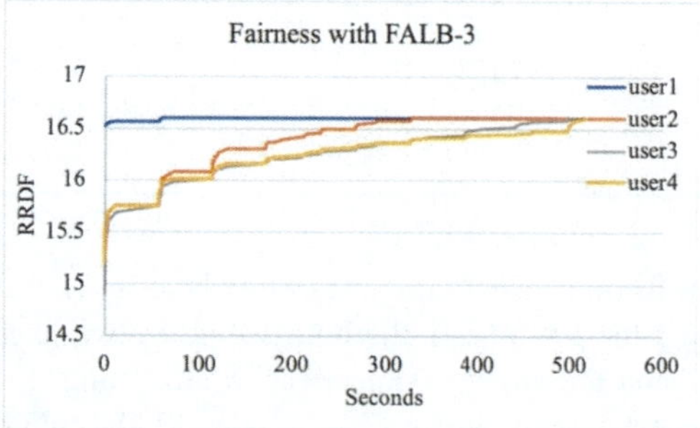

Fig. 4. Fairness in stream scenario (adopting FALB)

28, the total execution of FALB is better than that of the original one. The original Yunikorn without DDRF causes other tenants' applications to wait when a large number of applications from a tenant consume certain resources. Moreover, this also increase the resource waste.

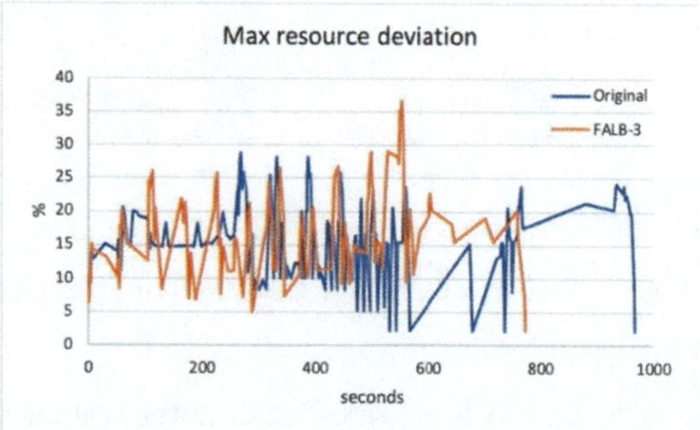

Fig. 5. Deviations in the stream scenario

Figure 6 shows the mean resource utilizations in the stream scenario. The memory utilization with FALB is stably higher than that of original Yunikorn after 270 s. FALB considering RRDF avoids scheduling a lot of same tenant's applications to exhaust specific resource. Comparing to the FALB, the original Yunikorn leads that other tenants' applications are waiting. The original Yunikorn increases the waiting time of application and the total execution time.

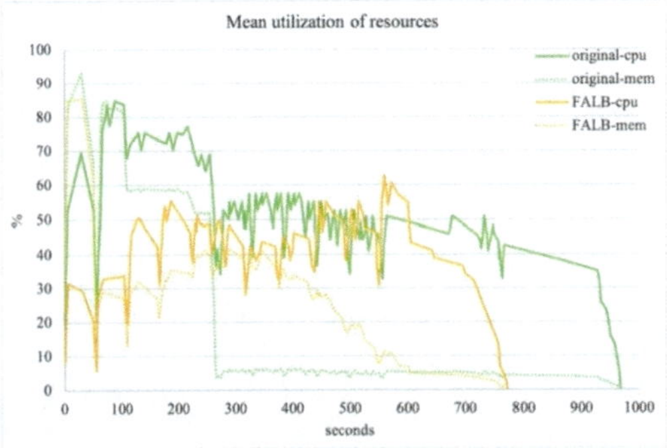

Fig. 6. Mean utilizations of nodes in stream scenario

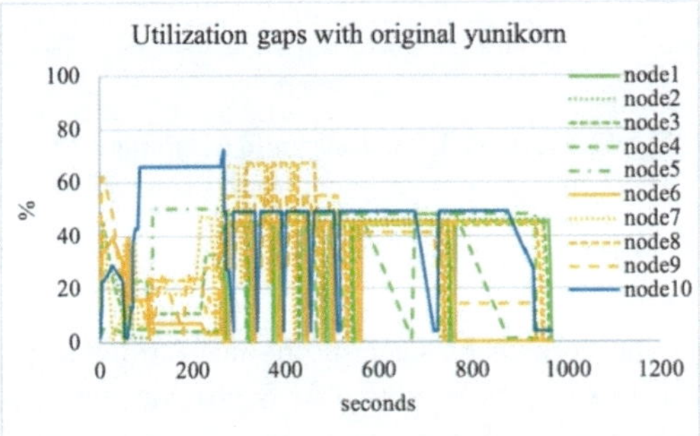

Fig. 7. Utilization gaps in stream scenario (original)

Figure 7 and Fig. 8 show the utilization gaps by different strategies. A small utilization gap of a node is better when the node can't run new applications. Figure 8 indicates average median value of utilization gaps is around 47. Compared to Fig. 7, the RRDF mechanism avoids the tenants' application waiting and some utilization gaps are reduced. Figure 9 shows that the FALB keeps the difference of every tenant's *RRDF* close. In hence, the curves of different tenants would be close each other. No tenant owns too many resource quantities to violate the fairness.

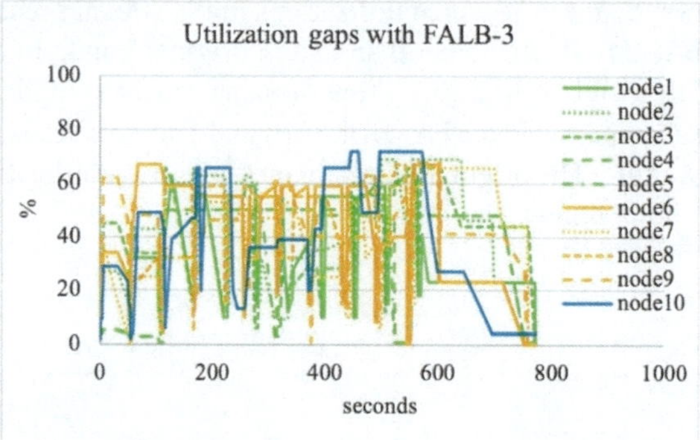

Fig. 8. Utilization gaps in stream scenario (adopting FALB)

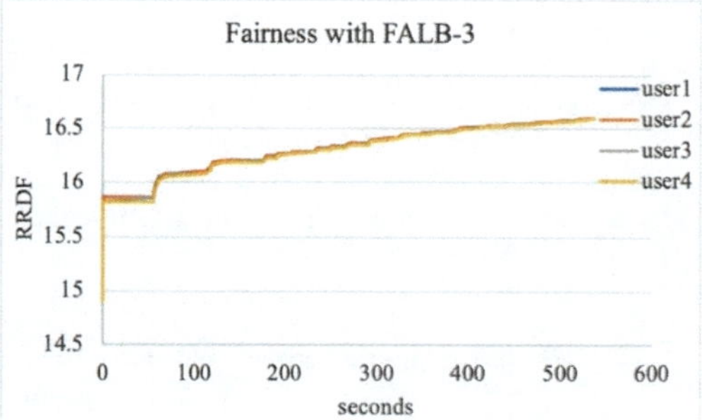

Fig. 9. Fairness in batch scenario (adopting FALB)

5 Conclusions

This study improves the fairness by keeping the utilizations of tenants' *RRDF* close. After keeping the fairness among tenants, FALB provides the reduction of utilization gaps in a node, the decreament of max. Deviation of resources in clusters to reduce the execution time. FALB evaluates nodes by TOPSIS in order to find a node to allocate an application. The results show that total execution time of the FALB is better than that of original Yunikorn.

Acknowledgement. This study was sponsored by the Ministry of Science and Technology, Taiwan, R.O.C., under contract numbers: MOST 111-2221-E-142-004-, and by the "Intelligent Manufacturing Research Center" (iMRC) from the Featured Areas Research Center Program within the framework of the Higher Education Sprout Project by the Ministry of Education, Taiwan, R.O.C.

References

1. Apache Yunikorn, 8 December 2022. https://yunikorn.apache.org
2. Carrión, C.: Kubernetes scheduling: taxonomy, ongoing issues and challenges. ACM Comput. Surv. **55**(7), 1–37 (2022)
3. Chung, W.C., Wu, T.L., Lee, Y.H., Huang, K.C., Hsiao, H.C., Lai, K.C.: Minimizing resource waste in heterogeneous resource allocation for data stream processing on clouds. Appl. Sci. **11**(1), 149 (2020)
4. Hilman, M.H., Rodriguez, M.A., Buyya, R.: Multiple workflows scheduling in multi-tenant distributed systems: a taxonomy and future directions. ACM Comput. Surv. (CSUR) **53**(1), 1–39 (2020)
5. Menouer, T.: KCSS: Kubernetes container scheduling strategy. J. Supercomput. **77**(5), 4267–4293 (2020). https://doi.org/10.1007/s11227-020-03427-3
6. Ramasamy, M., Balakrishnan, M., Thangaraj, C.: Priority queue scheduling approach for resource allocation in containerized clouds. In: Smys, S., Bestak, R., Rocha, Á. (eds.) ICICIT 2019. LNNS, vol 98, pp. 758–765. Springer, Cham (2020). https://doi.org/10.1109/TPDS.2014.2362139
7. Wang, W., Liang, B., Li, B.: Multi-resource fair allocation in heterogeneous cloud computing systems. IEEE Trans. Parallel Distrib. Syst. **26**(10), 2822–2835 (2015)
8. Pfreundschuh, S., Brown, P.J., Kummerow, C.D., Eriksson, P., Norrestad, T.: GPROF-NN: a neural network based implementation of the Goddard Profiling Algorithm. Atmos. Meas. Tech. Discuss. **15**(17), 5033–5060 (2022)

Comments on a Double-Blockchain Assisted Data Aggregation Scheme for Fog-Enabled Smart Grid

Pei-Yu Lin[1], Ya-Fen Chang[2], Pei-Shih Chang[2], and Wei-Liang Tai[3(✉)]

[1] Department of Electrical Engineering, National Kaohsiung University of Science and Technology, Kaohsiung, Taiwan

[2] Department of Computer Science and Information Engineering, National Taichung University of Science and Technology, Taichung, Taiwan

[3] Bachelor Degree Program of Artificial Intelligence, National Taichung University of Science and Technology, Taichung, Taiwan

twl@nutc.edu.tw

Abstract. To comply with the specific requirements of smart grids, Chen et al. proposed a data aggregation scheme by utilizing double blockchains and the Paillier cryptosystem that is an additive homomorphic encryption system for public-key cryptography. Chen et al. claimed that their scheme could resist various attacks and ensure data confidentiality, data integrity, validity, identity anonymity and authenticity. However, after thoroughly analyzing their scheme, we find that it suffers from five flaws. Firstly, anonymity is not ensured as claimed. Secondly, private keys of smart meters and fog nodes can be easily retrieved. Thirdly, after a smart meter or a fog node's private key is revealed, a malicious entity can impersonate it and generate a valid signature of the forged data's ciphertext. Fourthly, in both of the UA-blockchain generation phase and FA-blockchain generation phase, the signature verification will never succeed. Fifthly, some statements in Chen et al.'s scheme are inaccurate or missing such that their scheme cannot work as claimed. The details of how these flaws damage Chen et al.'s scheme are shown in this paper.

Keywords: Blockchain · Smart Grid · Authentication · Homomorphism · Anonymity · Fog Computing · Data Aggregation

1 Introduction

Smart grids are designed to manage the two-way flow of electricity and information between utilities and users. Digital technologies are used to improve the efficiency, reliability, and security of the grid and to provide users more information and control over their energy usage. For example, smart meters can send users' electricity consumption information to the control center, and the control center can analyze the obtained information to provide users with electricity consumption reports and suggestions to save energy. However, smart grids have special communication and computing requirements

J. C. Hung et al. (Eds.): FC 2023, LNEE 1134, pp. 234–245, 2024.
https://doi.org/10.1007/978-981-99-9342-0_25

such that they need to be deployed widely. In addition, users' electricity consumption information sent by smart meters is often collected and stored by electricity suppliers without the appropriate permission. The data may be accessed by a third party such that users' living habits and related economic status will be revealed. This leads to privacy breaches while users are not aware that their personal data is being collected and shared in this way. That is, how to protect user privacy is an important issue in smart grids. On the other hand, because of the properties of smart grids, a variety of attacks may result in catastrophic damage. Thus, how to ensure the security of smart grids is another essential issue that has to be taken into consideration, and proper security mechanisms need to be employed to protect smart grids from these various attacks such as eavesdropping, tampering, and counterfeiting. Besides, performance is also a key issue to determine whether a smart grid can work and offer desired services well or not. To sum up, performance, privacy, and security are the main issues in smart grids. And data aggregation can be regarded as a representative mechanism in smart grids because of its remarkable advantages. Thus, a plenty of data aggregation schemes preserving privacy are proposed [1–7].

Zhang et al. [8] took advantages of the superior features of blockchain such as non-repudiation, untamperability, decentralization, and easy-to-trace and proposed a keyless signature scheme based on the blockchain architecture for smart grids. In Zhang et al.'s scheme, a new consensus mechanism is designed to turn the blockchain into an automatic access control manager such that no trusted third party is needed. Because blockchain possesses superior features, several researches are proposed from then on. In 2020, Alcaraz et al. [9] proposed a smart grid structure by using the three-layer-based interconnection architecture and blockchain technology to manage connections among devices, resources, and processes while ensuring reliability and security. Li et al. [10] proposed a blockchain-based anomalous electricity consumption detection method for smart grids. In Li et al.'s method, electricity consumption data is from readings of sensors and smart meters, a trained machine learning model is adopted to detect electricity consumption anomalies, and the blockchain is used to record all processes.

Recently, Chen et al. [11] proposed a data aggregation scheme for smart grids with a double-blockchain structure, which is called the double-blockchain-assisted secure and anonymous data aggregation scheme, DA-SADA. DA-SADA presents a network model with three layers, user layer, fog computing layer, and service supporting layer. And, there are two types of blockchains, UA-blockchain and FA-blockchain. In the user layer, the whole area is divided into several subareas, and smart meters SM's are deployed to collect users' electricity consumption information. In a subarea, data collected by smart meters is encrypted and sent to a specific smart meter, namely an aggregation node in the user layer. An aggregation node in the user layer is responsible for aggregating data, generating the new block in UA-blockchain and sending the generated UA-blockchain to the subarea's corresponding fog node in the fog computing layer. In the fog computing layer, when a fog node receives information, it generate the corresponding digital signature and sends required data to a specific node, namely an aggregation node in the fog computing layer. The aggregation node in the fog computing layer is responsible for aggregating data, generating the new block in FA-blockchain and sending the generated FA-blockchain to the cloud server in the service supporting layer. When the cloud server receives the FA-blockchain, it can retrieve the needed information and make analysis to further determine strategies and improve the power utilization efficiency.

However, after thoroughly analyzing their scheme, DA-SADA, we find that it suffers from five flaws. Firstly, anonymity is not ensured as claimed. Secondly, private keys of smart meters and fog nodes can be easily retrieved. Thirdly, after a smart meter or a fog node's private key is revealed, a malicious entity can impersonate it and generate a valid signature of the forged data's ciphertext. Fourthly, in both of the UA-blockchain generation phase and FA-blockchain generation phase, the signature verification will never succeed. Fifthly, some statements in DA-SADA are inaccurate or missing such that it cannot work as claimed.

The rest of this paper is organized as follows: Sect. 2 reviews DA-SADA. The security analysis of DA-SADA is made in Sect. 3. At last, some conclusions are made in Sect. 4.

2 Review of DA-SADA

In this section, we review the double-blockchain assisted secure and anonymous data aggregation scheme, DA-SADA, proposed by Chen et al. DA-SADA consists of four phases, system initialization phase, UA-blockchain generation phase, FA-blockchain generation phase, and service supporting phase. Notations commonly used in DA-SADA are listed in Table 1, and the details are as follows.

Table 1. Notations used in DA-SADA.

Notation	Definition
TA	a trust authority that generates parameters for all devices
p, q	two large prime numbers
κ	a system security parameter denotes the length of prime numbers
N/λ	the system public/private key
SM_{ij}	the i-th smart meter in the j-th subarea
X_{ij}	SM_{ij}'s public key
Y_{ij}	SM_{ij}'s private key
$Pseu_{ij}$	SM_{ij}'s pseudonym
fog_j	the fog node responsible for the j-th subarea in the user layer
X_j	fog_j's public key
Y_j	fog_j's private key
$H(.)$	a cryptographic hash function
$\|\|$	the concatenation operator

2.1 System Initialization Phase

In DA-SADA, *TA* is responsible for system initialization. First, *TA* generates all system parameters including the system's public and private keys, all smart meters and fog nodes' public and private keys, and pseudonyms for all smart meters and fog nodes. Then, *TA* distributes system parameters. At last, *TA* generates Bloom filters for all subareas and a Bloom filter in the fog computing layer. The details are as follows:

Step 1. *TA* selects the security parameter κ and two prime numbers p and q of length κ bits.

Step 2. *TA* computes the system public key $N = p \times q$ and the system private key $\lambda = lcm\ (p\text{-}1, q\text{-}1)$ for the homomorphic encryption algorithm.

Step 3. *TA* chooses a number $r \in \mathbb{Z}_N^*$ randomly, computes $s = r^N \bmod N^2$, and defines a function $L(u) = \frac{u-1}{N}$, where u denotes the input of the function $L(.)$.

Step 4. For each smart meter SM_{ij}, *TA* randomly chooses a prime number X_{ij} as SM_{ij}'s public key and computes SM_{ij}'s private key $Y_{ij} = X_{ij}^{-1} \bmod N^2$ and SM_{ij}'s pseudonym $Pseu_{ij} = X_{ij} \bmod N^2$.

Step 5. For each fog node fog_j, *TA* randomly chooses a prime number X_j as fog_j's public key and computes fog_j's private key $Y_j = X_j^{-1} \bmod N^2$ and fog_j's pseudonym $Pseu_j = X_j \bmod N^2$.

Step 6. *TA* chooses a cryptographic hash function $H(.)$: $\{0, 1\}^* \rightarrow Z_N{}^*$.

Step 7. After λ, N, s, $H(.)$, X_{ij}, Y_{ij}, X_j, and Y_j are generated, N and $H(.)$ are published online while (X_{ij}, Y_{ij}, s), (X_j, Y_j), and λ are assigned to SM_{ij}, fog_j, and the cloud server through a secure channel, respectively.

Step 8. For the *j*-th subarea, *TA* collects pseudonyms of all corresponding smart meters and generates a Bloom filter that is a θ-bit array and the element's value is set to one when its index is equal to $H(Pseu_{ij}) \bmod \theta$. Similarly, *TA* collects pseudonyms of fog nodes to generate a Bloom filter in the fog computing layer.

Step 9. At last, *TA* sends Bloom filters generated for subareas to the corresponding smart meters and the Bloom filter in the fog computing layer to fog nodes.

2.2 UA-Blockchain Generation Phase

After the system is initialized, each smart meter encrypts a user's power consumption data and generates the corresponding signature for integrity. Then, one of smart meters in the same subarea is chosen as the aggregation node in the user layer because it possesses the most remaining computational resources. The aggregation node collects reports of smart meters, verifies signatures, obtains the aggregated ciphertext, and generates the transaction, where the aggregation node and these smart meters are in the same subarea. The aggregation node records the transaction in a block and broadcasts this block in the subarea for authentication. When the number of positive verification results reaches the threshold, the new block is added to the UA-blockchain in the user layer. Figure 1 depicts the structure of the UA-blockchain and how the corresponding Merkle root is generated. The details of this phase are as follows:

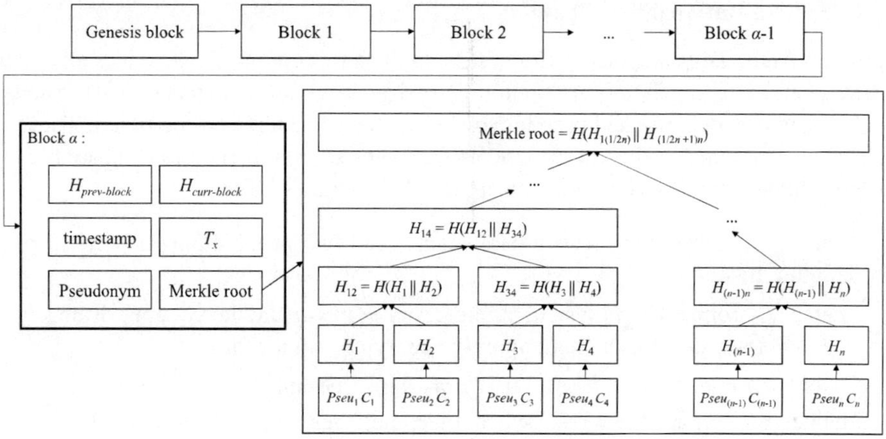

Fig. 1. The structure of UA-blockchain and how the Merkle root is generated in DA-SADA.

Step 1. In the j-th subarea in a certain time slot t_s, SM_{ij} first sets a parameter $g = N + 1$ and encrypts the collected data d_{ij} by computing $C_{ij} = (1 + d_{ij}N) \times s \bmod N^2$ instead of $g^{d_{ij}} \times r^N \bmod N^2$ to improve computation efficiency, where $C_{ij} = g^{d_{ij}} \times r^N \bmod N^2 = (N+1)^{d_{ij}} \times r^N \bmod N^2 = (1 + d_{ij}N) \times s \bmod N^2$, i is in $[1, n]$ and n denotes the number of smart meters in the j-th subarea.

Step 2. SM_{ij} uses the encrypted data C_{ij}, its pseudonym $Pseu_{ij}$, its private key Y_{ij}, and the timeslot t_s to generate the signature $\sigma_{ij} = H\left(u_{ij}||Pseu_{ij}\right)^{Y_{ij}} \bmod N^2$, where $u_{ij} = H(C_{ij} \parallel t_s)$.

Step 3. SM_{ij} sends the report $\{Pseu_{ij}, t_s, C_{ij}, \sigma_{ij}\}$ to the aggregation node.

Step 4. After receiving the report $\{Pseu_{ij}, t_s, C_{ij}, \sigma_{ij}\}$, the aggregation node checks the effectiveness of $Pseu_{ij}$ with the Bloom filter and the validity of the report with the timestamp.

Step 5. The aggregation node uses batch verification to verify these received signatures by checking if $\prod_{i=1}^{n} \sigma_{ij}^{X_{ij}} \bmod N^2 = \prod_{i=1}^{n} H(H(C_{ij}||t_s)||Pseu_{ij}) \bmod N^2$.

Step 6. If it holds, the aggregation node obtains the aggregated ciphertext C_j for the j-th subarea by computing $C_j = \prod_{i=1}^{n} C_{ij} \bmod N^2$ and generates the transaction $T_x = (C_j, Pseu_{ij}, t_s)$.

Step 7. Then the aggregation node records the transaction $T_x = (C_j, Pseu_{ij}, t_s)$ in a new block that also includes the Merkle root, the hash value of the previous block $H_{prev\text{-}block}$, and the hash value of the current block $H_{curr\text{-}block}$, where the Merkle root is obtained by setting a leaf node's value with the ciphertext and the pseudonym and hashing them and the corresponding hash results as shown in Fig. 1, and $H_{curr\text{-}block} = \text{SHA256}(index + H_{prev\text{-}block} + Pseu_{ij} + timestamp + C_j + \sum_{ij} transactions_{ij})$.

Step 8. After generating the new block, the aggregation node broadcasts it in the j-th subarea. Then, each smart meter SM_{ij}, an ordinary node, verifies records in the new block and its related data only by checking whether it is identical to the original data or not. If the verification is successful, the smart meter SM_{ij}, an ordinary node, broadcasts the verification result in the j-th subarea.

Step 9. If the number of the correctness confirmation messages sent by other distinct smart meters in the j-th subarea is equal to or more than $2n/3+1$, the new block is considered to be valid and added to the UA-blockchain, where n denotes the number of smart meters in the j-th subarea.

2.3 FA-Blockchain Generation Phase

The process to generate the FA-blockchain is similar to that to generate the UA-blockchain. In the FA-blockchain generation phase, each fog node first receives encrypted data from the corresponding UA-blockchain and generates the corresponding signature of the encrypted data for integrity. Then, one of fog nodes is chosen as the aggregation node in the fog computing layer because it possesses the most remaining computational resources. The aggregation node in the fog computing layer collects reports of fogs, verifies signatures, obtains the aggregated ciphertext for all subareas, and generates the transaction. The aggregation node in the fog computing layer records the transaction in a block and broadcasts this block to other fog nodes for authentication. When the number of positive verification results reaches the threshold, the new block is added to the FA-blockchain in the fog computing layer. The details of the FA-blockchain generation phase are as follows:

Step 1. When fog_j that is responsible for the j-th subarea gets the aggregated power consumption ciphertext C_j, fog_j uses the encrypted data C_j, its pseudonym $Pseu_j$, its private key Y_j, and the timeslot t_s to generate the signature $\sigma_j = H\left(u_j||Pseu_j\right)^{Y_j} \bmod N^2$, where j is in $[1, m]$, m denotes the number of fog nodes in the fog computing layer and $u_j = H(C_j \parallel t_s)$.

Step 2. fog_j sends the report $\{Pseu_j, t_s, C_j, \sigma_j\}$ to the aggregation node in the fog computing layer.

Step 3. After receiving the report $\{Pseu_j, t_s, C_j, \sigma_j\}$, the aggregation node in the fog computing layer checks the effectiveness of $Pseu_j$ with the Bloom filter and the validity of the report with the timestamp.

Step 4. The aggregation node in the fog computing layer uses batch verification to verify these received signatures by checking if $\prod_{j=1}^{m} \sigma_j^{X_j} \bmod N^2 = \prod_{j=1}^{m} H(H\left(C_j||t_s\right)||Pseu_j) \bmod N^2$.

Step 5. If it holds, the aggregation node in the fog computing layer obtains the aggregated ciphertext C_{AS} for all subareas by computing $C_{AS} = \prod_{j=1}^{m} C_j \bmod N^2$ and generates the transaction $T_x{}' = (C_{AS}, Pseu_j, t_s)$.

Step 6. Then the aggregation node in the fog computing layer records the transaction $T_x' = (C_{AS}, Pseu_j, t_s)$ in a new block that also includes the Merkle root, the hash value of the previous block $H_{prev\text{-}block}'$, and the hash value of the current block $H_{curr\text{-}block}'$, where $H_{curr\text{-}block}' = \text{SHA256}(index + H_{prev\text{-}block}' + Pseu_j + timestamp + C_{AS} + \sum_j transactions_j)$.

Step 7. After generating the new block, the aggregation node in the fog computing layer broadcasts it to other fog nodes. Then, each fog node fog_j, an ordinary node in the fog computing layer, verifies records in the new block and verifies its related data only by checking whether it is identical to the original data or not. If the verification is successful, the fog node fog_j, an ordinary node in the fog computing layer, broadcasts the verification result to other fog nodes in the fog computing layer.

Step 8. If the number of the correctness confirmation messages sent by other distinct fog nodes in the fog computing layer is equal to or more than $2m/3+1$, the new block is considered to be valid and added to the FA-blockchain, where m denotes the number of fog nodes in the fog computing layer.

2.4 Service Supporting Phase

When the cloud server receives the FA-blockchain of the fog computing layer, it gets the aggregated power consumption ciphertext C_{AS} for all subareas and decrypts it to get the aggregated plaintext $M = L({C_{AS}}^{\lambda} \bmod N^2) / L(g^{\lambda} \bmod N^2)$ by using the decryption procedure mentioned in the Paillier cryptosystem. Then the cloud server recovers subareas' data UA_j's with their proposed Horner rule-based analytical algorithm, where $UA_j = \sum_{i=1}^{n} d_{ij}$ and $M = \sum_{j=1}^{m} UA_j$. When the cloud server obtains the power consumption of each subarea, the future power usage of each subarea can be predicted, and decision support for power dispatch and price adjustment can be provided.

3 Security Analysis of DA-SADA

Chen et al. claimed that DA-SADA could resist various attacks and ensure data confidentiality, data integrity, validity, identity anonymity and authenticity. However, after thoroughly analyzing DA-SADA, we find that it suffers from five flaws. Firstly, anonymity is not ensured as claimed because a smart meter's pseudonym is fixed. Secondly, private keys of smart meters and fog nodes can be easily retrieved. Thirdly, after a smart meter or a fog node's private key is revealed, a malicious entity can impersonate it and generate a valid signature of the forged data's ciphertext. Fourthly, in both of the UA-blockchain generation phase and FA-blockchain generation phase, the signature verification will never succeed such that legal signatures are always regarded as invalid. Fifthly, some statements in DA-SADA are inaccurate or missing such that DA-SADA cannot work as claimed. The details are as follows:

3.1 Failure to Ensure Anonymity

In the system initialization phase, *TA* randomly chooses a prime number X_{ij} as the smart meter SM_{ij}'s public key and computes SM_{ij}'s private key $Y_{ij} = X_{ij}^{-1} \bmod N^2$ and SM_{ij}'s pseudonym $Pseu_{ij} = X_{ij} \bmod N^2$. And, *TA* randomly chooses a prime number X_j as the fog node fog_j's public key and computes fog_j's private key $Y_j = X_j^{-1} \bmod N^2$ and fog_j's pseudonym $Pseu_j = X_j \bmod N^2$. Each smart meter SM_{ij}'s pseudonym $Pseu_{ij}$ and each fog node fog_j's pseudonym $Pseu_j$ are always fixed because they are not updated in other phases. Moreover, these pseudonyms are not concealed when transmitted or included in blocks. Thus, a specific smart meter or fog node will be traced or monitored. According to the above, anonymity is not ensured in Chen et al.'s scheme.

3.2 Disclosure of the Private Key

In the system initialization phase, a smart meter SM_{ij} is assigned with the private key Y_{ij} and the public key X_{ij}, where $Y_{ij} = X_{ij}^{-1} \bmod N^2$ and SM_{ij}'s pseudonym $Pseu_{ij} = X_{ij} \bmod N^2$. SM_{ij}'s pseudonym $Pseu_{ij}$ will be transmitted in the UA-blockchain generation phase or included in blocks of the UA-blockchain, so $Pseu_{ij}$ can be easily obtained. In addition, the parameter N is published online. Thereupon, SM_{ij}'s private key Y_{ij} can be retrieved by computing $Y_{ij} = Pseu_{ij}^{-1} \bmod N^2 = X_{ij}^{-1} \bmod N^2$. Similarly, a fog node fog_j is assigned with the private key Y_j and the public key X_j, where $Y_j = X_j^{-1} \bmod N^2$ and fog_j's pseudonym $Pseu_j = X_j \bmod N^2$. Because fog_j's pseudonym $Pseu_j$ will be transmitted in the FA-blockchain generation phase or included in blocks of the FA-blockchain, $Pseu_j$ can be easily obtained. Thereupon, fog_j's private key Y_j can be retrieved by computing $Y_j = Pseu_j^{-1} \bmod N^2 = X_j^{-1} \bmod N^2$. As a result, private keys of smart meters and fog nodes can be easily retrieved in Chen et al.'s scheme.

3.3 Generation of a Valid Signature of the Forged Data's Ciphertext

After a smart meter's private key is revealed, a malicious entity can impersonate it and generate a valid signature of the forged data's ciphertext. The details are as follows. In the system initialization phase, a smart meter SM_{ij} is assigned with the private key Y_{ij}, the public key X_{ij}, and s securely, where $s = r^N \bmod N^2$ and $r \in \mathbb{Z}_N^*$. In the UA-blockchain generation phase, SM_{ij} computes the ciphertext C_{ij} of the data d_{ij} by computing $C_{ij} = (1 + d_{ij}N) \times s \bmod N^2$. Actually, the Paillier homomorphic cryptosystem allows a user to compute his/her personal ciphertext with an arbitrary random number, and the receiver who gets the aggregated ciphertext can retrieve the aggregated data without knowing what the involved random numbers are. That is, the malicious entity can chooses a random number $r_{ij} \in \mathbb{Z}_N^*$, generates the forged data d_{ij}', and computes the corresponding ciphertext C_{ij}' by computing $C_{ij}' = g^{d_{ij}'} \times r_{ij}{}^N \bmod N^2$. After retrieving a smart meter SM_{ij}'s private key Y_{ij} by computing $Y_{ij} = Pseu_{ij}{}^{-1} \bmod N^2$, the malicious entity computes $u_{ij}' = H(C_{ij}' \| t_s)$ and the signature $\sigma_{ij}' = H\left(u_{ij'} \| Pseu_{ij}\right)^{Y_{ij}} \bmod N^2$ and sends the report $\{Pseu_{ij}, 't_s, C_{ij}', \sigma_{ij}'\}$ to the aggregation node. Thus, after retrieving a smart meter SM_{ij}'s private key Y_{ij}, a malicious entity can impersonate SM_{ij} and further generate a valid signature of the forged data's ciphertext. Similarly, because a fog node fog_j's private key Y_j can be easily retrieved, a malicious entity can impersonate fog_j and generate a valid signature of the forged data's ciphertext as well.

3.4 Failed Signature Verification

In both of the UA-blockchain generation phase and FA-blockchain generation phase, an aggregation node verifies signatures with batch verification. In the UA-blockchain generation phase, the aggregation node in the user layer verifies these received signatures σ_{ij}'s by checking if $\prod_{i=1}^{n} \sigma_{ij}^{X_{ij}} \bmod N^2 = \prod_{i=1}^{n} H(H(C_{ij}||t_s)||Pseu_{ij}) \bmod N^2$. That is, the equality of $\sigma_{ij}^{X_{ij}} \bmod N^2$ and $H(H(C_{ij} \parallel t_s) \parallel Pseu_{ij}) \bmod N^2$ is checked with a batch approach, where i is in [1, n] and n denotes the number of smart meters in the j-th subarea. Because $\sigma_{ij} = H\left(u_{ij}||Pseu_{ij}\right)^{Y_{ij}} \bmod N^2$ and $u_{ij} = H(C_{ij} \parallel t_s)$, $\sigma_{ij}^{X_{ij}} \bmod N^2$ $= H\left(H\left(C_{ij}||t_s\right)||Pseu_{ij}\right)^{Y_{ij}X_{ij}} \bmod N^2$. Unfortunately, $\sigma_{ij}^{X_{ij}} \bmod N^2 \neq H(H(C_{ij} \parallel t_s) \parallel Pseu_{ij}) \bmod N^2$. In the system initialization phase, the smart meter SM_{ij} is assigned with the private key Y_{ij} and the public key X_{ij}, where $Y_{ij} = X_{ij}{}^{-1} \bmod N^2$ and SM_{ij}'s pseudonym $Pseu_{ij} = X_{ij} \bmod N^2$. To show that $\sigma_{ij}^{X_{ij}} \bmod N^2 \neq H(H(C_{ij} \parallel t_s) \parallel Pseu_{ij})$ $\bmod N^2$, an example is given. First, we set $N = 3 \times 5$ and randomly select a prime number $X_{ij} = 227$, the private key $Y_{ij} = 227^{-1} \bmod 225 = 113$. Suppose $H(u_{ij} \parallel Pseu_{ij})$ $= H(H(C_{ij} \parallel t_s) \parallel Pseu_{ij}) = 2$. Then the signature $\sigma_{ij} = H\left(u_{ij}||Pseu_{ij}\right)^{Y_{ij}} \bmod N^2 = 2^{113}$ $\bmod 225 = 109$, and $\sigma_{ij}^{X_{ij}} \bmod N^2 = 109^{227} \bmod 225 = 121 \neq H(H(C_{ij} \parallel t_s) \parallel Pseu_{ij})$. As a result, valid signatures will be never verified successfully in the UA-blockchain generation phase. Similarly, in the FA-blockchain generation phase, the aggregation node in the fog computing layer verifies these received signatures σ_j's by checking if $\prod_{j=1}^{m} \sigma_j^{X_j} \bmod N^2 = \prod_{j=1}^{m} H(H(C_j||t_s)||Pseu_j) \bmod N^2$. That is, the equality of $\sigma_j^{X_j} \bmod N^2$ and $H(H(C_j \parallel t_s) \parallel Pseu_j) \bmod N^2$ is checked with a batch approach, where j is in [1, m] and m denotes the number of fog nodes in the fog computing layer. Because $\sigma_j = H\left(u_j||Pseu_j\right)^{Y_j} \bmod N^2$ and $u_j = H(C_j \parallel t_s)$, $\sigma_j^{X_j} \bmod N^2 = H\left(H\left(C_j||t_s\right)||Pseu_j\right)^{Y_jX_j} \bmod$ N^2. Unfortunately, $\sigma_j^{X_j} \bmod N^2 \neq H(H(C_j \parallel t_s) \parallel Pseu_j) \bmod N^2$ because the fog node foj_j is assigned with the private key Y_j and the public key X_j, where $Y_j = X_j{}^{-1} \bmod$ N^2. As a result, valid signatures will be never verified successfully in the FA-blockchain generation phase.

3.5 Inaccurate and Missing Statements

Some statements in the proposed scheme are inaccurate or missing such that Chen et al.'s scheme cannot work as claimed. The details are as follows.

How to Obtain the Related Public Keys. In the system initialization phase, after TA generates λ, N, s, $H(.)$, X_{ij}, Y_{ij}, X_j, and Y_j, N and $H(.)$ are published online while (X_{ij}, Y_{ij}, s), (X_j, Y_j), and λ are assigned to SM_{ij}, fog_j, and the cloud server through a secure channel, respectively. In the UA-blockchain generation phase, the aggregation node in the user layer needs each smart meter SM_{ij}'s public key X_{ij} to verify the received signatures σ_{ij}'s by checking if $\prod_{i=1}^{n} \sigma_{ij}^{X_{ij}} \bmod N^2 = \prod_{i=1}^{n} H(H(C_{ij}||t_s)||Pseu_{ij}) \bmod N^2$. In the FA-blockchain generation phase, the aggregation node in the fog computing layer needs each fog node fog_j's public key X_j to verify the received signatures σ_j's by checking if $\prod_{j=1}^{m} \sigma_j^{X_j} \bmod N^2 = \prod_{j=1}^{m} H(H(C_j||t_s)||Pseu_j) \bmod N^2$. However, how these aggregation nodes obtain the related public keys X_{ij}'s and X_j's is missing.

Inaccurate Block Structure. Aggregation nodes in the user layer and fog computing layer store the transactions in blocks of the UA-blockchain and FA-blockchain, respectively. In the UA-blockchain generation phase, the aggregation node in the user layer generates the transaction $T_x = (C_j, Pseu_{ij}, t_s)$ and records the transaction $(C_j, Pseu_{ij}, t_s)$ in a new block that also includes the Merkle root, the hash value of the previous block $H_{prev\text{-}block}$, and the hash value of the current block $H_{curr\text{-}block}$, where the Merkle root is obtained by setting a leaf node's value with the ciphertext and the pseudonym and hashing them and the corresponding hash results as shown in Fig. 1, and $H_{curr\text{-}block} =$ SHA256($index + H_{prev\text{-}block} + Pseu_{ij} + timestamp + C_j + \sum_{ij} transactions_{ij}$). And, the aggregation node broadcasts the new block in the j-th subarea. Then, each smart meter SM_{ij}, an ordinary node, verifies records in the new block and verifies its related data only by checking whether it is identical to the original data or not. If the verification is successful, SM_{ij} broadcasts the verification result in the j-th subarea.When the number of the correctness confirmation messages sent by other distinct smart meters in the j-th subarea is equal to or more than $2n/3 + 1$, the new block is considered to be valid and added to the UA-blockchain. However, parameters or symbols *index*, $Pseu_{ij}$, *timestamp*, $\sum_{ij} transactions_{ij}$, and " +" are not defined accurately. And, only one transaction $T_x = (C_j, Pseu_{ij}, t_s)$ is recorded in the new block while all involved C_{ij}'s are absent. This approach makes it impossible for each smart meter SM_{ij}, an ordinary node, to verify its related data only by checking whether it is identical to the original data or not. Moreover, SM_{ij}, an ordinary node, cannot verify whether C_{ij} is indeed aggregated to generate C_j. Thus, as shown in Fig. 2, a block should be modified to consist of a block header and a block body. The block head should include the sequence number of the block *index*, all involved transactions $(C_{ij}, Pseu_{ij}, t_s)$'s, the pseudonym of the aggregation node in the user layer, the aggregated ciphertext C_j, the Merkle root, the hash value of the previous block $H_{prev\text{-}block}$, and the hash value of the current block $H_{curr\text{-}block}$, where $H_{curr\text{-}block} =$ SHA256(*index* || $H_{prev\text{-}block}$ || the pseudonym of the aggregation node in the user layer || t_s || C_j || transactions || the Merkle root) and the Merkle root is obtained by setting a leaf node's value with the ciphertext C_{ij} and the corresponding pseudonym $Pseu_{ij}$ and hashing them to get the corresponding hash results hierarchically. The corresponding Merkle tree is stored in the block body.

On the other hand, the similar problems will be encountered in the FA-blockchain generation phase. To overcome these problems, some modifications should be made. For simplicity, the differences between blocks of the UA-blockchian and those of the FA-blockchain are listed. First, transactions $(C_j, Pseu_j, t_s)$'s, the pseudonym of the aggregation node in the fog computing layer and the aggregated ciphertext C_{AS} instead of $(C_{ij}, Pseu_{ij}, t_s)$'s, the pseudonym of the aggregation node in the user layer and C_j are stored in the block header of the FA-blockchain. Second, $H_{curr\text{-}block} =$ SHA256(*index*

|| $H_{prev\text{-}block}$ || the pseudonym of the aggregation node in the fog computing layer || t_s || C_{AS} || transactions || the Merkle root), and the Merkle root is obtained by setting a leaf node's value with the ciphertext C_j, the corresponding pseudonym $Pseu_j$, and the time slot t_s, and hashing them to get the corresponding hash results hierarchically.

Failure to Retrieve Subareas' Data. When the cloud server receives the FA-blockchain of the fog computing layer, it gets the aggregated power consumption ciphertext C_{AS} for all subareas and decrypts it by using the Paillier homomorphic decryption algorithm to get the aggregated plaintext $M = L({C_{AS}}^{\lambda} \bmod N^2) / L(g^{\lambda} \bmod N^2)$. Chen et al.'s proposed a Horner rule-based analytical algorithm and attempted to recovers subareas' data UA_j's, where $UA_j = \sum_{i=1}^{n} d_{ij}$ and $M = \sum_{j=1}^{m} UA_j$. However, the theoretical derivations to show why their proposed Horner rule-based analytical algorithm can recovers subareas' data are not correct. Meanwhile, the designed algorithm cannot work, either. In their designed algorithm, UA_j is computed with a modulus R that a product of all random numbers r_j's. This makes UA_j is in $[0, R]$ while the range of UA_j should be in $[0, N^2–1]$. On the other hand, in the FA-blockchain generation phase, the aggregation node in the fog computing layer generates the transaction $T_x = (C_{AS}, Pseu_j, t_s)$ and records the transaction $(C_{AS}, Pseu_j, t_s)$ in a new block. Because only one transaction $T_x = (C_{AS}, Pseu_j, t_s)$ is recorded in the new block while all involved C_j's are absent in the FA-blockchain. Thus, it is impossible for the cloud server to retrieves subareas' data UA_j's because C_j's are unknown.

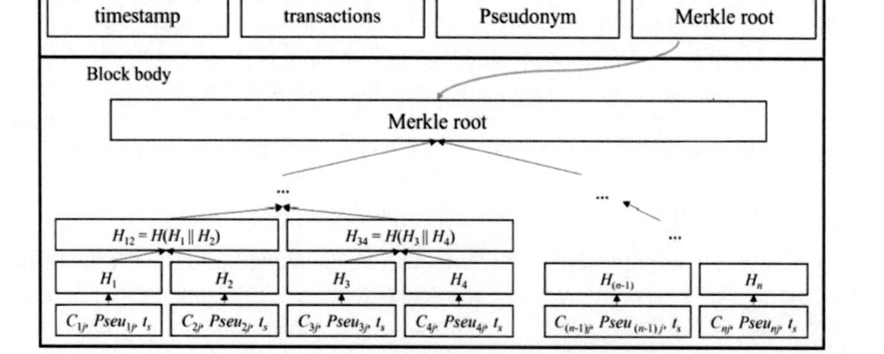

Fig. 2. The structure of the modified blockchain.

4 Conclusions

Chen et al. proposed a data aggregation scheme by using double blockchains to satisfy the specific requirements of smart grids. They claimed that their scheme could defend against various attacks and guarantee data confidentiality, data integrity, validity, anonymity of identity, and authenticity. However, after thoroughly analyzing their scheme, we find that it suffers from five flaws. Firstly, because the pseudonyms of the smart meters and fog nodes are fixed, anonymity is not guaranteed as claimed. Secondly, private keys of smart meters and fog nodes can be easily obtained. Thirdly, after a smart meter or fog node's private key is revealed, a malicious entity can impersonate it and generate a valid signature of the forged data's ciphertext. Fourthly, in both of the UA-blockchain generation phase and FA-blockchain generation phase, the signature verification will never succeed such that legal signatures are always regarded as invalid. Fifthly, in Chen et al.'s scheme, how to obtain the related public keys is absent, the block structure is inaccurate, and the cloud server cannot retrieve subareas' data. Due to the above analysis, proper modification is needed; otherwise, Chen et al.'s scheme can neither work nor preserve the claimed superior properties.

Acknowledgement. This work was supported in part by National Science and Technology Council under the Grants MOST 110–2221-E-992–097-MY3, MOST 110–2221-E-025–012-, MOST 111–2221-E-025–007-, and MOST 110–2221-E-025–014-MY2.

References

1. Lu, R., Liang, X., Li, X., Lin, X., Shen, X.: An efficient and privacy-preserving aggregation scheme for secure smart grid communications. IEEE Trans. Parallel Distrib. Syst. **23**(9), 1621–1631 (2012)
2. Jia, W., Zhu, H., Cao, Z., Dong, X., Xiao, C.: Human-factor-aware privacy-preserving aggregation in smart grid. IEEE Syst. J. **8**(2), 598–607 (2014)
3. Liang, K., Susilo, W.: Searchable attribute-based mechanism with efficient data sharing for secure cloud storage. IEEE Trans. Inf. Forensics Secur. **10**, 1981–1992 (2015)
4. Wan, Z., Zhu, W.T., Wang, G.: PRAC: efficient privacy protection for vehicle-to-grid communications in the smart grid. Comput. Secur. **62**, 246–256 (2016)
5. Lu, R., Heung, K., Lashkari, A.H., Ghorbani, A.A.: A lightweight privacy-preserving data aggregation scheme for fog computing-enhanced IoT. IEEE Access **5**, 3302–3312 (2017)
6. Mahmood, K., et al.: Pairing based anonymous and secure key agreement protocol for smart grid edge computing infrastructure. Futur. Gener. Comput. Syst. **88**, 491–500 (2018)
7. Eltayieb, N., Elhabob, R., Hassan, A., Li, F.: An efficient attribute-based online/offline searchable encryption and its application in cloud-based reliable smart grid. J. Syst. Architect. **98**, 165–172 (2019)
8. Zhang, H., Wang, J., Ding, Y.: Blockchain-based decentralized and secure keyless signature scheme for smart grid. Energy **180**, 955–967 (2019)
9. Alcaraz, C., Rubio, J.E., Lopez, J.: Blockchain-assisted access for federated smart grid domains: coupling and features. J. Parallel Distrib. Comput. **144**, 124–135 (2020)
10. Li, M., Zhang, K., Liu, J., Gong, H., Zhang, Z.: Blockchain-based anomaly detection of electricity consumption in smart grids. Pattern Recogn. Lett. **138**, 476–482 (2020)
11. Chen, S., Yang, L., Zhao, C., Varadarajan, V., Wang, K.: Double-Blockchain assisted secure and anonymous data aggregation for fog-enabled smart grid. Engineering **8**, 159–169 (2020)

Pavement Distress Detection Using YOLO and Faster RCNN on Edge Devices

Chen-Kang Chiu[1(✉)], Jung-Chun Liu[1], Yu-Wei Chan[2], and Chao-Tung Yang[1,3]

[1] Department of Computer Science, Tunghai University, Taichung 407224, Taiwan, ROC
aganggggg@gmail.com, {jcliu,ctyang}@thu.edu.tw

[2] Department of Information Management, Providence University, Taichung 43301, Taiwan, ROC
ywchan@gm.pu.edu.tw

[3] Research Center for Smart Sustainable Circular Economy, Tunghai University, Taichung 407224, Taiwan, ROC

Abstract. In this study, transfer learning techniques will be used for model training, using edge computing [1] and deep learning object detection technology, combined with image road pothole detection applications, and deploying devices and tools that accelerate neural network operations, including DeepStream [2] and Intel NCS2. The performance and accuracy of model recognition will be compared, and finally, real-time streaming video technology will be used to present the results on the web. According to the experimental results, the best model achieved an mAP of 70.% in YOLOv4-tiny-3l, and in terms of operating efficiency, deployment on Jetson Xavier NX using DeepStream for acceleration can achieve 30FPS. Finally, the deep learning model recognizes the screen presented on the web. This application can improve the accuracy of Pavement Distress identification and help road maintenance units improve the efficiency of repairing roads.

Keywords: Transfer Learning · Deep Learning · Edge Computing · Object Detection · DeepStream

1 Introduction

Traditional road survey work usually requires manual visual inspection and written records [3], but this method has some problems. Factors such as the pressure of driving on the road, rain, and high temperatures caused by sun exposure can cause various potholes in the road. Passing through potholes can reduce the service life of vehicle parts, increase vehicle maintenance rates, and seriously affect the safety of road users. We can develop more efficient and accurate image-processing methods for various application scenarios based on this approach [4].

On the other hand, You Only Look Once (YOLO) [5] belongs to the one-stage object detection algorithm, in which the entire architecture consists of only convolutional and fully connected layers. After inputting an image, the

J. C. Hung et al. (Eds.): FC 2023, LNEE 1134, pp. 246–252, 2024.
https://doi.org/10.1007/978-981-99-9342-0_26

algorithm can quickly obtain the positions and categories of objects much faster than other methods like R-CNN [6] that require obtaining candidate regions before classification. Yang et al. [7] design a management system that uses drone-mounted cameras combined with object identification and live broadcast systems to assist disaster relief.

2 System Design and Implementation

2.1 System Architecture

In this study, we used Ubuntu 18.04 as the operating system for the research. We conducted deep learning and object detection experiments using Tensorflow and PyTorch, respectively, with Python as the primary programming language. We collected videos of pavement distress in Taiwan using a dashcam and used Python OpenCV to extract frames from the videos to augment the dataset. Various neural network architectures were constructed as different machine learning models for training, followed by model evaluation and performance comparison. The most suitable model was selected and deployed on different edge devices, utilizing acceleration tools to improve FPS, ensure stability, and reduce data transmission latency and network issues. The generated data was transmitted to the client-side through a web server to enable real-time viewing functionality. Figure 1 illustrates the complete system architecture.

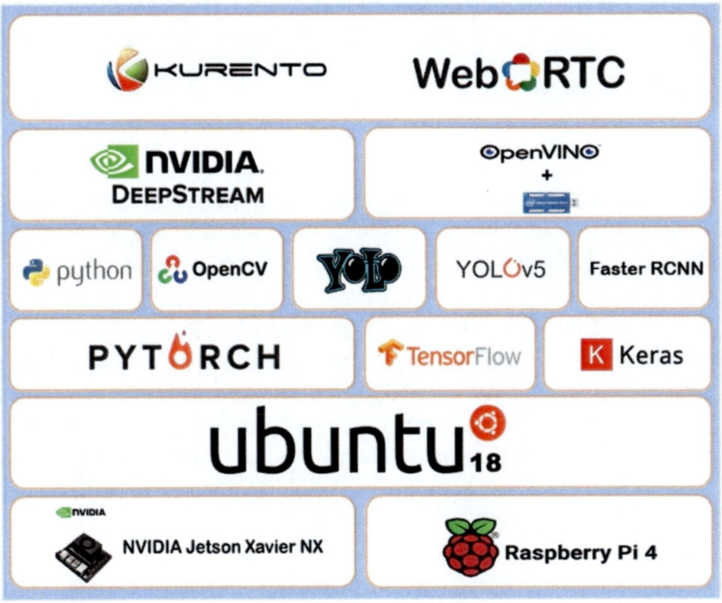

Fig. 1. System architecture diagram.

3 Type of Pavement Distress and Data Collection

3.1 Type of Road Damage

Considering that different countries and regions have varied types of road damage, this study takes into account the limited availability of self-labeled data and also refers to a pavement distress dataset from six countries in 2022: Japan, India, the Czech Republic, Norway, the United States, and China. [8] This dataset consists of 47,420 road images and categorizes pavement distress into four classes, as shown in Table 1.

Due to the limited availability of pavement distress data currently provided in Taiwan, it is not possible to segment the data and conduct training, we will utilize the pavement distress dataset from 2022 to conduct model training and performance testing.

Table 1. 2022 Pavement Distress Dataset

Damage type/Eng	Class Name
Longitudinal crack	D00
Transverse crack	D10
Alligator cracks	D20
Pothole	D40

3.2 Data Collection

To train a deep learning model, a sufficient amount of dataset is required, and it takes time to train a good model. In this study, we used the pavement distress dataset provided by RDD2022 for our data. This dataset includes categories such as potholes, alligator cracks, transverse cracks, and longitudinal cracks. It was collected from roads in six different countries and exhibits excellent complexity. However, only the test set of this dataset is annotated. Since our experiments are conducted in Taiwan, we selected a dataset from Japan that is more representative of Taiwanese roads to ensure the accuracy of the model. The Japanese road dataset consists of a total of 16,470 images. Additionally, we collected 2,000 images of pavement distress in Taiwan through car recorders, resulting in a current total of 18,470 images. Figure 2 shows the pavement distress dataset.

3.3 Data Augmentation

The training process of deep learning involves extracting meaningful image features from the training data, such as edges, colors, orientations, positions, and textures. However, since the collected data cannot cover all scenarios, this study used a total of 18,470 images as the training dataset. Data organization and augmentation were performed using the online data management platform,

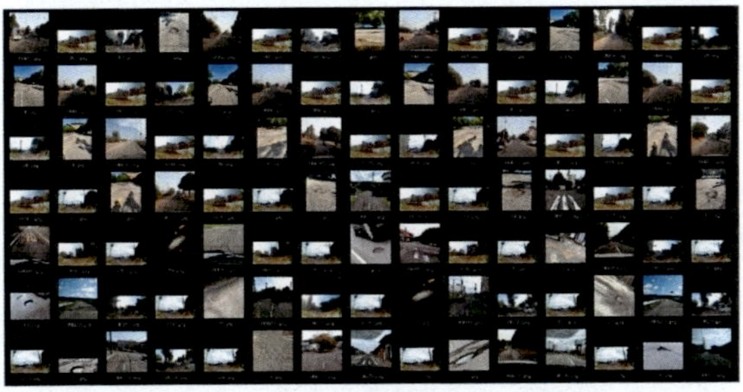

Fig. 2. Pavement Distress Dataset.

Roboflow. Each image was augmented every 30°, and additional transformations like translation, flipping, and distortion were applied to account for rotational variations in the shooting angles. This expanded the dataset to a total of 6,867 images, resulting in a final dataset of approximately 25,337 images. Furthermore, all images were resized to a dimension of 416×416. In terms of data distribution, the datasets were divided into 60% for training, 20% for validation, and 20% for testing. Figure 3 shows the After data augmentation.

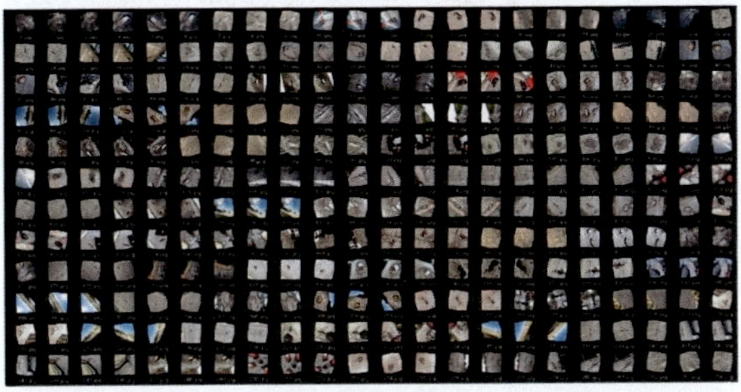

Fig. 3. After data augmentation.

3.4 Feature Label and CSV File Generation

In this study, LabelImg [9] will be used for feature labeling. This tool is open-source software that allows setting the locations of the original and storage files and annotating specific positions and classes of images. The generated XML files provide detailed information about the labeled images, including the image name, image type, and x, y position values of the potholes, among other relevant information. Figure 4 display the content of the CSV files, and Fig. 5 illustrates the number of labels for each category.

Fig. 4. Data annotation.

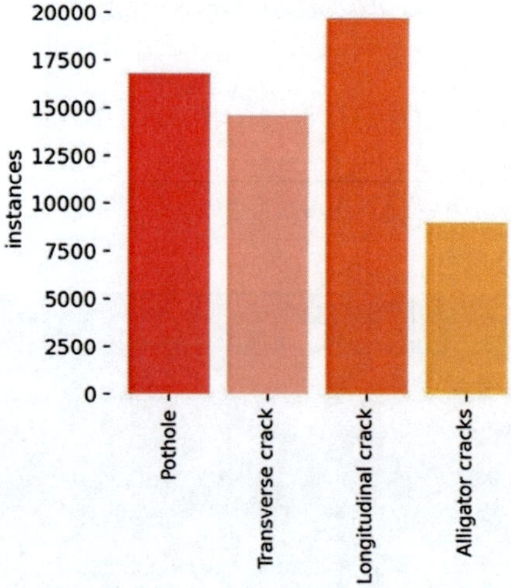

Fig. 5. Amount of annotation

4 Pavement Distress Model Training Results

According to the comparison results in Table 2, YOLOv4-tiny-3l is more suitable for road pothole detection. This study trained models using six different neural network architectures. Since the images were resized to a size of 416×416 during the data augmentation process, all six models were trained with an input size of 416×416. It can be observed that YOLOv4-tiny-3l achieved the highest Precision at 75% while having a relatively lower Recall of 63%. This indicates that the models have conservative predictions, resulting in high Precision but lower Recall. On the other hand, YOLOv5m generated higher Recall but led to a decrease in Precision. This implies that if a model is too greedy and aims to predict more positive examples as Ground Truth, it may result in false positives. In such a trade-off situation, another metric, F1-score, which is the harmonic mean of Precision and Recall, is derived. It can be seen that YOLOv4-tiny-3l, YOLOv5s, and YOLOv5m all achieved the same F1-score of 68%. Considering the same value, further comparison of the models is needed. Therefore, this study adopted mAP0.5 as the final evaluation metric. The results showed that YOLOv4-tiny-3l achieved the highest mAP0.5 at 70.5% among the six models, while YOLOv5s and YOLOv5m achieved 68.4% and 69% respectively. On the contrary, due to the complex architectures of VGG16 and ResNet101, they required more training time and exhibited relatively poorer performance in terms of evaluation metrics compared to YOLOv4-tiny-3l. Meanwhile, YOLOv4-tiny has only two yolo layers, reducing the computational complexity and model size compared to YOLOv4-tiny-3l.

Table 2. Model prediction result table

	F1-score	mAP0.5	Precision	Recall	Training Time
VGG16	66%	66.5%	65%	66%	168H
ResNet101	66%	67.5%	65%	64%	192H
YOLOv4-tiny	65%	66.0%	66%	65%	36H
YOLOv4-tiny-3l	68%	70.5%	75%	63%	48H
YOLOv5s	68%	68.4%	67%	66%	26H
YOLOv5m	68%	69.0%	65%	68%	46H

5 Web Page Presentation

In terms of real-time streaming, this study utilized Kurento and WebRTC [10] technologies to establish a web application for live video streaming. To enable

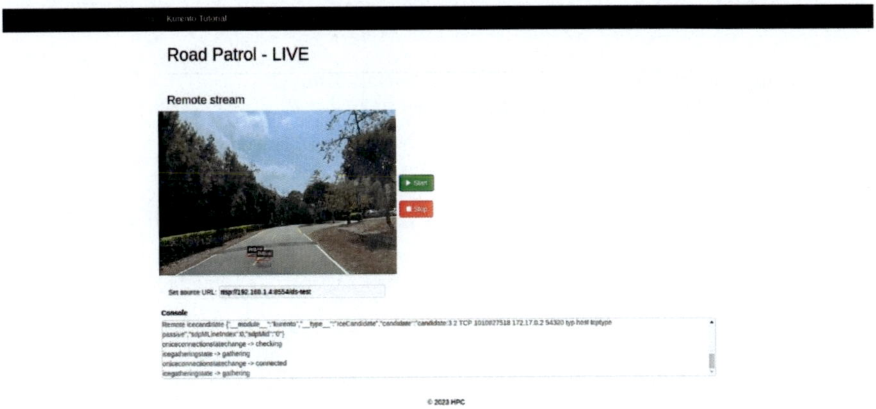

Fig. 6. Web results screen

communication between WebRTC and IP cameras, their media formats must be compatible. This encoding format conversion task is handled by a WebRTC gateway such as Kurento. The web application allows direct viewing in the browser, while the server side is capable of performing video streaming recognition. In addition to providing the edge device page, the web application also offers a page for real-time video processing on the server side, supporting synchronized real-time streaming. Figure 6 showcases the interface of the real-time pavement distress recognition web application.

References

1. Huynh, L.N., Lee, Y., Balan, R.K.: DeepMon: mobile GPU-based deep learning framework for continuous vision applications. In: MobiSys 2017, pp. 82–95. Association for Computing Machinery, New York, NY, USA (2017)
2. Abdulghafoor, N.H., Abdullah, H.N.: A novel real-time multiple objects detection and tracking framework for different challenges. Alex. Eng. J. **61**(12), 9637–9647 (2022)
3. Lai, Y.-J.: Analysis of forward-looking plans improving the road quality (highway system) - first maintenance office, directorate general of highways, MOTC (2020). http://ir.lib.ncu.edu.tw:88/thesis/viewetd.asp?URN=107352005
4. Wang, Y.-C., Yu, C.-W., Lu, X.-Y., Chen, Y.-L.: Road semantic segmentation and traffic object detection model based on encoder-decoder CNN architecture. In: 2022 IEEE International Conference on Consumer Electronics - Taiwan, pp. 421–422 (2022)
5. Redmon, J., Divvala, S., Girshick, R., Farhadi, A.: You only look once: unified, real-time object detection (2016)
6. Girshick, R., Donahue, J., Darrell, T., Malik, J.: Rich feature hierarchies for accurate object detection and semantic segmentation (2014)
7. Yang, C.T., Lin, W.Y., Chen, Y.C., Wang, Z.Y., Lee, C.H.: Flame recognition system using YoLo. In: Chang, J.W., Yen, N., Hung, J.C. (eds.) Frontier Computing. FC 2020. LNEE, vol. 747, pp. 239–246. Springer, Singapore (2021). https://doi.org/10.1007/978-981-16-0115-6_23
8. Arya, D.: Crowdsensing-based road damage detection challenge (CRDDC-2022) (2022)
9. Labelimg. https://github.com/heartexlabs/labelImg
10. WebRTC architecture. https://webrtc.org/

The Application of Artificial Intelligence to Support Behavior Recognition by Zebrafish: A Study Based on Deep Learning Models

Yi-Ling Fan, Fang-Rong Hsu(✉), Jing-Yaun Lu, Min-Jie Chung, and Tzu-Ching Chang

Department of Information Engineering and Computer Science, Feng Chia University, Taichung 407, Taiwan
frhsu@fcu.edu.tw

Abstract. Zebrafish (Danio rerio) is an ideal model organism for biological research due to its ease of breeding, maintenance, observation, and complete genome sequencing. As a small aquatic organism with a body length of about 3–5 cm, zebrafish mainly exhibits its behavior through swimming in water. Therefore, trajectory tracking is crucial for a deep understanding of zebrafish behavior and physiological states, as well as for revealing its associations with specific diseases. In addition, zebrafish is widely used for drug screening and toxicology testing to explore the underlying neural and physiological mechanisms. Because of the high efficiency, accuracy, structural simplicity, and versatility of YOLO series models in object detection, they have become one of the preferred deep learning models for many researchers and developers. In this study, a model trained using YOLOv7 was proposed to track the movement trajectories of zebrafish and classify their behaviors into three categories: swimming, sinking, and static, through time-series sorting. According to experimental testing, our method exhibits excellent performance in detecting zebrafish movement trajectories. On a test set consisting of one frame per second, the model achieved a 100% accuracy rate and a 100% recall rate, demonstrating its potential in automated trajectory tracking.

Keywords: zebrafish · deep learning · object detection · behavior recognition

1 Introduction

Zebrafish is a small tropical fish native to Southeast Asia, known for its external fertilization, transparent embryos, and short life cycle. In addition, its genome was fully sequenced in 2013 [1], and its genetic structure shares up to 70% similarity with that of humans. Zebrafish offers numerous advantages as a model organism: (1) it is small in size and has a high survival rate, (2) it is less expensive to maintain than mice, (3) it can produce hundreds of eggs per week, providing a large number of embryos for research, (4) up to 84% of genes related to human diseases have corresponding genes in zebrafish, and (5) as a vertebrate with major organ systems that are similar to those of humans, it shares many similar characteristics with humans. Zebrafish has been widely

J. C. Hung et al. (Eds.): FC 2023, LNEE 1134, pp. 253–262, 2024.
https://doi.org/10.1007/978-981-99-9342-0_27

used as a model organism in laboratory research for many years [2], with its analysis extending to numerous research applications, including drug safety, behavioral genetics, ecotoxicology, circadian rhythms [3–7], and many others. Moreover, zebrafish has a unique ability to regenerate heart muscle, which makes it a valuable tool for studying heart-related diseases.

1.1 Zebrafish Applied to Biomedicine

Zebrafish (Danio rerio) has rapidly emerged as a promising tool for disease modeling and drug discovery in the field of biomedical research [2, 8]. As an experimental animal, zebrafish has a wide range of applications in the medical field. Their genome is simple, easy to manipulate, and they grow quickly, making them useful for studying the mechanisms, treatments, and drug screening of various diseases [9–11]. In recent years, the use of zebrafish in cancer research, neuroscience [12, 13], and cardiology has gradually increased. Moreover, zebrafish has also been used to study ecotoxicology [14] to assess the impact of certain chemicals on organisms, thus protecting public health.

As a research model organism, zebrafish (Danio rerio) has been widely used in scientific and medical fields [15, 16]. Therefore, objective, impartial, and reproducible data collection and analysis methods are necessary to ensure the credibility and accuracy of experimental results [17]. Such methods can eliminate subjectivity and bias, reduce uncertainty in experiments, and help scientists further understand the experimental results. In addition, objective, impartial, and reproducible data collection and analysis methods make it easier for other scientists to verify and replicate experimental results, thereby further enhancing the reliability and sustainability of research outcomes. Therefore, objective, impartial, and reproducible data collection and analysis methods are crucial for zebrafish experiments and other scientific research fields.

1.2 Deep Learning Applied to Zebrafish Research

The use of deep learning techniques and zebrafish imaging studies can provide more accurate tracking and analysis of zebrafish behavior and movement trajectories. In recent years, many research teams have applied deep learning techniques to zebrafish image analysis [18–20], such as using convolutional neural networks to detect zebrafish movement and track their trajectories [21–23]. With these techniques, valuable features can be extracted from large amounts of zebrafish image data, and accurate models can be established to predict zebrafish behavior and study their movement trajectories [24–26]. These research findings help to deepen our understanding of zebrafish movement and behavior, providing important foundational information for biological and medical research. Additionally, utilizing deep learning techniques and zebrafish imaging studies can also develop more advanced image analysis tools, improving the automated analysis of zebrafish images, and providing more possibilities for the study of zebrafish behavior and physiology [12].

1.3 Introduction to YOLOv7

Dr. Chien-Yao Wang, Alexey Bochkovskiy, and Director Hung-Yu Liao successively introduced YOLOv4, ScaledYOLOv4, and YOLOR in 2020–2021, and recently launched their latest masterpiece - YOLOv7 [27] in early July of this year. YOLOv7 outperforms all known object detectors in terms of speed and accuracy within the range of 5 FPS to 160 FPS and at 30 FPS on GPU V100. The YOLOv7-E6 object detector (56 FPS V100, 55.9% AP) outperforms the transformer-based detector SWIN-L Cascade-Mask R-CNN (9.2 FPS A100, 53.9% AP) by 509% in speed and accuracy and the convolution-based detector ConvNeXt-XL Cascade Mask R-CNN (8.6 FPS A100, 55.2% AP) by 551% in speed and 0.7% AP in accuracy. Moreover, YOLOv7 surpasses YOLOR, YOLOX, Scaled-YOLOv4, YOLOv5, DETR, Deformable DETR, DINO-5scale-R50, ViT-Adapter-B, and many other object detectors in terms of speed and accuracy. Additionally, they trained YOLOv7 from scratch only on the MS COCO dataset without using any other datasets or pre-trained weights (Fig. 1).

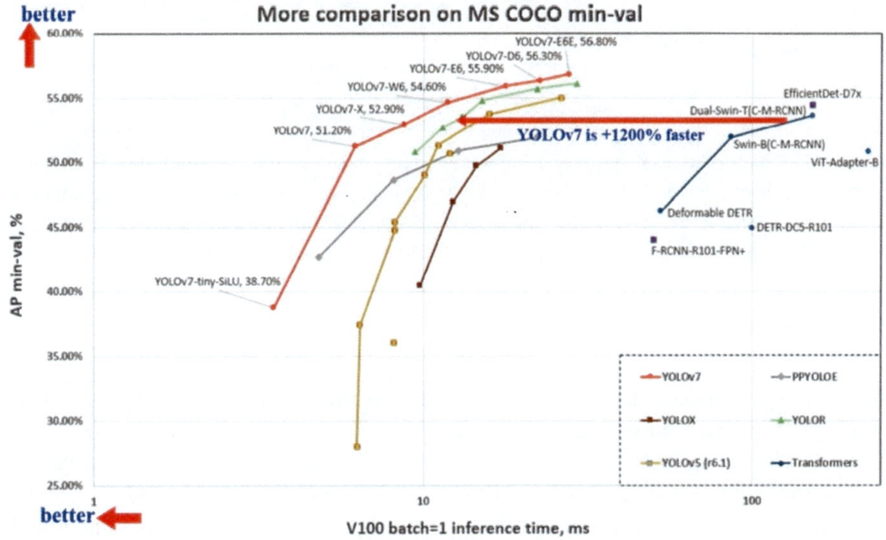

Fig. 1. Comparison with other object detectors, their proposed methods achieve state-of-the-arts performance [27].

YOLOv7 reduces the parameter count and computation cost of today's real-time object detection SOTA by about 40% and 50%, respectively. It mainly optimizes the model architecture and training process, proposing extended and scaling methods for effective utilization of parameters and computation costs in the model architecture optimization. As for the training process optimization, the authors proposed the "bag-of-freebies" method in YOLOv4, which increases accuracy at the cost of training, but does not increase inference cost, and in YOLOv7, they used re-parameterized techniques to replace the original modules and dynamic label assignment strategy to assign labels more efficiently to different output layers (Fig. 2).

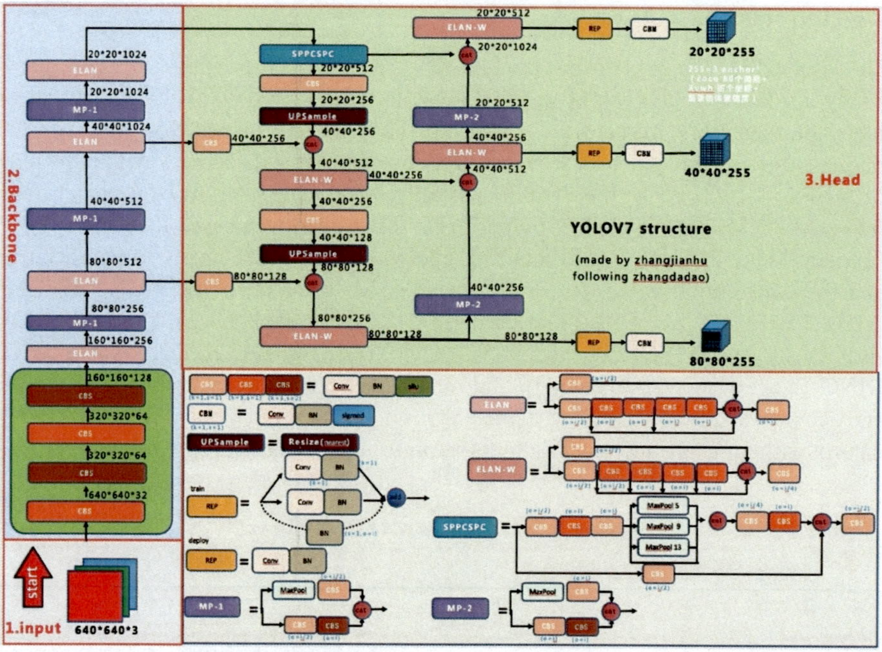

Fig. 2. The picture shows the overall network architecture of yolov7, which consists of three parts: input, backbone and head. Unlike yolov5, the neck layer and the head layer are combined as the head layer, which actually has the same function. The functions of each part are the same as yolov5, such as backbone is used to extract features, and head is used for prediction [27].

2 Method

In this study, self-recorded videos were first converted to images and labeled. These labeled data were then used to train a convolutional neural network model. A small amount of test samples were input to confirm the accuracy and usability of the model. The videos containing the desired behavior to be detected were then input, and the behavior category was determined by detecting the position of zebrafish and its timeline. Finally, the results were output on a self-designed UI interface (Fig. 3).

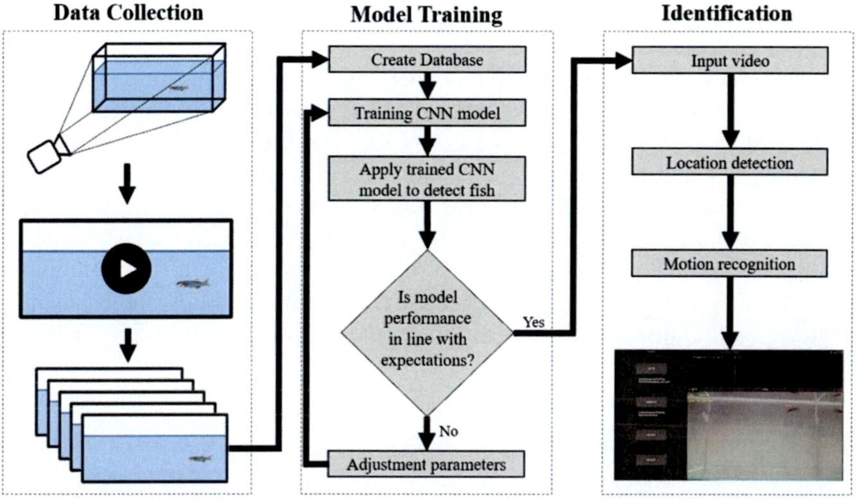

Fig. 3. Experimental flow chart of this study.

2.1 Data Collection and Processing

We collected our own video footage to build the dataset, accumulating a total of 1 h 59 min and 52 s of video material. Subsequently, the videos were edited using Potplayer, and the annotations of the area of interest were performed using the LabelImg software (Fig. 4).

Fig. 4. Mark software LabelImg actual operation screen.

2.2 Model Training and Behavioral Judgment

After preprocessing, the images were divided into training, validation, and test sets. YOLOv7.pt among the six initial weights of YOLOv7 was used as the weight for training the model. After detecting the position of the zebrafish, the behavior category was determined by comparing the current position with the previous position.

In this experiment, zebrafish behavior was classified into three categories: normal movement, bottom-dwelling, and stationary. A timing unit of 1.5 s was used, and if the coordinates appeared within the bottom 20% of the tank during the detection process, meaning within the range of 4 cm from the tank bottom upwards, the behavior was defined as bottom-dwelling. If the coordinates did not move within 1.5 s, the behavior was considered stationary.

2.3 Experimental Environment

All deep learning is trained and evaluated on a machine equipped with AMD R5 5600X, 32GM RAM, NVIDIA GeForce RTX 3070Ti, and the operating system uses Window10 x64. Use Python 3.9.13 for table creation and data encoding. Use LabelImg to label datasets. Network construction and training verification are all run on the virtual environment of Anaconda3 architecture. The videos required for the experiments in this study were all shot by iPhone 12, and the size of the fish tank used was 30*16*20 in length, width and height.

2.4 Evaluation Indicators

In terms of evaluating the classification network, the precision and recall (also known as sensitivity or true positive rate (TPR)) were used as the evaluation metrics. The calculation formula is as follows:

$$\text{Precision} = \frac{T_p}{T_p + F_p}$$

$$\text{Recall } (TPR) = \frac{T_p}{T_p + F_n}$$

3 Results

The first batch of training used a total of 1200 labeled data, trained nine times with different parameters, and selected two of them for detection testing. However, the accuracy and recall rate were not satisfactory in the testing videos. Therefore, the amount of training data was increased to 2000, and the results showed that both the accuracy and recall rate reached 97% (Fig. 5). In the confusion matrix, TP also achieved a good performance of 98% (Fig. 6). Furthermore, in the videos with a speed of one frame per second, the accuracy and recall rate reached 100%, and the behavior categories were successfully automatically determined (Fig. 7).

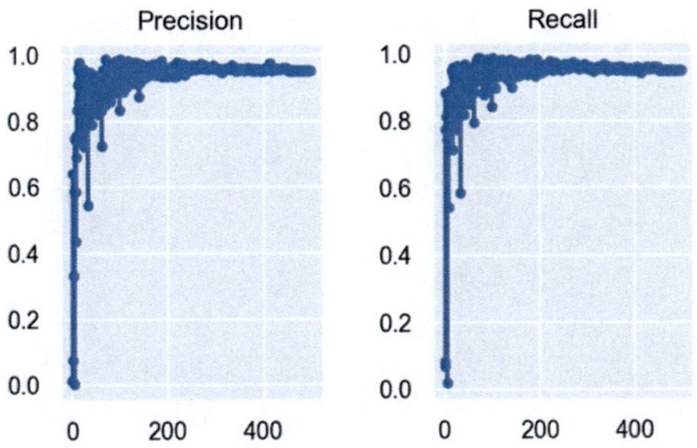

Fig. 5. After increasing the amount of data for model training, the ideal accuracy and recall rates were achieved.

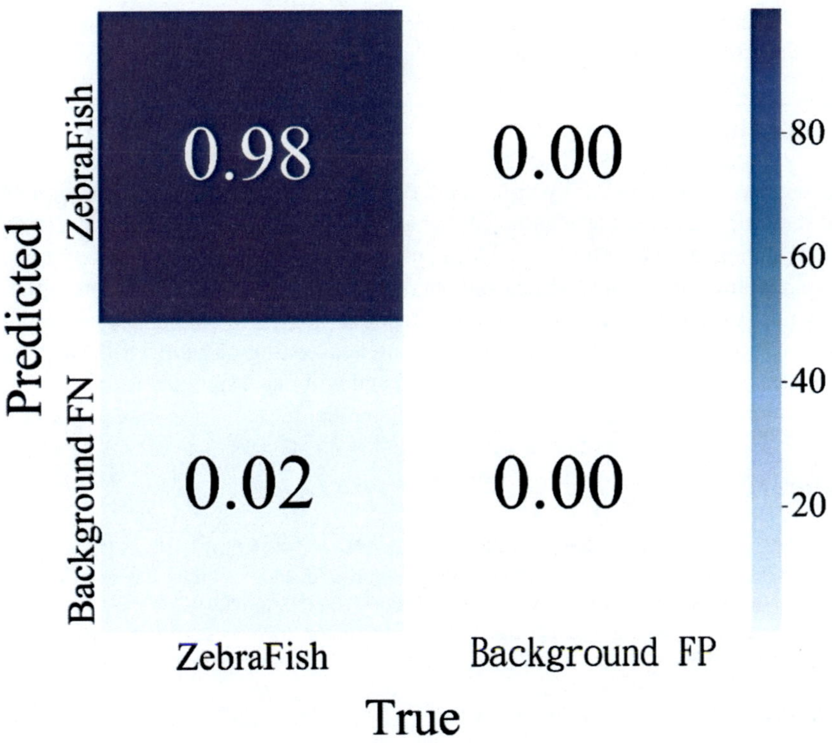

Fig. 6. After increasing the amount of data to train the model, the success rate of the model detecting zebrafish is as high as 98%.

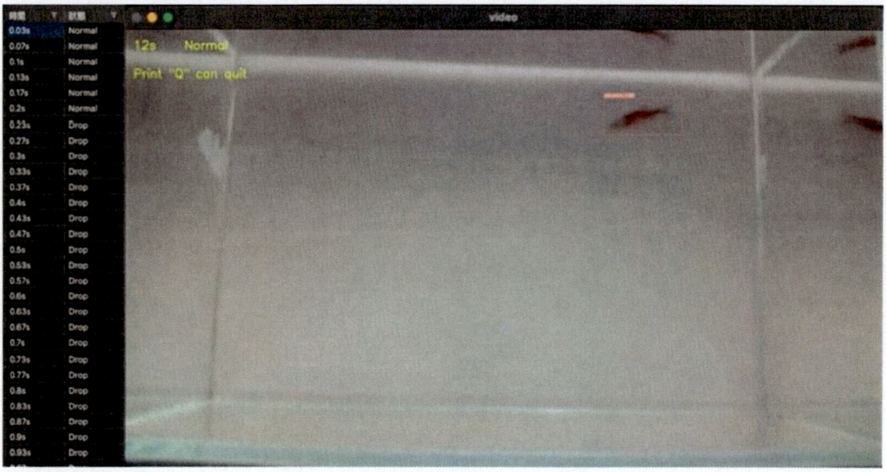

Fig. 7. Input the location detection and behavior recognition of the video, and the table on the left is the behavior recognition record.

4 Conclusion

The present study successfully utilizes the combination of object position identification and the time axis to achieve zebrafish behavior recognition, indicating great potential in automating trajectory tracking. Through this approach, we can accurately track the movement trajectories of zebrafish and obtain valuable data and information, which can further help us understand zebrafish behavior and ecological habits and play an important role in scientific research both in laboratory and field settings. Therefore, the application prospects of this method are extremely broad and will make significant contributions to the development of zebrafish research and related fields.

References

1. Schier, A.F.: Zebrafish earns its stripes. Nature **496**, 443–444 (2013)
2. Lieschke, G.J., Currie, P.D.: Animal models of human disease: zebrafish swim into view. Nat. Rev. Genet. **8**(5), 353–367 (2007). https://doi.org/10.1038/nrg2091
3. Darland, T., Dowling, J.E.: Behavioral screening for cocaine sensitivity in mutagenized zebrafish. Proc. Natl. Acad. Sci. USA **98**(20), 11691–11696 (2001). https://doi.org/10.1073/pnas.191380698
4. Gerlai, R., Lahav, M., Guo, S., Rosenthal, A.: Drinks like a fish: zebra fish (Danio rerio) as a behavior genetic model to study alcohol effects. Pharmacol. Biochem. Behav. **67**(4), 773–782 (2000). https://doi.org/10.1016/s0091-3057(00)00422-6
5. Guo, S.: Linking genes to brain, behavior and neurological diseases: what can we learn from zebrafish? Genes Brain Behav. **3**(2), 63–74 (2004). https://doi.org/10.1046/j.1601-183x.2003.00053.x
6. Levin, E.D., Chrysanthis, E., Yacisin, K., Linney, E.: Chlorpyrifos exposure of developing zebrafish: effects on survival and long-term effects on response latency and spatial discrimination. Neurotoxicol. Teratol. **25**(1), 51–57 (2003). https://doi.org/10.1016/s0892-0362(02)00322-7

7. Linney, E., Upchurch, L., Donerly, S.: Zebrafish as a neurotoxicological model. Neurotoxicol. Teratol. **26**(60), 709–718 (2004). https://doi.org/10.1016/j.ntt.2004.06.015
8. Fetcho, J.R., Liu, K.S.: Zebrafish as a model system for studying neuronal circuits and behavior. Ann. N. Y. Acad. Sci. **860**, 333–345 (1998). https://doi.org/10.1111/j.1749-6632.1998.tb09060.x
9. Rink, E., Wullimann, M.F.: Connections of the ventral telencephalon and tyrosine hydroxylase distribution in the zebrafish brain (Danio rerio) lead to identification of an ascending dopaminergic system in a teleost. Brain Res. Bull. **57**(3–4), 385–387 (2002). https://doi.org/10.1016/s0361-9230(01)00696-7
10. Demin, K.A., et al.: Developing zebrafish experimental animal models relevant to schizophrenia. Neurosci. Biobehav. Rev. **105**, 126–133 (2019). https://doi.org/10.1016/j.neubiorev.2019.07.017
11. Egan, R.J., et al.: Understanding behavioral and physiological phenotypes of stress and anxiety in zebrafish. Behav. Brain Res. **205**(1), 38–44 (2009)
12. Bozhko, D.V., et al.: Artificial intelligence-driven phenotyping of zebrafish psychoactive drug responses. Prog. Neuropsychopharmacol. Biol. Psychiatry **112**, 110405 (2022). https://doi.org/10.1016/j.pnpbp.2021.110405
13. Lillesaar, C., Stigloher, C., Tannhauser, B., Wullimann, M.F., Bally-Cuif, L.: Axonal projections originating from raphe serotonergic neurons in the developing and adult zebrafish, Danio rerio, using transgenics to visualize raphe-specific pet1 expression. J. Comp. Neurol. **512**(2), 158–182 (2009). https://doi.org/10.1002/cne.21887
14. Paganotto Leandro, L., et al.: Behavioral changes occur earlier than redox alterations in developing zebrafish exposed to Mancozeb. Environ. Pollut. **268**(Pt B), 115783 (2021). https://doi.org/10.1016/j.envpol.2020.115783
15. Yang, P., Takahashi, H., Murase, M., Itoh, M.: Zebrafish behavior feature recognition using three-dimensional tracking and machine learning. Sci. Rep. **11**(1), 13492 (2021). https://doi.org/10.1038/s41598-021-92854-0
16. Al-Imari, L., Gerlai, R.: Sight of conspecifics as reward in associative learning in zebrafish (Danio rerio). Behav. Brain Res. **189**(1), 216–219 (2008). https://doi.org/10.1016/j.bbr.2007.12.007
17. Stuart, G.W., Vielkind, J.R., McMurray, J.V., Westerfield, M.: Stable lines of transgenic zebrafish exhibit reproducible patterns of transgene expression. Development **109**(3), 577–584 (1990). https://doi.org/10.1242/dev.109.3.577
18. Qian, Z.M., Chen, Y.Q.: Feature point based 3D tracking of multiple fish from multi-view images. PLoS ONE **12**(6), e0180254 (2017). https://doi.org/10.1371/journal.pone.0180254
19. Gao, Y., et al.: A high-throughput zebrafish screening method for visual mutants by light-induced locomotor response. IEEE/ACM Trans. Comput. Biol. Bioinform. **11**(4), 693–701 (2014). https://doi.org/10.1109/TCBB.2014.2306829
20. Mikut, R., et al.: Automated processing of zebrafish imaging data: a survey. Zebrafish **10**(3), 401–421 (2013). https://doi.org/10.1089/zeb.2013.0886
21. Jalal, A., Salman, A., Mian, A., Shortis, M., Shafait, F.: Fish detection and species classification in underwater environments using deep learning with temporal information. Ecol. Inf. **57**, 101088 (2020). https://doi.org/10.1016/j.ecoinf.2020.101088
22. Wang, H., Zhang, S., Zhao, S., Wang, Q., Li, D., Zhao, R.: Real-time detection and tracking of fish abnormal behavior based on improved YOLOV5 and SiamRPN++. Comput. Electr. Agric. **192**, 106512 (2022). https://doi.org/10.1016/j.compag.2021.106512
23. Barreiros, M.D.O., Dantas, D.D.O., Silva, L.C.D.O., Ribeiro, S., Barros, A.K.: Zebrafish tracking using YOLOv2 and Kalman filter. Sci. Rep. **11**(1), 3219 (2021). https://doi.org/10.1038/s41598-021-81997-9
24. Xu, Z., Cheng, X.E.: Zebrafish tracking using convolutional neural networks. Sci. Rep. **7**, 42815 (2017). https://doi.org/10.1038/srep42815

25. Sun, M., Li, W., Jiao, Z., Zhao, X.: A multi-target tracking platform for zebrafish based on deep neural network. In: IEEE 9th Annual International Conference on CYBER Technology in Automation (2019)
26. Breier, B., Onken, A.: Analysis of video feature learning in two-stream CNNs on the example of zebrafish swim bout classification. In: ICLR 2020 Conference (2020)
27. Wang, C.-Y.: YOLOv7 Trainable bag-of-freebies sets new state-of-the-art for real-time object detectors. arXiv (2022)

A Survey of Speech Recognition for People with Cerebral Palsy

Yu-Ru Wu(✉), Jason C. Hung, and Jia-Wei Chang

National Taichung University of Science and Technology, Taichung City, Taiwan
wu870626@gmail.com

Abstract. This study aims to address the communication barriers related to speech for individuals with cerebral palsy, with the goal of using technological methods to assist or alleviate difficulties in oral communication. To achieve this, the study plans to analyze and test mainstream speech recognition services or platforms available in the market to understand their current speech recognition capabilities for individuals with cerebral palsy, and explore the possibility of assisting them in solving their communication problems, in order to enhance their quality of life and promote their social skills. As the author is a person with congenital cerebral palsy, the study is particularly meaningful to him because the congenital brain damage affecting the nervous system has made his speech unclear, seriously affecting his ability to express himself orally. Therefore, the author plans to record a dataset of speech samples from individuals with cerebral palsy, collecting conversations from various aspects of daily life. This dataset will be tested and analyzed using mainstream speech recognition services such as Google, Microsoft, and YaTing, among others, in order to infer the current difficulties in speech recognition technology for individuals with cerebral palsy and propose potential solutions for oral communication barriers, with the hope that the contribution of this research will promote the development of mature assistive technologies for individuals with communication difficulties in the near future.

Keywords: Congenital Cerebral Palsy · Speech Recognition · Speech Clarity

1 Introduction

This study intends to record speech files of cerebral palsy patients and collect the correct textual answers of their speech, in order to provide a self-made speech dataset for cerebral palsy patients. Meanwhile, using the speech recognition services provided by well-known artificial intelligence companies such as Google, Microsoft, and YaTing as the testing benchmark, the study aims to investigate and analyze whether the most advanced speech recognition technology can recognize the speech of cerebral palsy patients, with the goal of assisting or alleviating their speech communication barriers.

J. C. Hung et al. (Eds.): FC 2023, LNEE 1134, pp. 263–268, 2024.
https://doi.org/10.1007/978-981-99-9342-0_28

2 Related Works

In the early 1960s, speech recognition technology was primarily based on pattern matching methods. Pattern matching involved comparing input speech to pre-stored speech templates to determine speech content. Due to the complexity of speech variations, the accuracy of this method was limited. However, contemporary speech recognition systems can only recognize basic single-speech sentences. In the late 1970s, a method based on Hidden Markov Models (HMM) emerged. This method built state transition models and mixtures to achieve higher accuracy than pattern matching, but still had certain limitations. The quality of speech recognition may be affected by environmental factors, such as noise, due to the characteristics of the speech signal itself [1].

In the 1990s, neural network methods emerged as a type of speech model that is capable of learning and adapting. This method can be trained using backpropagation algorithm, and its accuracy is higher than the previous two methods. However, it requires more computing resources and data. For example, [2] proposed a time-reversal backpropagation neural network speech recognition method.

In the 2000s, with the development of deep learning, speech recognition methods based on deep learning emerged. Deep learning builds deep neural networks to learn speech features, which further improves speech recognition accuracy. Convolutional neural networks and recurrent neural networks are commonly used models, for example, [3] proposed a speech recognition method based on recurrent neural networks that converts speech signals into text sequences.

In recent years, with the continuous development of speech recognition technology, new methods based on deep learning have emerged, such as end-to-end learning, which directly maps speech signals to text sequences and avoids the complexity of intermediate steps, further improving the speech recognition performance [4]. In general, the technology of speech recognition has been advancing constantly, evolving from pattern matching, hidden Markov models, neural networks, to deep learning and end-to-end learning methods. These advancements have continuously improved the accuracy and application range of speech recognition.

In the field of speech, the speech model plays an important role. The knowledge base behind the speech model makes predictions based on the context of the knowledge base, providing appropriate sentences. By pre-training the model on a large amount of text, performance on many downstream tasks often improves with different model sizes and increasing amounts of unsupervised data using transfer learning. The model does not need to learn external knowledge, only to memorize, and can provide appropriate responses in a speech question-and-answer format [5].

In summary, early speech recognition technology was developed based on pattern matching, until the current speech recognition technology was developed using deep learning-related techniques.

3 Methodology

3.1 Metrics of Speech Recognition

For the analysis and ranking of speech recognition results for people with cerebral palsy, three speech models, Google, Microsoft, and YaTing, were used for recognition. Using the Character/Word Error Rate (CER/WER) scoring tool, the accuracy of the speech recognition models was analyzed based on the recognition results, and the other commonly used evaluation indicators for speech recognition, such as Match Error Rate (MER), Word Information Preserved (WIP), and Word Information Lost (WIL), were used to obtain the average, median, maximum, minimum, and standard deviation for each evaluation [6].

1. CER represents the indicators of word error rate. It calculates the total number of hits, insertions, deletions, and substitutions of words in the recognized sentence compared to the correct reference sentence, and divides it by the total number of words in the reference sentence to obtain the word error rate percentage.
2. MER is a measurement metric in speech recognition that represents the percentage of characters in the recognized speech that are missing compared to the correct speech. A lower MER value indicates better integrity in speech recognition, and the value range of MER is between 0 and 1.
3. WIP is an evaluation indicator in speech recognition that measures the percentage of correct word count in the recognized text by the speech model, compared to the total word count in the correct transcription. A higher WIP indicates better recognition performance of the speech recognition system. The WIP value ranges from 0 to 1.
4. WIL measures the text loss rate of a speech recognition model. It is the percentage of incorrectly recognized words in the total number of correct words. A lower WIL indicates better recognition performance of the speech recognition system. The WIL value ranges from 0 to 1.

3.2 Speech Recognition Performance of Google, Microsoft and YaTing

Speech recognition technology has come a long way in recent years, and major players like Google, Microsoft, and YaTing are leading the charge in delivering accurate and efficient speech recognition services. Therefore, we conduct the experiments with the three service providers. Their performance of CER, MER, WIL and WIP are shown in the Fig. 1.

1. The experimental results of Google speech recognition are as follows. The average CER is 104%, with a median of 93%, a maximum of 500%, a minimum of 0%, and a standard deviation of 83%. The average MER is 66%, with a median of 75%, a maximum of 100%, a minimum of 0%, and a standard deviation of 37%. The average WIP is 28%, with a median of 7%, a maximum of 100%, a minimum of 0%, and a standard deviation of 36%. The average WIL is 72%, with a median of 93%, a maximum of 100%, a minimum of 0%, and a standard deviation of 36%.
2. The experimental results of Microsoft speech recognition are as follows. The average CER is 93%, with a median of 89%, a maximum of 450%, a minimum of 0%, and

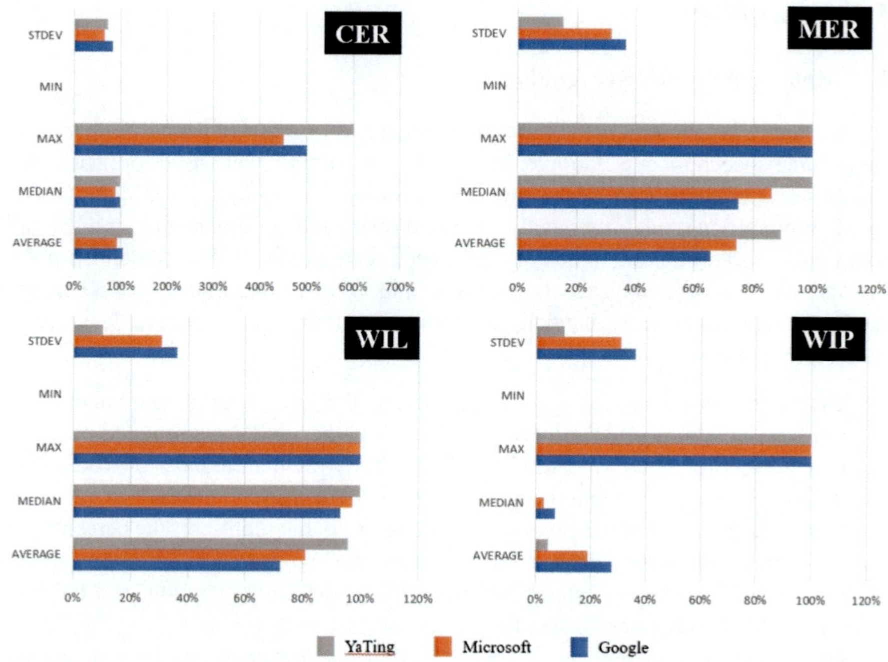

Fig. 1. Comparisons of YaTing, Microsoft and Google on CER, MER, WIL and WIP.

a standard deviation of 65%. The average MER is 74%, with a median of 86%, a maximum of 100%, a minimum of 0%, and a standard deviation of 32%. The average WIP is 19%, with a median of 3%, a maximum of 100%, a minimum of 0%, and a standard deviation of 31%. The average WIL is 81%, with a median of 97%, a maximum of 100%, a minimum of 0%, and a standard deviation of 31%.

3. The experimental results of YaTing speech recognition are as follows. The average CER is 126%, with a median of 100%, a maximum of 600%, a minimum of 0%, and a standard deviation of 73%. The average MER is 89%, with a median of 100%, a maximum of 100%, a minimum of 0%, and a standard deviation of 15%. The average WIP is 5%, with a median of 0%, a maximum of 100%, a minimum of 0%, and a standard deviation of 10%. The average WIL is 95%, with a median of 100%, a maximum of 100%, a minimum of 0%, and a standard deviation of 10%.

The CER represents the accuracy rate, and a lower rate indicates better recognition capability. Among the four evaluation indicators, the average CER of Microsoft's speech model is 93%, compared to Google's 104% and YaTing's 126%. The median CER of Google is 93%, Microsoft is 89%, and YaTing is 100%. The maximum CER of Google is 500%, Microsoft is 450%, and YaTing is 600%. The standard deviation of Google's CER is 83%, Microsoft's is 32%, and YaTing's is 73%. Therefore, Microsoft's speech model performs better in terms of CER recognition ability. Ranking of average CER is Microsoft > Google > YaTing.

The MER represents the word omission rate, and a lower rate indicates better speech recognition capability. Among the four evaluation indicators, the average MER of Google's speech model is 66%, while Microsoft's is 74%, and YaTing's is 89%. In comparison, the average MER of Google is much lower than that of Microsoft and YaTing. The median MER of Google is 75%, Microsoft is 86%, and YaTing is 100%. The median order is Google, Microsoft, and YaTing. The maximum and minimum values of the three speech models are all 100% and 0%, respectively. The standard deviation of Google's MER is 37%, Microsoft's is 32%, and YaTing's is 15%. The standard deviation order is YaTing, Microsoft, and Google. The order of average MER is Google > Microsoft > YaTing.

WIP represents the percentage of unidentified speech in the total amount of speech, and a higher value indicates better speech recognition ability. Among the four evaluation indicators, in terms of WIP, the average value for Google is 28%, for Microsoft is 19%, and for YaTing is 5%. The order of average values is YaTing, Microsoft, and Google, respectively. The median values for WIP are 7% for Google, 3% for Microsoft, and 0% for YaTing, and the order of median values is Google, Microsoft, and YaTing, respectively. The maximum and minimum values for the three speech models are 100% and 0%, respectively. The standard deviation for Google WIP is 36%, for Microsoft WIP is 31%, and for YaTing WIP is 10%. The order of standard deviation values is Google, Microsoft, and YaTing, respectively. The average WIP values are in the order of Google > Microsoft > YaTing.

WIL represents the percentage of identified speech in the total amount of speech, and a lower value indicates better speech recognition ability. Among the four evaluation indicators, in terms of WIL, the average value for Google is 72%, for Microsoft is 97%, and for YaTing is 95%.The order of average values is Google, Microsoft, and YaTing, respectively. The median values for WIL are 93% for Google, 97% for Microsoft, and 100% for YaTing, and the order of median values is Google, Microsoft, and YaTing, respectively. The maximum and minimum values for the three speech models are all 100% and 0%, respectively. The standard deviation for Google WIL is 36%, for Microsoft WIL is 31%, and for YaTing WIL is 10%. The order of standard deviation values is Google, Microsoft, and YaTing, respectively. The average WIL values are in the order of Google > Microsoft > YaTing.

Based on the above four evaluation indicators and the testing data of 500 speech files, 500 sentences were selected for the evaluation and the speech recognition ability of the 500 sentences was sorted. The order of speech recognition ability is Google > Microsoft > YaTing. Each of the three speech recognition services, Google, Microsoft, and YaTing, has its own advantages. Google performs better in MER, WIP, and WIL, while Microsoft performs better in CER. However, the speech recognition performance of the models may vary depending on the speech data used. The experiments only represent the results of this small-scale evaluation for people with cerebral palsy.

References

1. Rabiner, L.R.: A tutorial on hidden Markov models and selected applications in speech recognition. Proc. IEEE **77**(2), 257–286 (1989)

2. Lippmann, R.P.: Speech recognition by machines and humans. Speech Commun. **22**(1), 1–15 (1997)
3. Graves, A., Mohamed, A.R., Hinton, G.: Speech recognition with deep recurrent neural networks. In 2013 IEEE International Conference on Acoustics, Speech and Signal Processing, pp. 6645–6649 (2013)
4. Hannun, A., et al.: Deep speech: scaling up end-to-end speech recognition. arXiv preprint: arXiv:1412.5567 (2014)
5. Roberts, A., Raffel, C., Shazeer, N.: How much knowledge can you pack into the parameters of a language model? arXiv preprint: arXiv:2002.08910 (2020)
6. Xu, B., Tao, C., Feng, Z., Raqui, Y., Ranwez, S.: A benchmarking on cloud based speech-to-text services for French speech and background noise effect. arXiv preprint: arXiv:2105.03409 (2021)

Fire and Smoke Detection Using YOLO Through Kafka

Kai-Yu Lien[1], Jung-Chun Liu[1], Yu-Wei Chan[2], and Chao-Tung Yang[1,3(✉)]

[1] Department of Computer Science, Tunghai University, No. 1727, Sec. 4, Taiwan Boulevard, Xitun District, Taichung 407224, Taiwan, ROC
jcliu@thu.edu.tw

[2] Department of Information Management, Providence University, Taichung City, Taiwan
ywchan@gm.pu.edu.tw

[3] Research Center for Smart Sustainable Circular Economy, Tunghai University, No. 1727, Sec. 4, Taiwan Boulevard, Xitun District, Taichung 407224, Taiwan, ROC
ctyang@thu.edu.tw

Abstract. This study is based on deep learning techniques, which compare various detection algorithms and implement the suitable one for firework detection. The considered factors include streaming, speed, accuracy, and portability. Through a detection algorithm, it can simultaneously identify the positions of smoke and fire, providing subsequent control of fire or other applications. After comparison, we plan to perform detection results in a streaming manner, where only real-time detection of the captured scene is carried out. The system can notify people or teams in need of notification via the network.

Keywords: YOLO · Deep learning · Machine learning · Fire and Smoke detection · Kafka

1 Introduction

Global warming causes climate change, increasing forest fire risks in flammable areas like forests and grasslands. Forests and grasslands are essential to the ecosystem and habitats of animals and plants. Forest fires destroy vegetation and habitats, causing severe ecological impact. Smoke and fire detection systems detect smoke and fire quickly, improving response time to prevent forest fires, and allowing authorities to take measures to protect the environment and public safety. As a result, we plan to build a system based on vision to detect fire and smoke. With the development and advancements in manufacturing technology and the evolution of algorithms of artificial intelligence, components can be made smaller, faster, more accurate, and power-efficient advantages, even making them into portable devices, let image detection has already had a wide range of applications. In this paper, we use the YOLO model as our primary development environment. We will explore the efficiency and accuracy differences by comparing the differences and applicability of various algorithm versions. We will

J. C. Hung et al. (Eds.): FC 2023, LNEE 1134, pp. 269–275, 2024.
https://doi.org/10.1007/978-981-99-9342-0_29

challenge the detection of relatively complex situations - smoke and fire sources - as a system and combine it with real-time streaming, Kafka data access, and edge devices for further exploration.

2 Literature Review

2.1 Machine Learning

Machine learning is a rapidly evolving field within AI that involves training computers to learn from data inputs and optimize algorithms for accurate predictions on new, unknown data. It has subfields such as supervised learning, deep learning, and reinforcement learning, which involve training models on labeled datasets, training neural networks, and using reward systems to guide optimal decision-making in different environments.

2.2 Image Recognition (Detection)

Image recognition is a type of machine learning technology in artificial intelligence, which mainly enables machines to automatically recognize objects in images, such as objects, people, animals, scenes, texts, and so on. Currently, the common related technologies and methods for image recognition include convolutional neural networks, object detection, YOLO, image segmentation, feature extraction, PyTorch, and TorchVision.

2.3 Kafka

Kafka is an open-source distributed event streaming platform developed by Apache, with the ability to process and store data streams reliably and efficiently. It is designed to handle large-scale real-time data streams by providing a unified, high-throughput, low-latency platform for data processing. Kafka operates on a publish-subscribe model, where producers publish messages or events to topics, and consumers subscribe to those topics to receive the messages. The platform is designed to handle massive amounts of data and can scale horizontally by adding more brokers, which are the servers that store and manage the data, also Kafka can be used to manage data from multiple sources, including sensors, web applications, and databases. The platform can process these streams of data quickly and efficiently, making it ideal for real-time analytics and decision-making.

2.4 YOLO Algorithm

YOLOv5 has diverse applications and support like Pytorch, CUDA acceleration, and Tensor support. This paper focuses on comparing YOLO (You Only Look Once) and extending its functionality for future data applications. This includes comparing algorithmic differences between various versions and their accuracy and efficiency. YOLO is a fast and accurate object detection system that simultaneously performs comprehensive inferences on entire images through regions and bounding boxes. Unlike sliding window detection or region proposal methods, it can see the entire image during training and testing. Therefore, YOLO can conceal globally interrelated class information within its encoding.

In this study, we chose three models from the YOLO series: YOLO v5, v7, and v8. The architecture is based on Region-Based Convolutional Neural Network Layers and is composed of two parts: feature extraction and detection. The feature extractor is a convolutional neural network responsible for extracting features from the image, with down-sampling for feature extraction from local to global regions. The detection head is responsible for mapping these features to object positions and categories and restoring them to their original size.

3 Data Collection and Experimental Results

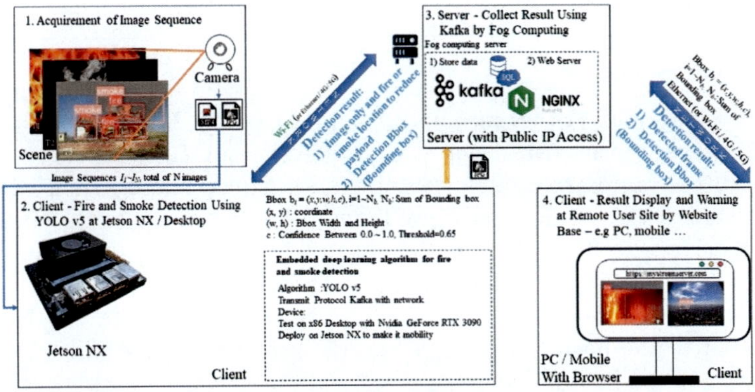

Fig. 1. System architecture

This research begins with the development of various algorithms on the x86 computing platform, and training and testing these algorithms in a Windows environment. The trained models are analyzed and compared to evaluate the differences between the algorithms. The advantages and disadvantages of each algorithm are compared quantitatively. The most suitable algorithm is selected for further development and integration with real-time streaming, data storage in Kafka, and the use of portable EDGE devices such as Jetson Xavier NX. The training process includes adjusting model size, training frequency, training content, parameter tuning, etc. to increase the accuracy of the model. The system architecture of this study is shown in Fig. 1.

We want to detect smoke and fire using image detection, as traditional methods are expensive and limited in range. In this study, we used a sample dataset consisting of two categories, fire, and smoke to perform feature extraction and labeling on it. By comparing the model performance using different feature extraction and labeling methods, we found their impact on model accuracy and generalization ability. Additionally, we analyzed the impact of data distribution on model performance, including the impact of different sample quantities and category ratios on model performance. The simplified research process is divided into data collection, processing, model training, and benchmarking the result of models using metrics in machine learning such as Accuracy, Loss, Confusion Matrix, etc.

3.1 Training

In this study, the various versions of standard-sized models were first trained using the recommended training frequency in the paper, and the accuracy of the initial training of each version was compared to determine which version would be used as the primary detection method for this study, then performed feature extraction and labeling on the dataset and investigated the impact of feature extraction and labeling on model performance and analyzed the impact of data distribution on model performance.

3.2 Server Setup

We have set up four virtual machines running Ubuntu 22.04 Server version on an ESXi server, each with 4 cores, 8 GB RAM, and 100G storage. Three of the virtual machines are used as containers for Kafka, and one is used for web hosting. During the setup process, we limited the disk speed of the three Kafka virtual machines to 100 IOPS to simulate the operation of Kafka on multiple machines and test its write performance. As we mainly transmit image data, there is almost no delay in reading and writing, and the data can be quickly transmitted to the host. The images can be displayed on the web page with a low response time (about 3 s).

3.3 Data Collection

We use Google Search, Kaggle datasets, and crawler to collect approximately 3,100 images of fire and smoke, then use data augmentation techniques to expand the dataset to 15,000 images for training, then split into training and testing sets, typically with 80% of the data for training and 20% for testing. The testing set may be further divided into either the entire validation set or a 10% validation set and the remaining testing set. However, due to the limited number of samples in the dataset, we enhanced the images and split them into training (85%), and validation (15%) sets with a pre-processing model size of 640×640, and keep the quantity of fire and smoke samples were balanced as much as possible, which is approximately 8000 images per class used for training on average. To meet the requirements, In the data augmentation process, we applied rotation enhancement to images taken from different angles to address the focus blur issue and then trained the dataset with default 300 iterations on YOLO v5, v7, and v8, and compared their difference.

3.4 Experimental Results

The actual training results were observed to be similar to our expectations, with better accuracy in fire detection and not-so-good results in smoke detection, although still better than expected. The reason may lie in the fact that smoke is more complicated, Moreover, its irregular shape and various factors such as color, lighting conditions during photography, and concentration (translucent), which may easily be confused with the environmental background can lead to uncertainty.

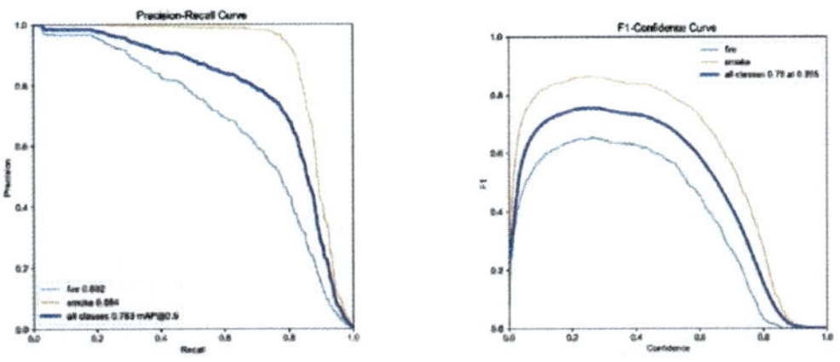

Fig. 2. F1 curve and PR curve.

In the YOLO v5 section, we trained the model an image size of 640. Our results showed that the precision (P) and recall rate (R) for smoke and fire were 0.9 and 0.7, respectively. Although the accuracy of smoke detection was 0.2 mAP higher than fire detection, the average accuracy of fire detection during actual testing was around 0.6, while the accuracy of smoke detection was around 0.4 when it was correctly classified. Unfortunately, it was often observed during actual testing that smoke was not detected and ignored by the model. In Fig. 2, we can see that even in the small fire or smoke is still detectable in the long distance of the small pixel.

We will use YOLOv5, an object detection model that is intended for real-time streaming video and can use in a real-time environment, It offers five models that come with differing depth and width multiple to control the network. These differences have a significant impact on the performance and suitability of the YOLOv5 model for different environments. In the following table, we can see the detection speeds of various sizes of YOLO v5 models. We tested these models on an x86 PC platform with an image processing core of RTX3090. As we are running our models on embedded platforms, we tested YOLO v5 on an Nvidia Jetson NX to compare the original processing speed without any accelerators. As a result, the speed of the YOLO v5m model starts to decrease rapidly but accuracy does not significantly improve. Since high resolution can lead to bandwidth waste and transmission delays, and a high frame rate is not necessary, we limit our image so the system can handle about one frame per second. This approach is suitable and sufficient for our experimental environment and practical use, where a difference of a second in fire status is not significant. For this

Table 1. SPEED BETWEEN YOLO V5 MODELS

Model	Image	Video	Video
	(RTX 3090)	(Stream @ RTX 3090)	(Native Jetson NX)
V5n	16 ms	7.8 ms (128FPS)	590 ms (1.7 FPS)
V5s	12 ms	7.0 ms (142FPS)	760 ms (1.3 FPS)
V5m	16 ms	8.0 ms (125FPS)	2000 ms (0.5 FPS)
V5l	15 ms	10.1 ms (99FPS)	3500 ms (0.3 FPS)
V5x	22 ms	13.5 ms (74FPS)	5000 ms (0.2FPS)

Table 2. ACCURACY BETWEEN YOLOV5 MODELS

Scenes	Fire			Smoke		
Model	Accuracy	S	B	Accuracy	S	B
YOLO v5m (Enforced)	0.75~0.98	V	O	0.7~0.98	V	O
YOLO v5n	0.3~0.72	O	O	0.3~0.89	O	V
YOLO v5s	0.2~0.77	O	O	0.3~0.92	O	O
YOLO v5m	0.3~0.85	O	O	0.4~0.95	O	X
YOLO v5l	0.4~0.89	V	V	0.4~0.96	V	V
YOLO v5x	0.4~0.88	O	V	0.4~0.95	O	V

reason, we implement the YOLO v5m mode and plan to optimize its detection speed in the future (Table 1).

$$Accuracy \quad = \quad (TP + \ TN)/(TP + FP + FN + TN). \tag{1}$$

We utilize the recall and precision rates as metrics to evaluate the proportion of correctly identified accuracy among all the objects present as the ground truth and estimate (1) and check the lowest and highest result in each test on validation, test, strange and live datasets, Table 2 shows that increasing the size of the model beyond YOLO v5m does not lead to a significant improvement in accuracy. Instead, it results in higher costs in terms of image processing time and memory resources, while also reducing the detection speed. We chose to optimize the YOLO v5m model to achieve an accuracy of 0.98 while maintaining a speed of around 1 FPS. This approach reduces the payload of transmission and processing and minimizes the risk of excessive data duplication, which would rapidly fill up our storage container and increase the load on the Kafka server. As a result, we can collect useful data continuously without requiring too many resources.

4 Conclusion and Future Works

With the continuous development of machine learning detection technology, we should not blindly pursue the latest version but rather conduct experiments and tests to compare and find the development environment and models that are suitable for our application system. In this experiment, the latest version achieved the best results for this task and its algorithm was optimized, but the resulting more closed code and relatively difficult code modification also bring challenges. Although this process may be time-consuming and may use older technology, it is necessary to ensure the practicality and real-time effects of the system. Afterward, we can focus on optimizing and applying the model, as well as studying the new versions.

Currently, this study has successfully implemented firework detection on the YOLO series model. And through Kafka, photos can be transmitted instead of text only. Although more time was spent on encoding and non-optimal encoders resulted in some wasted resources during transmission, this model can accurately recognize different forms of fires, including large flames and small fires, and can also detect the presence of smoke. Even occurs outdoors and the source is hidden or uncleared, the algorithm can provide different prompts through additional functions when it detects thick smoke by adding the smoke detection model function that prevents the system from misjudging some light sources similar to flames to improve the accuracy of detection results. The original images are also kept for subsequent modifications and verification.

References

1. Redmon, J., Divvala, S., Girshick, R., Farhadi, A.: You only look once: unified, real-time object detection. In: Computer Vision and Pattern Recognition, May 2016
2. https://github.com/ultralytics/yolov5
3. https://github.com/ultralytics/ultralytics
4. Kim, Y.-K., Jeong, C.-S.: Large scale image processing in real-time environments with Kafka. In: Proceedings of the 6th AIRCC International (2017)
5. Frizzi, S., Kaabi, R., Bouchouicha, M., Ginoux, J.-M., Moreau, E., Fnaiech, F.: Convolutional neural network for video fire and smoke detection. In: IECON 2016 - 42nd Annual Conference of the IEEE Industrial Electronics Society, October 2016
6. Namozov, A., Cho, Y.I.: An efficient deep learning algorithm for fire and smoke detection with limited data. Adv. Electric. Comput. Eng. **18**(4), 121–128 (2018)
7. Bradski, G., von Kleist-Retzow, F.T., Tiemerding, T., Elfert, P., Haenssler, O.C.: The OpenCV library. Dr. Dobb's journal of software tools. J. Comput. Commun. **4**(3), 25–33 (2016)
8. Viswanatha, V., Chandana, R.K., Ramachandra, A.: Real time object detection system with YOLO and CNN models: a review. In: Computer Vision and Pattern Recognition (cs.CV); Image and Video Processing (eess.IV), July 2022
9. Wang, C.-Y., Bochkovskiy, A., Liao, H.-Y.M.: YOLOv7: trainable bag-of-freebies sets new state-of-the-art for real-time object detectors. In: Computer Vision and Pattern Recognition, July 2022

Check for updates

mKIPS: A Lightweight Modular Kernel-Level Intrusion Detection and Prevention System

Yuan-Zheng Yi and Mei-Ling Chiang(✉)

Department of Information Management, National Chi Nan University, Nantou, Taiwan, R.O.C.
s105213529@mail1.ncnu.edu.tw, joanna@mail.ncnu.edu.tw

Abstract. With many research results and the development of related tools, user-level intrusion detection and prevention systems (IDPS) have been widely used to defend systems against network attacks. However, there are still bottlenecks in their high packet drop rate and low detection efficiency under heavy network traffic. In contrast, kernel-level IDPS has a higher packet detection rate and higher efficiency, whereas kernel-level design faces many challenges. The system designed with the monolithic architecture has high performance. The dynamically loadable module architecture design has higher flexibility and scalability; however, the increased operating costs lower system performance.

This paper explores the modular architecture of kernel-level IDPS that can expand or reconfigure system functions through dynamic plug-in modules and maintain the system's stability and high performance. We have developed a lightweight, high-efficiency, scalable, and highly modular kernel-level IDPS named mKIPS. This modular architecture divides the system into several kernel modules, in which functional components can be dynamically inserted or removed during runtime to adapt to changing demands. Therefore, administrators can control the IDPS's packet processing by mounting modules of different versions and functions for their needs. Besides, mKIPS dispatches packets to various cores for processing through software and hardware functions by properly setting the IRQ affinity and using Receive Packet Steering technology. As a result, the load of each core can be more balanced to utilize the multicores. Experimental results show that our mKIPS can achieve a high detection rate and efficiency.

Keywords: Intrusion Detection and Prevention System · Multicore Systems · Kernel Level · Linux Kernel

1 Introduction

As network attack events occur frequently, providing system security for information systems is an important issue. The intrusion detection and prevention system (IDPS) [1–5] has been proven effective against information security attacks.

IDPSs can be classified into several categories according to deployment, functionality, and detection methods. The network-based IDPS is deployed within the network structure to monitor network traffic in real-time, which analyzes network packets to detect and prevent intrusions. The host-based IDPS is deployed on the host node of the

J. C. Hung et al. (Eds.): FC 2023, LNEE 1134, pp. 276–288, 2024.
https://doi.org/10.1007/978-981-99-9342-0_30

information system, which analyzes the activity of the host to identify potential intrusions or unauthorized activities. The detection method can be misuse detection, which checks the signature by comparing the network traffic or host behavior against a predefined set of signatures to detect known threats. Another detection method can be anomaly detection, which monitors the system's operation to detect anomalies that could indicate an attack or suspicious activity. IDPSs can be implemented and run at the kernel level or user level.

Snort [6] and Suricata [7] are the famous open-source network IDPS widely used in related fields and research. However, our practical experience and previous research [8] observed that Snort operating at the user level could not handle packet inspection under heavy network traffic, and its packet drop rate is relatively high. Compared with Snort and Suricata operating at the user level, kernel-level IDPS can directly intervene in the kernel's processing flow of network packets and detect threats as soon as packets are received or sent. Furthermore, it can avoid the operating cost of copying packets to the user level for inspection, waiting for the kernel scheduler's scheduling to execute user-level IDPS, and switching the protection domain back and forth between kernel mode and user mode, significantly reducing the impact on network performance.

Although kernel-level IDPS has the advantages of high efficiency and packet detection rate, developers need to have an in-depth understanding of operating system (OS) operations because it is within the kernel and directly interacts with kernel operations. Furthermore, because it operates at the kernel level, if there is a problem with the IDPS system design or a program bug is generated due to the negligence of the developers, it is easy to degrade the stability of the OS or directly cause a kernel panic. In addition, the kernel-level IDPS is highly dependent on the OS kernel. If the source code of the OS kernel is greatly changed, the kernel-level IDPS may need to be modified accordingly. These factors make it challenging to design and implement a kernel-level IDPS.

Due to the increasing maturity of virtualization technology, its use is becoming increasingly common. Furthermore, data centers virtualize the original physical server farms due to their many advantages. Therefore, how to perform intrusion detection and defense in a virtualized environment is also the focus of our research [8]. As a result, we have developed VMM-IPS [8] operating at the kernel level. It implements a reaction mechanism to respond to attacks in terms of intrusion detection functions and blocks the possibility of subsequent attacks by interrupting the attacker's connection.

From our practical experience [8] and related research [9], it is observed that kernel-level IDPS has better operating performance and better detection rate than user-level IDPS. Nevertheless, in contrast, every network packet is inspected because it is involved in the kernel's processing of network packets. Therefore, when a large number of packets come in, if the kernel-level IDPS is blocked in the kernel processing flow, it will affect the system's stability when encountering a performance bottleneck. Therefore, if it is necessary to maintain stable operating performance, providing the admission control mechanism can maintain the system's quality of service.

On the other hand, some technologies for evading detection and defense systems (IDS Evasion) [10] have been developed, and the methods are constantly being updated. With the development of defense tools, network attacks continue to evolve, and every time a new attack method is discovered, new defense tools, technologies, or systems must be added. However, developing new systems is not easy, and the newly added functions may also be conflicts with the original system, coupled with the replacing developers, making the development of the new system difficult. How to effectively expand system functions to adapt to changing needs has become the research focus.

In view of the increasing diversification of network services and the importance of network security, we began to study the design and development of an IDPS that is scalable, highly modular, and located at the kernel level. We study the structure and dynamic plug-in modules with expandable system functions, as well as the modules for detecting IDS evasion technology, to make the defense function of the IDPS more complete so that the operation or system development can be more flexible and convenient.

In this paper, we explore the modular architecture of kernel-level IDPS. The critical work includes designing a lightweight modular architecture that can expand or reconfigure system functions through dynamic plug-in modules. The goal is to allow the system to maintain stability and good performance with a high detection rate even under heavy network traffic. So we developed a lightweight, high-efficiency, scalable, and highly modular IDPS run in the Linux kernel [11], named mKIPS.

At the same time, we also explored the processing flow of the Linux kernel in receiving and sending network packets under multicore systems. The Linux system is based on interrupt and subsequent processing for packet processing. If the same core is responsible for packet processing, although it can improve its cache usage, the multicore's parallel processing performance is not fully utilized. The Linux networking stack provides technologies [12–14] for the parallel processing of network packets under multicore systems, which can improve system performance by distributing packets to cores for processing through software and hardware setting. We employ these technologies and correctly set the IRQ affinity to distribute the processing of packets to different cores to avoid the performance bottlenecks caused by the centralized processing of packets on the same core. Experimental results show that the proposed mKIPS can have a very high detection rate and efficiency even under heavy network traffic.

2 Related Work

Snort [6] and Suricata [7] are well-known open-source network-based IDPS. The development of Snort is relatively mature. Snort has a large and active developer community and is widely used in related research. Since version 3.0, Snort has been developed from scratch with a new software framework, especially in multithreading, automatic configuration, and cross-platform support. It adopts misuse detection technology, operates at the user level, and captures live network packets from the kernel through the packet capture library – libpcap [15]. First, the Sniffer module of Snort determines the packet type and performs statistical analysis. Then the Preprocessor module performs operations such as decoding and reassembling the packet content. After that, the Detection Engine module compares the packet payload with the rules of the attack signature database. Finally, the Output module determines the packet's response method and returns the control to the original processing flow. Snort's attack signature database is constructed based on the threat reports by the Cisco Talos Intelligence Group [16].

Like Snort, Suricata [7] is also a user-level network-based IDPS initially developed using a multithreaded software framework. It is a relatively lightweight IDPS, and its detection engine also uses misuse detection technology. Its attack signature database is also built based on the threat reports by the Cisco Talos Intelligence Group. Even so, the attack signature databases of Snort and Suricata have unique features in their design, which limits their compatibility.

Many studies on IDPSs work on making IDPSs more effective, efficient, complete, and with more applications. Gaddam and Nandhini [17] analyzed various IDPSs against various types of attacks in different environments. They then proposed an architecture to improve Snort's detection rate and reduce the packet drops under heavy traffic. In the study [18] of Yuan et al., Snort is used to form a distributed IDPS. By building multiple detection nodes, the detection performance is improved. However, there may be problems with repeated detection of packets. The research suggests that distributed IDPS should strengthen the communication ability between nodes.

Shah and Issac [19] reported that Suricata could have a lower packet drop rate than Snort under high network traffic but consumed higher computational resources. Whereas, Snort had higher detection accuracy. They explored using machine learning (ML) technology to improve the efficiency and detection rate of IDPS. This study developed a Snort adaptive plug-in that implements ML algorithms to determine attacks, and this module runs parallel with Snort's original detection engine. Shah and Bendale [20] surveyed related research on using AI technology in IDPS and anomaly detection.

Chin et al. [9] proposed the kernel-level IDS built under the SDN network and discussed the advantages and challenges of building the IDS at the kernel level. They pointed out that the kernel-level IDS obtains better performance than the user-level IDS, and network packets can be immediately examined when packets are received or sent. The disadvantage is that the kernel-level IDS does not have a rich library of functions available, and developers must have complete knowledge of the kernel to implement the system. They implemented a kernel-level IDS and constructed it in the SDN network, and tried to ensure that the network function of the system would not be affected when there were problems with the functional components of the IDS.

The Linux networking stack provides technologies for parallel processing of network packets and improving performance in multicore systems by distributing packets to different cores for processing through software and hardware, such as Receive Side Scaling [12–14], Receive Packet Steering [12–14], Receive Flow Steering [12–14], Accelerated Receive Flow Steering [12–14], and Transmit Packet Steering [13, 14]. However, they require experienced administrators to correctly enable and perform settings.

3 System Design and Implementation

3.1 System Overview

The proposed kernel-level IDPS named mKIPS is lightweight and modular. It is implemented as a set of kernel modules that can be dynamically inserted/removed into/from the Linux kernel during runtime. Its implementation employs the netfilter [21], a packet-filtering framework built into the Linux kernel, to intercept all network packets entering or leaving the system. The netfilter framework allows developers to register functions to the hook point to intervene in the kernel's packet processing flow. Therefore, the mKIPS module is registered on the PRE_ROUTING hook point of the netfilter. When the packet enters the system, the packet can be sent to the mKIPS for inspection. After completing the packet inspection, a corresponding response will be given according to the detection result. If outcoming packets need to be inspected, the mKIPS module must also be registered on the LOCAL_OUT hook point of the netfilter.

3.2 Lightweight Modular Architecture

This research aims to strike a balance between the system performance and the flexibility of the system architecture. Because too much processing will degrade the system performance, which affects the packet detection rate under heavy network traffic. Therefore, we develop a lightweight dynamic modular architecture, which can avoid excessive processing due to the design providing high flexibility.

The modular architecture of mKIPS divides the system into different Linux kernel modules. As shown in Fig. 1, the calling module is called a demand module, and the called module is called a supply module. In the implementation, the demanding function pointers point to the target functions implemented by other modules, and functions of the mKIPS modules can communicate with each other. The demand module must declare function pointers and implement an empty function that only returns the default value. The demand module exports the access authority through Linux kernel macro EXPORT_SYMBOL. The supply module needs to declare an extern modifier to obtain the access authority of the demand module and implement the supply function.

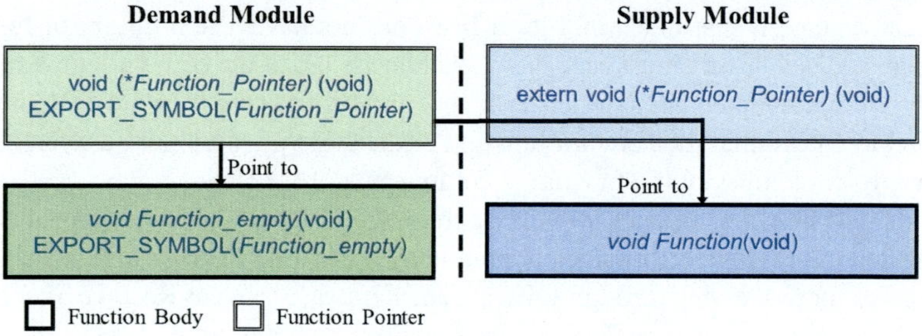

Fig. 1. Modular architecture design

When the demand module is inserted into the kernel, its demand function pointer will point to the empty function. When the supply module is inserted into the kernel, the demand function pointer will be changed to point to the implemented function of the supply module. Finally, when the supply module is removed from the kernel, it will redirect the demand function pointer to the empty function again. Under this framework, the mKIPS' kernel modules can be dynamically inserted or removed during system runtime. The administrator can also control the packet processing flow of IDPS by inserting modules of different versions and functions. This lightweight modular architecture has less impact on the system operations and performance.

Under this lightweight modular architecture, the mKIPS system is divided into three kernel modules: the Manager module responsible for intervening in the kernel to process network packets, the Detection module for inspecting packet content, and the Response module for defending against attacks and giving responses. The modular architecture and system processing flow are shown in Fig. 2.

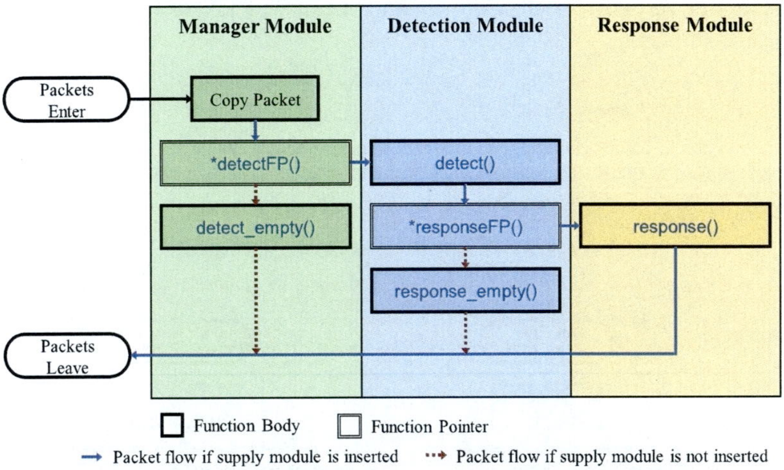

Fig. 2. mKIPS modular architecture and system processing flow.

As a network packet enters the system, the kernel invokes the Manager module to begin the packet inspection flow. It then executes the function component (i.e., detect()) of the Detection module to detect attacks through the invocation of the function pointer (i.e., *detectFP()). The Detection module manages the rule database that stores detection rules and compares the packet payload with the detection rules. When an attack is detected, it will call the function component (i.e., response()) of the Response module to respond to attacks through the invocation of function pointer (i.e., *responseFP()). It will then call the corresponding response method to deal with the attack according to the type of attack. Finally, the control returns to the kernel's original packet handling flow to continue subsequent processing.

3.3 Detection Module and Detection Rules

The Detection module implements misuse detection and uses the detection rules in the signature database for threat detection. It compares the payload of the network packet with the attack signature database to determine the type of external attack. Our detection rules are derived from Snort [6]. After converting Snort's detection rules into our dedicated detection rule format, these rules are imported to construct all the AC Trees used for threat detection when the mKIPS system is initialized. The Detection module uses the AC-BM algorithm [22] for fast string comparison, which combines the advantages of the Boyer-Moore [23] and Aho-Corasick [24] string comparison algorithms.

The mKIPS-specific rule format is shown in Fig. 3. Each rule includes the response method, transport layer protocol type, TCP/IP information, warning message, and attack string. The design of the attack signature database built based on this is shown in Fig. 4. The attack signature database is constructed as rule trees, and the corresponding rule tree is established according to the type of transport layer protocol. Each rule tree comprises a TCP/IP Tree Node containing TCP/IP information and several Rule Tree Nodes consisting of a response mechanism, warning message, and attack string, as shown in Fig. 5.

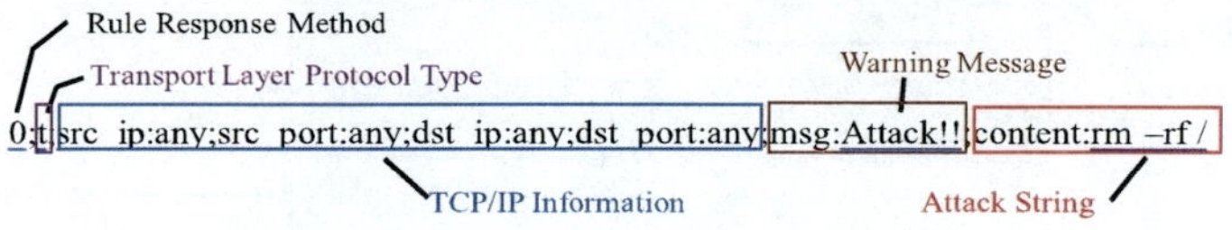

Fig. 3. The mKIPS-specific rule format.

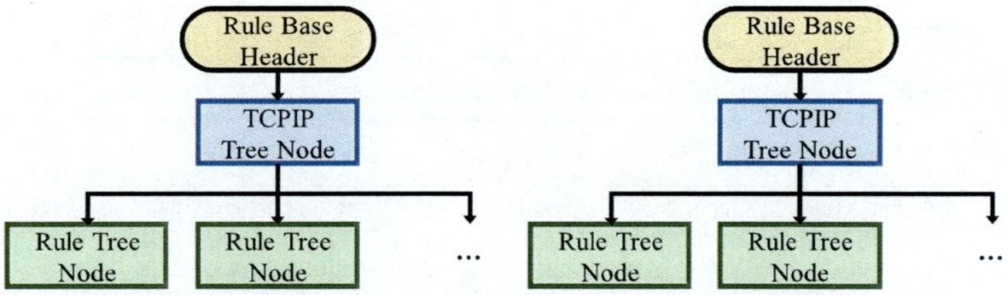

Fig. 4. The structure of the attack signature database.

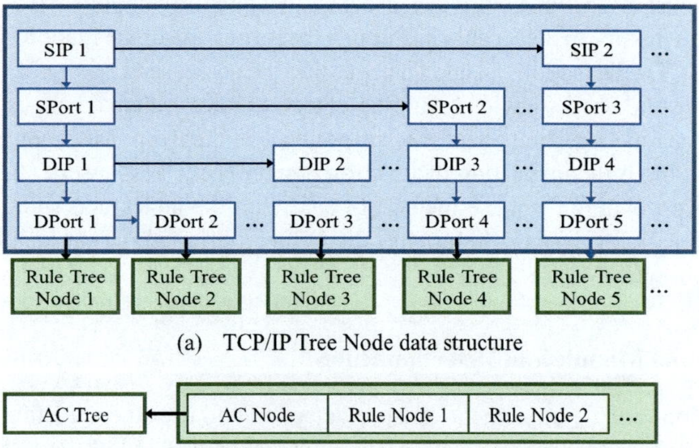

(a) TCP/IP Tree Node data structure

(b) Rule Tree Node data structure

Fig. 5. The structure of TCPIP Tree Node and Rule Tree Node

The TCP/IP Tree Node structure shown in Fig. 5(a) is composed of several SIP (Source IP), Sport (Source Port), DIP (Destination IP), and DPort (Destination Port). The SIP will point to the SIP node on the right, the SPort node on the lower layer, and so on. The DPort at the end will point to the right DPort node and the Rule Tree Node that matches the TCP/IP information.

Figure 5(b) shows the Rule Tree Node structure consisting of an AC Node and several Rule Nodes. The AC Node records the number of Rule Nodes and points to the AC Tree jointly constructed by all Rule Nodes. Moreover, each Rule Node records the response mechanism, warning message, and attack string in a single rule.

When the Detection module examines a packet, it will start visiting from the corresponding rule tree according to the TCP/IP information of the packet. First, it will visit SIP, SPort, DIP, and DPort in the TCP/IP Tree Node. Then, the AC-BM algorithm is executed to search for attack strings by comparing the packet payload with the AC Tree of Rule Tree Nodes conforming to that TCP/IP Node.

3.4 Response Module and Reaction Mechanisms

When the Detection module detects a malicious packet, it will record the matching Rule Node and send it to the Response module to execute the response mechanism. The response mechanism that conforms to the rule is recorded in the Rule Node.

The data structure of the response mechanism is shown in Fig. 6, and each response method occupies 1 bit as a switch. Four response methods are implemented: Alert (display warning message), Drop (discard packet), Reset Connection (interrupt the TCP/IP connection between the attacking end and the receiving end), and Logfile (record threat information). This design allows administrators to combine response mechanisms to form the required response mechanism for different detection rules. For example, the administrator can set the response mechanisms for a specific rule as Reset Connection and Logfile. The Unused bits are reserved for future expansion.

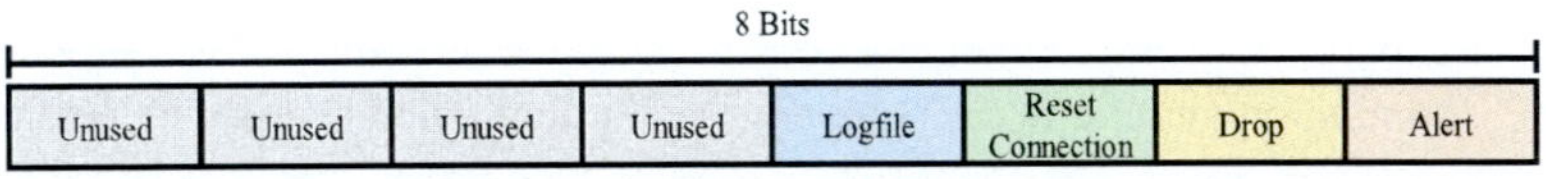

Fig. 6. Response mechanisms

3.5 Receive Packet Steering (RPS) Enabling

In addition to the lightweight and modular design, to allow the system to maintain stability and high performance when the network traffic is heavy, we study the Linux kernel's handling of network packets and utilize mechanisms to distribute packets to different cores for processing under a multicore system. For example, the administrator must correctly set the IRQ affinity setting to dispatch network packets to different cores to fully utilize the multicores performance under Linux.

Receive Packet Steering (RPS) [12–14] is a technology the Linux kernel selects for subsequent processing when Softirq is triggered. Since low-level NICs usually do not implement the function of hardware distribution of packets, resulting in the performance bottleneck problem of centralized processing of packets on a single core. Therefore, the Linux kernel triggers Softirq in the final stage of top-half interrupt execution and distributes packets to different cores for processing. RPS bases on the NIC queues to set the affinity setting that can trigger Softirq on a specific core. The default setting does not turn on the RPS, that is, to prioritize triggering Softirq on the local core. Therefore, we use RPS to distribute packets to different cores for processing through software and hardware setting so that the loading of each core can be balanced as much as possible to make the most use of multicore system performance.

4 Experimental Results

This section evaluates the system performance of the proposed kernel-level IDPS named mKIPS and compares the effectiveness of the experimental system using mKIPS and user-level IDPS (i.e., Snort [6]).

4.1 Experimental Environment

In order to measure the impact of IDPS on system performance, this research takes the native system without IDPS as the base system and compares the performance of running the Web server in three system environments, including the base system, the system runs Snort IDPS, and the system runs the mKIPS IDPS. The same detection rules (i.e., snortrules-snapshot-2983) are used to compare the performance fairly. A total of 5432 rules are selected, each containing only one attack string for detection.

In the experimental environment we constructed, a server provides Web services, and an IDPS running on another machine operates in bridge mode and is connected to the router and the Web server. The client's request to the Web server will be sent to IDPS first and then forwarded to the Web server after being examined by IDPS. Likewise, the Web server's response to the client will also be sent to IDPS first and then forwarded to the client after being examined by IDPS. Five clients were used to generate a large number of requests to the Web server to obtain Web data for measuring the system performance. As more packets are received or sent by the Web server, more packets are inspected by IDPS. Therefore, IDPS's performance does affect system performance. The detailed software and hardware specifications are shown in Table 1.

Table 1. Software and hardware specifications of the experimental environment

	Clients	Web Server	IDPS
Processor	Intel i5–3470 3.2GHz	Intel i7 -7700 3.6GHz 4C8T	Intel i7–9700 3GHz 8C8T
Memory	Kingston 2G DDR3–1333 * 2	Kingston 16G DDR4–2666 * 2	Kingston 16G DDR4–2666 * 2
NIC	RTL8111/8168/8411	I219V	EXPI9301CTBLK * 2
OS	Ubuntu Server 18.04.1 LTS	Ubuntu Server 18.04.1 LTS	Ubuntu Server 18.04.1 LTS
Kernel	Linux Kernel 4.15.0	Linux Kernel 4.15.0	Linux Kernel 4.15.0
Benchmark	ApacheBench 2.3	-	-
Web Server	-	Apache 2.4.29 [26]	-
IDPS			mKIPS / Snort 2.9.15.1

ApacheBench [25] is a performance testing tool measuring Web server performance. Each client used ApacheBench during the experiment to measure the Web server performance by sending 100,000 requests with 1,000 concurrent connections. A total of 5,000 concurrent connections and 500,000 requests were sent. Each experiment was tested ten times, and the average value was calculated.

4.2 Experimental Results

We first measured the Softirq distribution for sending and receiving packets under different RPS settings. The default setting is off RPS. Table 2 shows that in the experimental environment, with the default off RPS setting, sending and receiving packets are wholly concentrated on Core 5 to trigger Softirq. After using the RPS technology, sending and receiving packets can be effectively distributed to each core for processing.

We then measured system performance under different RPS settings. The experimental results of the transfer rate measurement are shown in Fig. 7. The results show that using the RPS mechanism can solve the performance bottleneck problem. When RPS is not enabled, due to inspecting each packet for threat detection, mKIPS will cause 52.73–53.35% performance loss compared with the base system. The base system stands for the native Linux system without running any IDPS. When RPS is turned on, since sending and receiving packets can be effectively distributed to each core for processing, system performance with running mKIPS can be significantly improved. Besides, in the bridging environment, the sending and receiving of packets between the Web server and clients for IDPS are to receive and then send packets, and the kernel of the IDPS simply forwards packets. Running mKIPS will not significantly affect the overall system performance when the RPS setting is enabled.

Table 2. Softirq distribution for sending and receiving packets under different RPS settings.

	RPS Configuration	
	Default (Disable)	RPS Enable
Core 0	0	896168
Core 1	0	880016
Core 2	0	880984
Core 3	0	893637
Core 4	0	894027
Core 5	6802114	890352
Core 6	0	882078
Core 7	0	904947
Total	6802114	7122207

The experimental results of the detection rate measurement in Fig. 8 show that even if the RPS setting is turned on, the user-level Snort still cannot perform packet detection well under such heavy network traffic, and the packet drop rate reaches 66.44–69%. Whereas the kernel-level mKIPS can have a very high detection rate and performance. Therefore, it can effectively utilize the function of IDPS to protect the system.

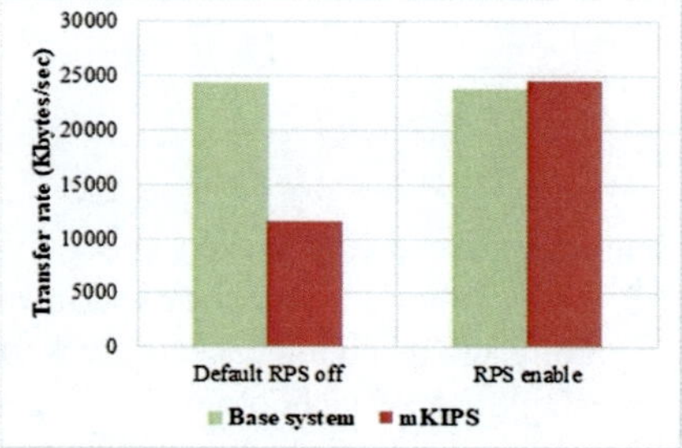

Fig. 7. Transfer rate comparison.

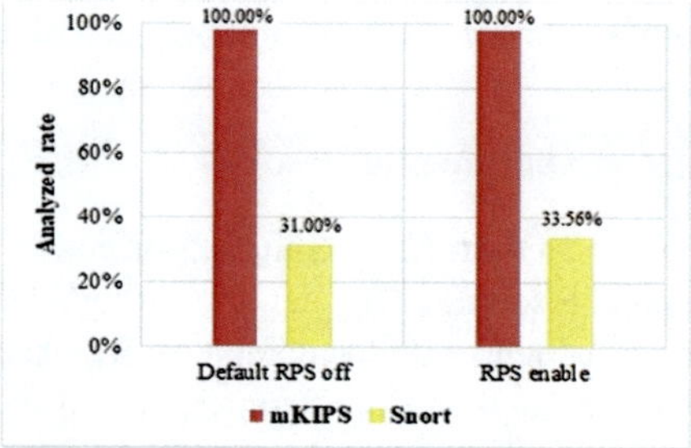

Fig. 8. Packet analyzed rate comparison.

5 Conclusions

We have designed and implemented a lightweight modular kernel-level IDPS named mKIPS. It adopts signature-based detection, inspecting each network packet to find malicious patterns in known attacks. mKIPS is implemented as a loadable kernel module that can be dynamically loaded into the Linux kernel during run time. Its modular architecture can support dynamic addition/deletion/replacement of functional components. Its implementation employs the netfilter framework, and all packets entering or leaving the system can be examined with in-place packet inspection. Furthermore, this work distributes packets to different cores for processing through software and hardware setting to make the most use of multicore performance.

Compared with user-level IDPS, mKIPS operating in the kernel can detect threats immediately after receiving packets. It needs not the overhead of copying packets to the user buffer for inspection. Furthermore, it does not need to wait for the scheduling to execute user-level IDPS and switch the protection domain back and forth between kernel mode and user mode for processing. These significantly reduce the impact on system performance. Experimental results show that mKIPS incurs less overhead on system performance and effectively ensures system safety with a high detection rate.

References

1. Liao, H.J., Lin, C.H.R., Lin, Y.C., Tung, K.Y.: Intrusion detection system: a comprehensive review. J. Netw. Comput. Appl.Netw. Comput. Appl. **36**(1), 16–24 (2013)
2. Patel, A., Taghavi, M., Bakhtiyari, K., Júnior, J.C.: An intrusion detection and prevention system in cloud computing: a systematic review. J. Netw. Comput. Appl.Netw. Comput. Appl. **36**(1), 25–41 (2013)
3. Modi, C., Patel, D., Patel, H., Borisaniya, B., Patel, A., Rajarajan, M.: A survey of intrusion detection techniques in Cloud. J. Netw. Comput. Appl.Netw. Comput. Appl. **36**(1), 42–57 (2013)
4. Kumar, U., Gohil, B.N.: A survey on intrusion detection systems for cloud computing environment. Int. J. Comput. Appl.Comput. Appl. **109**(1), 6–15 (2015)
5. Araújo, J.D., Abdelouahab, Z.: Virtualization in intrusion detection systems: a study on different approaches for cloud computing environments. Int. J. Comput. Sci. Network Secur. **13**(11), 135–142 (2013)
6. Snort. https://www.snort.org. Accessed 30 Apr 2023
7. Suricata. https://suricata-ids.org. Accessed 30 Apr 2023
8. Chiang, M.L., Wang, J.K., Feng, L.C., Chen, Y.S., Wang, Y.C., Kao, W.Y.: Design and implementation of a lightweight kernel level network intrusion prevention system for virtualized environment. In: 13th International Conference on Information Security Practice and Experience, 13–15 Dec, 2017, Melbourne, Australia, December 2017
9. Chin, T., Xiong, K., Rahouti, M.: Kernel-space intrusion detection using software-defined networking. Secur. Saf. **5**(15), 155–168 (2018)
10. Cheng, T.H., Lin, Y.D., Lai, Y.C., Lin, P.C.: Evasion techniques: sneaking through your intrusion detection/prevention systems. IEEE Commun. Surv. Tutorials **14**(4), 1011–1020 (2012)
11. The Linux Kernel Archives. http://www.kernel.org. Accessed 30 Apr 2023
12. Linux Network Scaling: Receiving Packets. https://garycplin.blogspot.com/2017/06/linux-network-scaling-receives-packets.html. Accessed 30 Apr 2023
13. Scaling in the Linux Networking Stack. https://www.kernel.org/doc/Documentation/networking/scaling.txt. Accessed 30 Apr 2023
14. Performance Tuning Guide. https://access.redhat.com/documentation/en-us/red_hat_enterprise_linux/6/html/performance_tuning_guide/index. Accessed 30 Apr 2023
15. libpcap. https://www.tcpdump.org. Accessed 30 Apr 2023
16. Cisco Talos Intelligence Group. https://www.talosintelligence.com. Accessed 30 Apr 2023
17. Gaddam, R.T., Nandhini, M.: Analysis of various intrusion detection systems with a model for improving snort performance. Indian J. Sci. Technol. **10**(20). https://doi.org/10.17485/ijst/2017/v10i20/108940, May 2017
18. Yuan, W., Tan, J., Le, P.D.: Distributed snort network intrusion detection system with load balancing approach. In: Proceedings of the International Conference on Security and Management (SAM), Athens (2013)
19. Shah, S.A.R., Issac, B.: Performance comparison of intrusion detection systems and application of machine learning to Snort system. Futur. Gener. Comput. Syst.. Gener. Comput. Syst. **80**, 157–170 (2018)

20. Shah, S., Bendale, S.P.: An intuitive study: intrusion detection systems and anomalies, how ai can be used as a tool to enable the majority. In: 5G era," 2019 5th International Conference on Computing, Communication, Control and Automation (ICCUBEA), Pune, India, Sept. 19–21, 2019, pp. 1–8. https://doi.org/10.1109/ICCUBEA47591.2019.9128786
21. Netfilter. https://www.netfilter.org. Accessed 30 Apr 2023
22. Coit, C.J., Staniford, S., McAlemey, J.: Towards faster string matching for intrusion detection or exceeding the speed of Snort. In: Proceedings of DARPA Information Survivability Conference & Exposition II, vol. 1, 2001, pp. 367–373 (2001)
23. Boyer, R.S., Moore, J.S.: A fast string searching algorithm. Commun. ACM. ACM **20**(10), 762–772 (1977)
24. Aho, A.V., Corasick, M.J.: Efficient string matching: An aid to bibliographic search. Commun. ACM. ACM **18**(6), 333–340 (1975)
25. ApacheBench. https://d.apache.org/docs/2.4/programs/ab.html. Accessed 30 Apr 2023
26. Apache Server. https://d.apache.org. Accessed 30 Apr 2023

SIAR: An Effective Model for Predicting Game Propagation

Tianyi Wang, Guodong Ye, Xin Liu(✉), Rui Zhou(✉), Jinke Li, and Tianzhi Wang

School of Information Science and Engineering, Lanzhou University, Lanzhou, China
xliu2019@lzu.edu.cn, zr@lzu.edu.cn
http://dslab.lzu.edu.cn/

Abstract. The COVID-19 pandemic has revitalized focus on predictive models, but scant research has been devoted to modeling game transmission, and current models are inadequate in this regard. To predict the spread of games within the population, this paper proposes the "addiction individuals", a new group based on the three groups of the SIR model. We applied the SIAR model, designed based on differential equations, to predict game transmission within this population. The SIAR model was validated on an existing dataset and compared with the traditional SIR model, demonstrating its greater accuracy.

Keywords: game propagation · addicted individuals · differential equations · BFGS algorithms

1 Introduction

Recently, there have been significant improvements in the quality of life of individuals through the use of predictive models [2,7]. Notably, models developed using machine learning and artificial intelligence techniques have become increasingly popular. Such models can learn and reveal patterns hidden in large datasets to anticipate future trends and behaviors. These include intelligent voice assistants, smart home controls, and personalized recommendation systems, all of which are built on predictive modeling technology. They aim to enhance people's lives by making them more convenient, efficient, and comfortable. Predictive models are increasingly expanding into diverse fields, such as healthcare, finance, and education, enhancing people's services and providing more accurate decision support. Therefore, predictive models are playing an increasingly important role in enhancing the quality of life for individuals. In particular, the application of predictive models in controlling infectious diseases has been under the spotlight during the COVID-19 pandemic. For instance, Cooper et al. made significant contributions to the COVID-19 management using their predictive models for control and prevention [6].

Although some models may perform well under specific circumstances, they tend to be ineffective in predicting the spread of games due to the characteristics of game propagation they overlook. Avid gamers commonly experience game

J. C. Hung et al. (Eds.): FC 2023, LNEE 1134, pp. 289–299, 2024.
https://doi.org/10.1007/978-981-99-9342-0_31

addiction, leading to their affinity towards specific games even after prolonged periods. As a result, the number of players likely remains stable once the game has been in the market for a while, contrasting the spread of infectious diseases. Game development is resource-intensive; both human and material resources are required to create successful games. Additionally, the number of active players is the key revenue driver for game companies. Thus, it is vital for game companies to predict the spread of games among the population. Regrettably, most existing models do not account for or clarify the game addiction phenomenon. The current challenge is to design models that incorporate this phenomenon of game addiction.

The problem solved by predictive models actually involves time series [1]. Various techniques have been proposed, one of which is based on deep learning techniques such as LSTM (Long Short Time Memory) [3] and RNN (Recurrent Neural Network) [20], which has shown good promise in fitting time series data. However, for smaller datasets (only a few hundred points), performance may degrade significantly due to over-fitting [21]. In addition, these models have poor explanatory power, especially for fluctuations in predicted outcomes. Another technique is based on mathematical models, such as the SIR (Susceptible, Infectious, or Recovered) model, which is considered a superior method for predicting the propagation of contagious phenomena like COVID-19 [6] and games. In real game scenarios, there are usually players who are very enthusiastic with a particular game and keen to promote their favourite game to those around them. However, these specific features are usually not captured by SIR models, which are often crucial when predicting the spread of games.

To tackle these challenges mentioned above, this paper proposes a new population classification based on the SIR model to elucidate the addiction phenomenon and introduces its own SIAR model, a prediction system that employs a system of differential equations. This model has demonstrated high accuracy in forecasting game spread among the population. Briefly, the contributions of this paper can be summarised as follows:

1. Through an analysis of the communication characteristics of games, we have identified a new group of individuals, referred to as the "addicted ones." This group is essential in explaining the observed phenomenon of a game's player count stabilizing after the fervor for the game has subsided.
2. We propose a novel extension of the SIR model that incorporates a new group of 'Addicted individuals' into our system of differential equations, leading to an improved model's ability to predict the spread of games in the population and provide additional explanatory power. The proposed model has coined the SIAR model.
3. Our proposed SIAR model has been tested on existing datasets and compared against current models, which has demonstrated its superior performance in predicting the spread of games among the population.

2 Related Works

Kermack and McKendrick were the first to propose a mathematical model for describing the spread and control of infectious diseases in a population [9–11]. The model categorizes the population into three groups: susceptible, infected, and recovered, and describes the transmission of infectious diseases in the population. Mathematical tools like calculus and difference equations [8] are used to derive the fundamental equations and basic laws of infectious disease spread in the population.

The COVID-19 pandemic prompted the widespread use of disease transmission models, including SIR models, to predict the spread and control of the virus. Cooper et al. applied SIR models to predict the spread of COVID-19 [6], while Mwalili et al. used SEIR models to predict the spread of COVID-19 propagation [14]. B Shayak et al. considered the lag between asymptomatic infected individuals and COVID-19 symptoms, added them to the SIR model, and derived predictions and analytical results for the spread and prevalence of the virus by numerical simulation and fitting to actual data. The results highlight the importance of planning and allocating resources for epidemic management [17]. Benjamin F. Maier et al. used the SIR-X model based on SIR models to explain the phenomenon of sub-exponential growth in mainland China during the early stages of the COVID-19 epidemic and to show that this growth was a direct consequence of epidemic control policies [13]. AK Singh et al. introduce a algorithm that uses the differential evolution algorithm in combination with Adam-Bashforth-Moulton method to learn the parameters in a system of variable-order fractional SIR model, which can predict the confirm COVID-19 cases in India considering the effects of nationwide lockdown and the possible estimate of the number of infliction inactive cases after the removal of lockdown on June 1, 2020 [18]. And Chen et al. utilized the α-path-based approach to determine the uncertainty distributions and expected values of the solutions. They also applied the method of moments to estimate the parameters and developed a numerical algorithm to solve the model. The proposed model was then used to describe the development trend of COVID-19 in Hubei province by analyzing infected and recovered data [4]. Ram et al. developed a customized age-structured SIR model by considering the social contact and distancing measures in Washington, USA [15].

Previous studies indicate that a variety of SIR-based models have significantly aided in forecasting the transmission of communicable illnesses. Nonetheless, there has been little research done so far on anticipating the propagation of games in societies, and current models are inadequate in projecting the spread of games.

3 The SIR Model

The SIR model is frequently utilized to forecast the spread of infectious diseases in epidemics and has shown success in modeling the spread of COVID-19 [6]. The model categorizes individuals into three groups:

a) Susceptible individuals **(S)**: Defined as individuals lacking immunity who do not currently have the disease but are at risk of contracting it when exposed to infected individuals.
b) Infected individuals **(I)**: Individuals who have contracted the disease and can pass it on to susceptible individuals.
c) Removed individuals **(R)**: Individuals who have recovered from the disease and, consequently, have developed immunity.

Then SIR model is controlled by the following ODE systems:

$$\left.\begin{aligned} \frac{\mathrm{d}S(t)}{\mathrm{d}t} &= -\frac{\beta IS}{N}, \\ \frac{\mathrm{d}I(t)}{\mathrm{d}t} &= \frac{\beta IS}{N} - \gamma I, \\ \frac{\mathrm{d}R(t)}{\mathrm{d}t} &= \gamma I, \end{aligned}\right\} \quad \text{SIR Model's ODE System} \tag{1}$$

where S(t), I(t), and R(t) represent the number of susceptible, infected, and removed individuals at time t, respectively. Here, β represents the rate of infection of susceptible persons by infected individuals per unit of time, while γ represents the rate of recovery of each infected individual per unit of time.

What's more, Fig. 1 provides a graph depicting the conversion of the three populations in the SIR model.

Fig. 1. Conversion of the Three groups in SIR model

By utilizing the SIR model for game prediction, we will redefine the meanings of the three populations:

a) Susceptible individuals **(S)**: People who have yet to be introduced to the game and who are likely to be recommended and subsequently become players.
b) Infected individuals **(I)**: Current players of the game who have a chance of losing interest, in addition to recommending the game to non-players.
c) Removed individuals **(R)**: Past players of the game who have lost interest and will no longer play.

4 Methods

4.1 SIAR Model

Although the SIR model is effective in modeling the spread of infectious diseases, it does not perform well when predicting the diffusion of non-medical phenomena, such as the popularization of a new game.

Therefore, in our modified model, we introduce a new category of individuals:

- Addicted individuals (**A**): People excessively hooked on the game. They have a low probability of losing interest in the game and a high probability of recommending it to others.

The modified model is called the SIAR model, and its differential equations are as follows:

$$\left.\begin{aligned} \frac{\mathrm{d}S(t)}{\mathrm{d}t} &= -\frac{\beta IS + \epsilon AS}{N}, \\ \frac{\mathrm{d}I(t)}{\mathrm{d}t} &= \frac{\beta IS + \epsilon AS}{N} - \gamma I - \alpha I, \\ \frac{\mathrm{d}R(t)}{\mathrm{d}t} &= \gamma I + \theta A, \\ \frac{\mathrm{d}A(t)}{\mathrm{d}t} &= \alpha I - \theta A \end{aligned}\right\} \quad \text{SIAR Model's ODE System} \tag{2}$$

Here, β and γ have the same meanings as in the previous SIR model. The parameter ϵ represents the probability that each addicted individual will promote the game to others. The parameter α represents the probability of a regular player becoming addicted, while θ represents the probability of an addicted person losing interest in the game.

In order for the model to be better understood, the relationship between the four groups in the model is shown in Fig. 2.

Fig. 2. Conversion of the four groups in SIAR model

4.2 The Solution of SIAR Model

To achieve our aim of predicting the spread of the game across the population, we first defined a loss function to quantify the accuracy of our predictions:

$$Loss(X_{pred}, X_{real}) = \sum_{i=1}^{n} (x_{pred,i} - x_{real,i})^2 \tag{3}$$

where x_{pred} is a vector indicating the predicted number of individuals who will play the game per day, and x_{real} is a vector indicating the real number of individuals who will play the game per day.

To determine the optimal parameters of our model, we utilize the BFGS algorithm [12], a numerical method to minimize multivariate functions. The BFGS algorithm is a Newton-like method that estimates the Hessian matrix of the objective function so that the search direction of the method can be progressively updated. This method has higher accuracy and a faster convergence rate when compared to the gradient descent algorithm [16].

The central concept of the BFGS algorithm involves an iterative method that approximates the Hessian matrix in the following algorithmic form:

$$B_{k+1} = B_k + \Delta B_k, \quad k = 0, 1, 2, ... \tag{4}$$

To initiate the iterative process, we choose B_0 to be the identity matrix, I. The rate of BFGS convergence can be increased by appropriately selecting ΔB_k as:

$$\Delta B_k = \alpha \mathbf{u}\mathbf{u}^T + \beta \mathbf{v}\mathbf{v}^T \tag{5}$$

Incorporating Newton's condition results in the following expression:

$$\mathbf{y}_k = B_k \mathbf{s}_k + \left(\alpha \mathbf{u}^T \mathbf{s}_k\right) \mathbf{u} + \left(\beta \mathbf{v}^T \mathbf{s}_k\right) \mathbf{v} \tag{6}$$

Setting $\alpha \mathbf{u}^T \mathbf{s}_k = 1, \beta \mathbf{v}^T \mathbf{s}_k = -1, \mathbf{u} = \mathbf{y_k}, \mathbf{v} = B_k s_k$, yields:

$$\alpha = \frac{1}{\mathbf{y}_k^T \mathbf{s}_k}, \quad \beta = -\frac{1}{\mathbf{s}_k^T B_k \mathbf{s}_k} \tag{7}$$

Incorporating all the above-step results in the computation formula for ΔB_k:

$$\Delta B_k = \frac{\mathbf{y}_k \mathbf{y}_k^T}{\mathbf{y}_k^T \mathbf{s}_k} - \frac{B_k \mathbf{s}_k \mathbf{s}_k^T B_k}{\mathbf{s}_k^T B_k \mathbf{s}_k} \tag{8}$$

After obtaining ΔB_k, as an additional step, we can obtain B_{k+1} by applying the recursive formula:

$$B_{k+1} = B_k + \frac{\mathbf{y}_k \mathbf{y}_k^T}{\mathbf{y}_k^T \mathbf{s}_k} - \frac{B_k \mathbf{s}_k \mathbf{s}_k^T B_k}{\mathbf{s}_k^T B_k \mathbf{s}_k} \tag{9}$$

Here, $\mathbf{y}_k = B_k^{-1} \Delta x_k$ is the change in gradient, $\mathbf{s}_k$ represents the change in the optimization variables and B_k is the approximation of the Hessian of the gradient.

5 Experiment Results

To further evaluate the validity of the model, we performed a series of tests using the dataset to assess its accuracy and reliability. We compared the model's performance to other established models, as well as analyzed its ability to generalize to new data and make accurate predictions. Our testing methodology involved a rigorous and comprehensive approach, to ensure the model's soundness was thoroughly assessed. The results of these tests confirmed the robustness of the model, indicating its suitability for use in real-world applications.

5.1 Dataset

The dataset we used is provided by MCM [5], which contains the number of people playing the game "Woddle" from Jan 07 2022 to Dec 31, 2022, which is shown in Fig. 3.

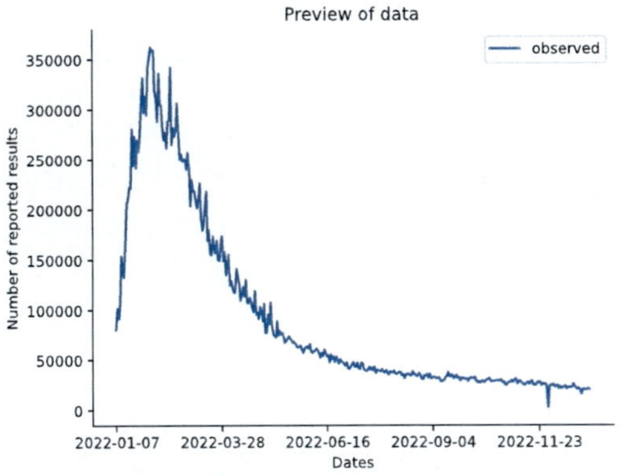

Fig. 3. The preview of dataset

In order to uphold the scientific validity of the experiment, two partitioning methods were implemented on the dataset.

- The first method involved dividing the 359-day dataset into two parts: the initial 299 d were assigned for training the model's parameters, while the last 60 d were reserved for testing the model's effectiveness. This method was utilized to assess the model's predictive performance.
- The second method employed all 359 d of the dataset for both training and testing purposes. This segmentation approach evaluated the model's ability to accurately fit the dataset.

5.2 Experiment Implementation

For the implementation of the model, we use the `optimize.minimize` function and `integrate.odeint` function of the scipy library [19]. The `optimize.minimize` function is used to optimize the parameters of the model and `integrate.odeint` is used to solve the differential equations.

Upon completion of the model training phase, the following formula was utilized to determine the model's level of error:

$$Error(Y_{pred}, Y_{real}) = 2\sum_{i=1}^{n} \frac{|y_{pred,i} - y_{real,i}|}{y_{pred,i} + y_{real,i}} \tag{10}$$

5.3 Analysis

Results in the Case of Spliting Dataset. After analyzing the outcomes of the SIR and SIAR models with the split dataset, as illustrated in Figs. 4 and 5 and Table 1, it is evident that the SIAR model offers significantly better predictions of the game's transmission phenomenon compared to the SIR model. Through the examination of both SIR and SIAR models, as depicted in Figs. 4 and 5 and Table 1, it is evident that the SIAR model provides a better representation and prediction of the game's transmission patterns compared to the SIR model.

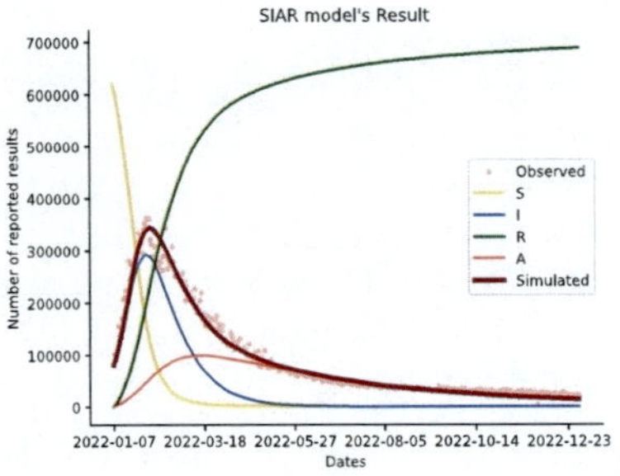

Fig. 4. Results of the SIAR model after splitting the training and test sets

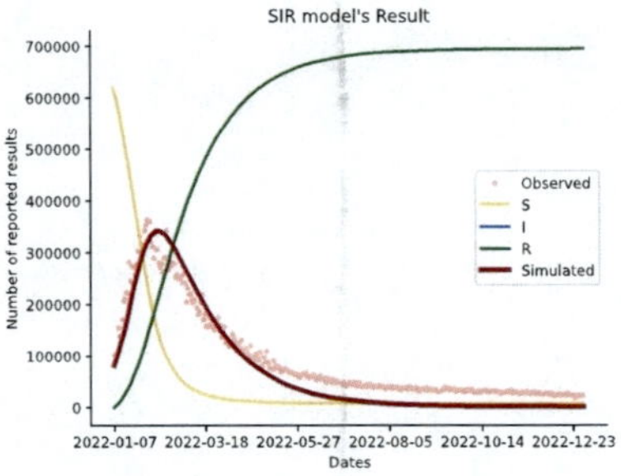

Fig. 5. Results of the SIR model after splitting the training and test sets

Table 1. Error table for SIR and SIAR model

	SIR	SIAR
Train Set Error	0.045	0.004
Test Set Error	0.48	0.099

Results in the Case of All Dataset. When we trained and tested the performance using the full dataset, the results are shown in Figs. 6 and 7 and Table 2. Analysis of the results shows that our SIAR model has a better ability to fit the propagation data of the game compared to the SIR model.

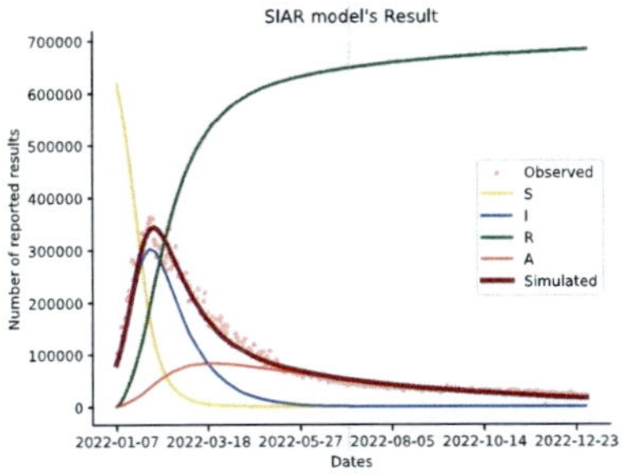

Fig. 6. Results of the SIAR model using all the dataset

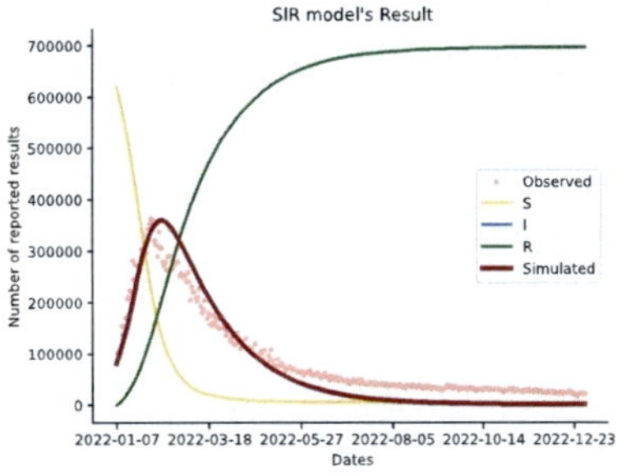

Fig. 7. Results of the SIR model using all the dataset

Table 2. Error table for SIR and SIAR model

	SIR	SIAR
All dataset Error	0.214	0.004

Under both scenarios, it became apparent that the SIR model exhibits a poor fit with the data during the latter half of the time period. We attribute this result to the fact that during this time frame, the majority of players demonstrated strong loyalty towards the game, making the likelihood of becoming disenchanted with the game extremely low. The SIR model is unable to accommodate this framework, as it does not account for loyal players, thereby impeding its ability to predict and fit the game's transmission in this condition. On the other hand, our SIAR model demonstrated outstanding success in accommodating prevalent loyal players, fitting the dataset consistently well.

6 Conclusion

This paper introduces a SIAR model that predicts the propagation of games among the population. Our model is constructed based on the SIR model while devoted to the gaming-specific attributes in the population. The implementation of our model can lead to more accurate and comprehendible predictions of gaming patterns among crowds. Additionally, we conducted multiple comprehensive tests to establish the dependability and validity of our model.

References

1. Adhikari, R., Agrawal, R.K.: An introductory study on time series modeling and forecasting. arXiv preprint arXiv:1302.6613 (2013)
2. Brynjolfsson, E., McAfee, A.: The second machine age: Work, progress, and prosperity in a time of brilliant technologies. WW Norton & Company (2014)
3. Cao, J., Li, Z., Li, J.: Financial time series forecasting model based on CEEMDAN and LSTM. Phys. A **519**, 127–139 (2019)
4. Chen, X., Li, J., Xiao, C., Yang, P.: Numerical solution and parameter estimation for uncertain sir model with application to COVID-19. Fuzzy Optim. Decis. Making **20**(2), 189–208 (2021)
5. COMAP: Predicting wordle results (2023). https://www.mathmodels.org/Problems/2023/MCM-C/index.html
6. Cooper, I., Mondal, A., Antonopoulos, C.G.: A sir model assumption for the spread of COVID-19 in different communities. Chaos, Solitons Fractals **139**, 110057 (2020)
7. Domingos, P.: The Master Algorithm: How the Quest for the Ultimate Learning Machine will Remake Our World. Basic Books (2015)
8. Hethcote, H.W.: The mathematics of infectious diseases. SIAM Rev. **42**(4), 599–653 (2000)
9. Kermack, W.O., McKendrick, A.G.: Contributions to the mathematical theory of epidemics. ii.-the problem of endemicity. Proc. Royal Soc. London A Math. Phys. Charact. **138**(834), 55–83 (1932)
10. Kermack, W.O., McKendrick, A.G.: Contributions to the mathematical theory of epidemics. iii.-further studies of the problem of endemicity. Proc. Royal Soc. London A Math. Phys. Charact. **141**(843), 94–122 (1933)
11. Kermack, W.O., McKendrick, A.G.: Contributions to the mathematical theory of epidemics-i. 1927. Bull. Math. Biol. **53**(1–2), 33–55 (1991)
12. Liu, D.C., Nocedal, J.: On the limited memory BFGS method for large scale optimization. Math. Program. **45**(1–3), 503–528 (1989)
13. Maier, B.F., Brockmann, D.: Effective containment explains subexponential growth in recent confirmed COVID-19 cases in China. Science **368**(6492), 742–746 (2020)
14. Mwalili, S., Kimathi, M., Ojiambo, V., Gathungu, D., Mbogo, R.: SEIR model for COVID-19 dynamics incorporating the environment and social distancing. BMC. Res. Notes **13**(1), 1–5 (2020)
15. Ram, V., Schaposnik, L.P.: A modified age-structured sir model for COVID-19 type viruses. Sci. Rep. **11**(1), 15194 (2021)
16. Ruder, S.: An overview of gradient descent optimization algorithms. arXiv preprint arXiv:1609.04747 (2016)

17. Shayak, B., Sharma, M., Rand, R.H., Singh, A.K., Misra, A.: Transmission dynamics of COVID-19 and impact on public health policy. medRxiv 2020.03.29.20047035 (2020). https://doi.org/10.1101/2020.03.29.20047035
18. Singh, A.K., Mehra, M., Gulyani, S.: A modified variable-order fractional sir model to predict the spread of COVID-19 in India. Math. Methods Appl. Sci. **46**(7), 8208–8222 (2023)
19. Virtanen, P., et al.: SciPy 1.0 contributors: SciPy 1.0: fundamental algorithms for scientific computing in Python. Nat. Methods **17**, 261–272 (2020). https://doi.org/10.1038/s41592-019-0686-2
20. Zaremba, W., Sutskever, I., Vinyals, O.: Recurrent neural network regularization. arXiv preprint arXiv:1409.2329 (2014)
21. Zhang, C., Bengio, S., Hardt, M., Recht, B., Vinyals, O.: Understanding deep learning (still) requires rethinking generalization. Commun. ACM **64**(3), 107–115 (2021)

Symbolic Regression Using Genetic Programming with Chaotic Method-Based Probability Mappings

Pu Cao[1], Yan Pei[1](✉), and Jianqiang Li[2]

[1] Graduate School of Computer Science and Engineering, University of Aizu, Aizuwakamatsu, Fukushima 965-8580, Japan
{m5252101,peiyan}@u-aizu.ac.jp

[2] Faculty of Information Technology, Beijing University of Technology, Beijing 100124, China
lijianqiang@bjut.edu.cn

Abstract. In this study, we propose a novel pre-learning approach for genetic programming (GP) that aims to investigate the effect of the probability of being selected for each operator. Furthermore, we present a technique that combines chaos theory and searches for a relatively good possibility mapping for each operator using one-dimensional chaotic mapping. We conducted several sets of comparative experiments on real-world data to test the viability of the proposal. These experiments included comparisons with conventional GP, examination of the impact of various chaotic mappings on the proposed algorithm, and implementation of different optimization strategies to find the relative optimal probability mapping. The experimental results demonstrate that the proposed method can achieve better results than conventional GP in the tested dataset, without considering the total quantitative calculation amount. Through statistical tests, it has been proven that the proposed method is significantly different from the conventional method. However, the discussion regarding the circumstances under which the proposed method can obtain better results when the total calculation amount is limited is not yet fully explored due to the small-scale nature of the experiments. Our future studies will focus on improving and fully discussing this idea.

Keywords: Genetic programming · Chaos theory · Symbolic regression · Evolutionary computation · Pre-learning

1 Introduction

Symbolic regression (SR) is a relatively specialized regression problem that requires consideration not only of the results but also of the structure. The goal of symbolic regression is to generate a set of formulas that can be used to fit the target data. The result is visible and understandable, such as generating an equation equivalent to $\sqrt{x_0^5 + x_1^2 - x_2^3}$. Genetic programming (GP) is a method that can generate explicit solutions without prior assumptions and only requires

J. C. Hung et al. (Eds.): FC 2023, LNEE 1134, pp. 300–312, 2024.
https://doi.org/10.1007/978-981-99-9342-0_32

knowledge of the basic components of the problem. Therefore, it is well-suited for solving symbolic regression problems [2].

In conventional tree-based GP, there are two categories of nodes: operator nodes(e.g. addition, subtraction, multiplication, division, etc.) and termination nodes, each with several types of operators and terminal nodes, respectively. During initialization and mutation, a random selection is made to decide what type of node the new node should be. At this point, the probability of each operator node being selected is equally likely.

However, there has not been extensive discussion on the selection of GP operators. When exploring non-prior knowledge using Symbolic Regression (SR), we do not know what basic components should be included in the target system. Therefore, we should provide enough choices for the initial operator set. But if the operator set is too large, the solution space will also be too large, which affects the algorithm's convergence performance. On the other hand, if the operator set is too small, the fitting of the target problem will not be accurate [8]. Therefore, this paper further discusses the selection of GP operators.

The motivation for this study is that certain operators should not be employed for specific problems as they can make the solution more complex and difficult to converge to. Therefore, it would be effective to propose a scheme that provides different probabilities of the operator being selected for different problems and data.

In response to this view, we propose a pre-learning method based on chaos theory. The aim is to find one or several sets of probability mappings that make it easier to obtain better answers from the searching space of different operator node possibilities. To provide more realistic and applicable results, this paper uses real-world data for experimentation and research.

Furthermore, we explore the use of chaotic systems to optimize our proposed algorithm. We conduct three sets of comparative experiments, including comparing the performance of our proposed algorithm with the original algorithm, exploring how different chaotic systems affect the algorithm's performance, and analyzing the algorithm's learning performance under controlled computational conditions. These experiments provide insights into the optimal conditions for the proposed algorithm and its sensitivity to the chaotic system used.

After the introduction, we present related studies in Sect. 2. In Sect. 3, we provide a detailed description of our proposed algorithm and two different implementation methods. Section 4 describes three sets of experiments conducted from different perspectives to analyze the proposed algorithm, and we discuss the experimental results. Finally, we summarize the contributions of this study and highlight current issues and future directions.

2 Related Works

2.1 Genetic Programming

Genetic Programming (GP) [2] is an evolutionary algorithm that utilizes biological phenomena such as heredity, mutation, selection, and crossover to generate solutions to problems. Unlike other optimization algorithms, GP resembles more of a machine learning method [1] as it uses hierarchical data structures to generate solutions. In recent years, some studies have focused on optimizing algorithms by leveraging external information.

For instance, Ying Bi, Bing Xue, and Mengjie Zhang decomposed the genetic programming problem into multiple sub-problems, which were solved separately, and their final results were combined [3]. The advantage of this approach is that it can reduce the search space's size and speed up the algorithm's running time, leading to improved performance. While various means exist to improve performance outside the algorithm, we focus on optimizing the algorithm itself by exploring the impact of operators on its overall performance.

Hengrui Xing, Ansaf Salleb-Aouissi, and Nakul Verma explored the use of convolutional neural networks to convert data into visual representations and then discussed the probability of operators being selected too [8]. However, they did not utilize real data for experiments. Moreover, we believe that converting data into images and pre-training neural networks for images is not an efficient approach. Thus, we propose a different implementation idea.

Overall, our research focuses on optimizing the genetic programming algorithm's performance by analyzing the impact of operators on its overall performance. We believe that our approach can lead to significant improvements in the algorithm's performance, thereby enhancing its effectiveness and efficiency in solving problems.

2.2 Chaos Theory for Optimization Algorithm

The chaotic system refers to a class of dynamical systems that demonstrate intricate, unpredictable behavior in both time and space. Such systems exhibit behavioral characteristics, including high sensitivity to initial conditions, deterministic chaos, and adaptability, which render them fascinating objects of study in various fields of science and engineering [9].

Numerous studies have investigated the utilization of the ergodicity and non-repetition of chaotic systems to enhance the performance of optimization algorithms. For instance, recent research has proposed an approach that enhances the performance of JADE by utilizing chaotic systems to generate more diverse initial populations [4]. In addition, CE is a new algorithm based on differential evolution that leverages the ergodic motion in the search space to improve performance by exploiting the ergodicity of chaos [5]. These studies suggest that the characteristics of chaos can enhance search efficiency to a certain extent for random-based optimization algorithms. By introducing chaos into the optimization process, the search space is explored more efficiently, leading to better solutions. Based on past research, we have reason to believe that the utilization of the characteristics of chaos can enhance the performance of optimization algorithms, especially for random-based optimization algorithms.

3 Different Probability Mapping for Operators are Used in Genetic Programming

As mentioned previously, a novel approach called "Different Probability Maps for Operators in Genetic Programming" (DPMOGP) has been proposed to investigate the impact of different operator selection probabilities on GP performance. The key idea behind DPMOGP is to use an appropriate optimization strategy to perform pre-learning and generate one or more probabilistic mappings for different operators, which are then saved to initiate the formal GP process. During the initialization and mutation phases, when a new operator node is generated, one of the stored probability mappings is selected using a weighted random search via the roulette algorithm to determine which operator the new node should be assigned to. The algorithmic implementation of DPMOGP is shown in Algorithm 1.

It can be seen that this will be an optimization problem to find some probability mappings that can relatively easier to generate the higher fitness individual from the searching space. Thus, DPMOGP is a versatile concept that can be implemented in various ways. At present, we have implemented two versions of DPMOGP: random search-based DPMOGP (RS-DPMOGP) and genetic algorithm-based DPMOGP (GA-DPMOGP).

Algorithm 1. DPMOGP

PS: population; **PM**: probability mapping; **CP**: cumulative probability

```
/* pre-learning for search the relatively optimal PM */
initialize PM = [], CP = []
PM = preLearning() {discuss in detail later}
CP = rouletteAlgorithm(PM)
/* start conventional GP */
/* Once the new node of tree is generated */
num = random.double() { random number between 0-1}
count = 0
for count = 0 to num.size() do
    if num <CP[i] then
        break
    end if
end for
newNode = PM[count]
```

3.1 Random Search-Based DPMOGP

In Random Search-based As mentioned previously, a novel approachrs used in Genetic Programming (RS-DPMOGP), the primary objective is to generate a set of probability mappings using a random number generator, which is explained in detail later. Each probability mapping is used to create a fixed number of solutions (trees) of a specified size, and the average fitness of all solutions is calculated as the final fitness. Finally, several relatively good mappings are selected among all the generated probability mappings, and it is saved as the selection basis when a new operator node is generated in the conventional GP. The algorithm for RS-DPMOGP is shown in Algorithm 2.

Algorithm 2. RS-DPMOGP

PM: probability mapping; **AN**: amount of random mapping; **AT**: amount of trees for each random mapping; **AP**: amount of PM

```
/* random searching for relatively optimal PM */
randomMappings =[]
for int i = 0 to AN do
    randomMappings.add(randomNumGenerator())
end for
sortingByFitness(randomMappings){each random mapping need to generate AT
tress then take the average fitness}
PM = randomMappings[0:AP]{pick top AP relatively better probability mapping
as PM}
/* The rest follow Algorithm 1*/
```

3.2 Genetic Algorithm-Based DPMOGP

Given that the solution space of the optimization problem presented above involves permutations and combinations of N arbitrary numbers ranging from 0 to 1, it is evident that it belongs to the class of NP problems. To solve this type of problem, metaheuristic algorithms are commonly employed [6], with the genetic algorithm (GA) being a popular choice due to its simplicity and efficiency.

Algorithm 3. GA-DPMOGP

PM: probability mapping;
AP: amount of PM

```
/*genetic algorithm for relatively optimal PM */
randomMappings =[]
for int i = 0 to AP do
    PM.add(genetic algorithm())
end for
/* The rest follow Algorithm 1*/
```

In genetic algorithm-based Different Probability Mapping for Operators is used in Genetic Programming (GA-DPMOGP) Similar to the RS-DPMOGP approach, the genetic algorithm utilizes the probability mapping generated to produce one or more trees. The trees are then utilized to assess the fitting performance of the generated formula using the least square method. The iteration process is stopped after a specified generation, and the best individuals are selected. Additionally, chaotic mapping is used as a random number generator when generating the initial population and mutations.

4 Experiments and Discussion

4.1 Dataset

Real-world data is utilized as a sample problem in this study. Specifically, the Yacht Hydrodynamics Data Set, which is a relevant dataset in fluid mechanics for exploring various ship hulls, is used for conducting experiments. The dataset is well-suited for applying symbolic regression to discover physical formulas. Refer to Table 1 for further details.

Table 1. Yacht Hydrodynamics Data Set

Date Donated	***Instances:***	***features:***	***Area:***
2013-01-03	308	6	Physical

4.2 Chaotic Mapping

In the experiments, one-dimensional chaotic maps are employed as the random number generator. If not explicitly stated, the logistic mapping is assumed to be the default choice for the random number generator.

A logistic mapping [7] is a classical and straightforward chaos model whose modeling expression is presented in (1). The distribution of the model is largely determined by the variable K. As shown in Fig. 1, when K equals four, the system is in a state of complete chaos, and the final long-term behavior is uniformly distributed in the interval [0,1]. In this study, all chaotic maps utilized are based on the scenario when the system is in a state of complete chaos.

$$X_{n+1} = X_n \cdot K(1 - X_n),\ K \in [0, 4], X \in [0, 1]. \tag{1}$$

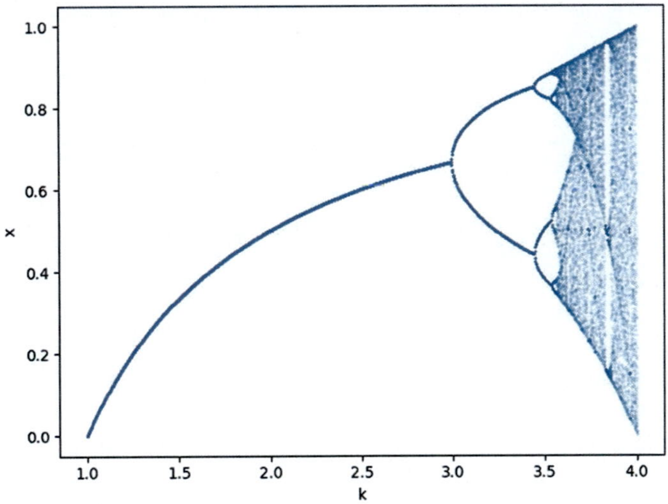

Fig. 1. Bifurcation diagram of the logistic mapping. When K is equal to 4, we consider the system to be in a chaotic state.

4.3 Results and Discussion

We conducted sets of comparative experiments to investigate the performance of DPMOGP in various aspects and the impact of using chaotic mapping for each experiment. Specifically, we performed 50 runs for each experimental group and then computed the R-squared as the result. Given that the division may be zero during calculations, we used protected division to replace the division with $\frac{y}{\sqrt{1+x^2}}$.

Initially, we compared the performance of DPMOGP and conventional GP with identical parameter settings. The parameter settings used for GP are presented in Table 2.

Table 2. Parameter for GP

Parameter	***Description***
Population size	100
Iterations	100
Initialize tree size	3
Mutation rate	0.1
Operator set	+,−,*, / protected, tan,sin,cos, reciprocal, remainder
Terminal set	x_0,x_1,x_2,x_3,x_4,x_5
Selection	Tournament selection of size 4

(1) Initially, we compared the performance of conventional GP and RS-DPMOGP, this group of experiments also evaluated the algorithm's performance in scenarios involving different numbers of chaotic mapping iterations (i.e., the total number of random searches). The parameter settings for this experiment are presented in Table 3.

Table 3. Parameter for RS-DPMOGP

Parameter	***RS-DPMOGP A***	***RS-DPMOGP B***
Number of chaos mapping	10000	15000
Trees generated per mapping	1	1
Number of saved mapping	5	5
Depth of tree	7	7

From the results depicted in Fig. 2, we observe that RS-DPMOGP B exhibits the best performance, while conventional GP yields relatively poor results. These results suggest that, at least for the dataset used in the experiment, DPMOGP can enhance the algorithm's performance. By comparing RS-DPMOGP A and RS-DPMOGP B, we note that, in general, increasing the amount of computation in DPMOGP pre-learning leads to better performance before reaching the

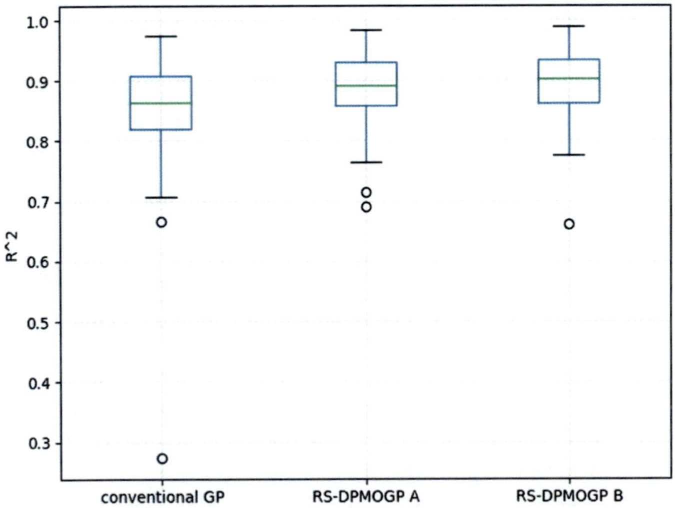

Fig. 2. The R^2 of results for conventional GP and DPMOGP. R^2 ranges from 0 to 1, and a value closer to 1 indicates a better fit of the model to the data. Typically, an R^2 value greater than 0.7 is considered a good fit.

threshold. Prior to reaching the global optimum, enhancing the optimization algorithm's amount of calculation in pre-learning can enhance the final GP's performance.

In this experiment, we adopt Wilcoxon signed-rank test to compare the performance of different algorithms. The use of the Wilcoxon signed-rank test in algorithm comparison has been well established in the literature [10]. Specifically, we use this non-parametric test to assess whether there is a statistically significant difference in the median performance of the two algorithms. By comparing the p-value of the Wilcoxon signed-rank test with a predetermined significance level (usually set at 0.05), we can determine whether the observed difference in performance is statistically significant or not. In this experiment, a p-value less than 0.05 indicates a statistically significant difference between the two algorithms with a high probability. The detailed results are presented in Table 4. Through the results of the statistical test, we have observed a significant difference between the conventional GP and the proposed method, which greatly strengthens the persuasiveness of our research.

Table 4. Results of Wilcoxon signed-rank test for experiment (1). The results of the conventional GP and those obtained with different parameters of the proposed algorithm exhibit significant differences.

Comparison	**GP**	**RS-DPMOGP A**
GP	–	0.023592951063958378
RS-DPMOGP B	0.01688572946162479	0.4961523833965189

(2) After confirming the effectiveness of DPMOGP, we conducted comparative experiments to investigate the influence of different chaotic mappings on the algorithm. One-dimensional chaotic mappings such as tent mapping and sine mapping were added to the experiment, and the distribution of mappings is shown in Fig. 3.

Upon observing the experimental results, we found that when logistic mapping is used as a random number generator, the probability of retention results falling within the range of 0–0.1 and 0.9–1 is higher than when random numbers are used directly. This phenomenon has a significant impact on the probability that the corresponding operator is selected, as we use these 0–1 numbers as the weight of each operator and then use the roulette algorithm to select. In other words, the larger the weight of the operator, the easier it is to be selected. If the weight is a tiny value, the probability of its corresponding operator being selected will obey exponential decay.

Therefore, we hypothesize that the special distribution of logistic mapping may lead to better results. To test this hypothesis, we constructed a set of pseudo-random models with similar distributions to logistic mapping and conducted experiments. The results are shown in Fig. 4. We can observe that logistic mapping and sine mapping follow the distribution as shown in Fig. 3 with more points located at the earlier and end stages, while the distribution of pseudo-random looks similar to logistic mapping and sine mapping. For tent mapping, the points are evenly distributed in the coordinate system.

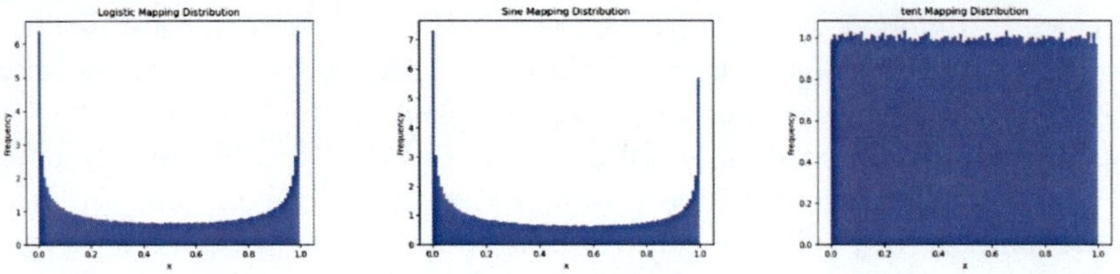

(a) logistic mapping (b) Sine mapping (c) Tent mapping

Fig. 3. Distribution histogram for chaotic mappings. The horizontal axis represents the generated results, and the vertical axis represents the frequency of occurrence of the generated numbers within that range.

The results of the comparative experiments are presented in Fig. 5. It is observed that logistic mapping, sine mapping, and the pseudo-random model outperformed random and tent mapping. As mentioned earlier, logistic mapping, sine mapping, and the pseudo-random model have the characteristic of generating more tiny and huge numbers, whereas the distribution of tent mapping closely resembles a random distribution. This suggests that our hypothesis that the distribution of the random number generator influences the performance of the algorithm is validated.

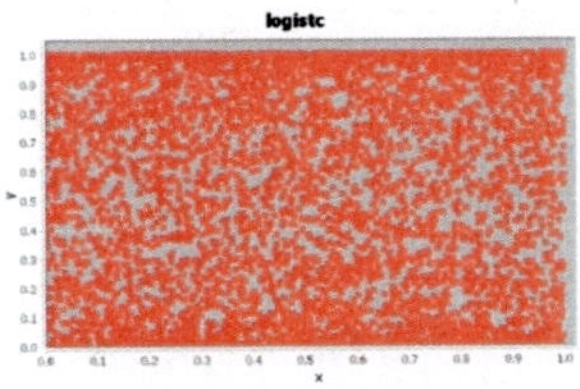

(a) logistic mapping

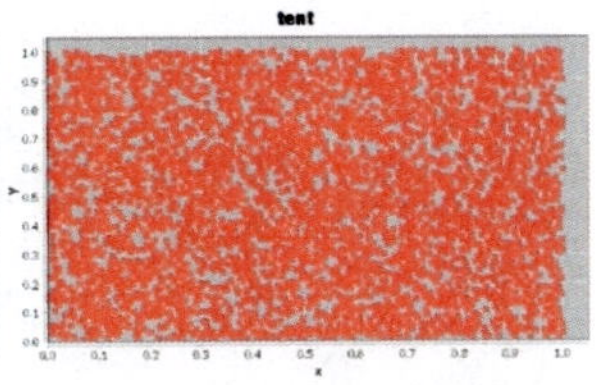

(b) tent mapping

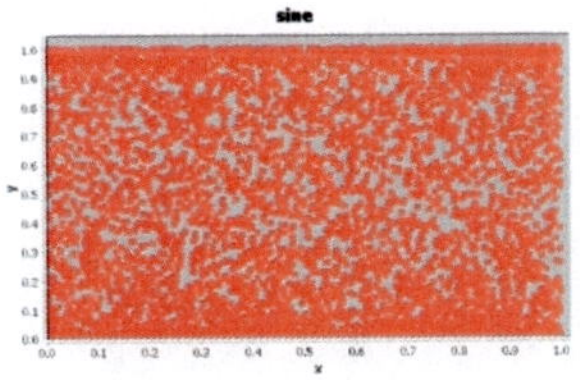

(c) sine mapping

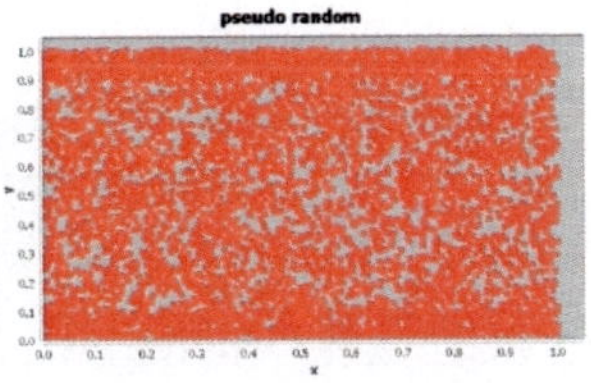

(d) pseudo random

Fig. 4. Distribution Scatter Plot for each model. Through the distribution of the lattice, we can see that (a), (b), and (d) are similar: more points are located near the vertical axis of 0 and 1.

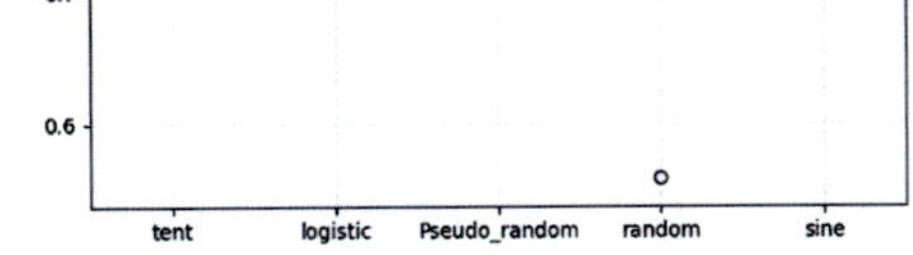

Fig. 5. The R^2 result for comparing the different random number generator.

Table 5. Parameter for experiment 3. Due to different algorithms, the parameters are not the same, but they have the same computational cost.

Parameter	*GP*	*RS-DPMOGP*	*GA-DPMOGP*
Population size	100	100	100
Iteration	100	90	90
Number of chaos mapping (RS)	–	1000	–
Trees generated per map (RS)	–	1	–
Number of saved mapping (RS)	–	20	–
Depth of tree (RS)	–	11	–
Population size (GA)	–	–	10
Depth of tree (GA)	–	–	11
Mutation (GA)	–	–	0.1
Iteration (GA)	–	–	10
Number of saved mapping (GA)	–	–	20
Total computation	10000	10000	10000

Additionally, analyzing the data from Fig. 5, it is observed that the pseudo-random model generates more outlier data than other models. We hypothesize that this could be due to the ergodicity of chaotic mappings, which we plan to investigate in our future study.

(3) In the final stage of our research, we examined the differences between conventional GP, RS-DPMOGP, and GA-DPMOGP when the amount of computation is limited. To ensure that we control the amount of computation for each algorithm, we redesigned the parameters as shown in Table 5. The total computation was set to 10,000 fitness evaluations, with 9,000 divided by formal GP for both RS-DPMOGP and GA-DPMOGP and only 1,000 reserved for pre-learning.

The experimental results are presented in Fig. 6. When the computation amount was limited to 10,000 times, our proposed method did not achieve significant advantages over conventional GA. This was due to the pre-learning process not being fully executed. However, we believe that when larger amounts of data are used in the experiments and sufficient runs are given for pre-learning, DPMOGP will achieve better results, even if the amount of computation is controlled to be the same as conventional GP. We will discuss this in more detail in future work.

From the figure, we can see that the results of the three methods are similar, but the results of GA-DPMOGP are the most constricted because all the results will be optimized in one direction due to the characteristics of the genetic algorithm. However, it is unclear whether fitness can fully account for the performance of the generated probability mapping, so we cannot determine whether the converging direction of the algorithm's result is correct or incorrect.

We found in our experiments that, although GA-DPMOGP has a faster convergence speed for pre-learning than RS-DPMOGP, the final results are generally better for RS-DPMOGP, especially when there is a large amount of computa-

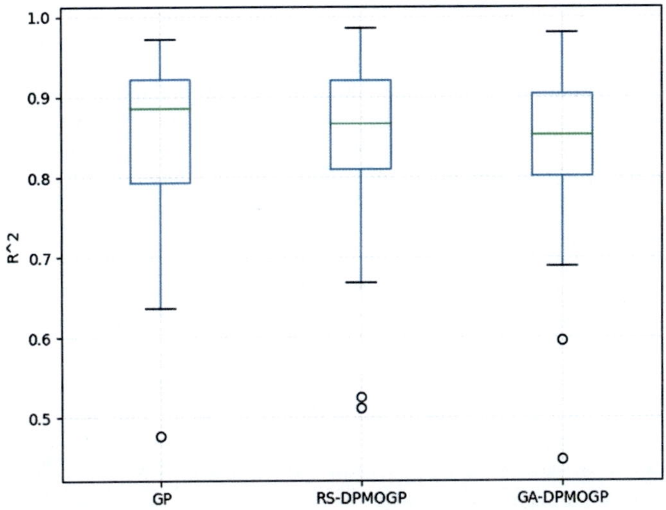

Fig. 6. The R^2 result for comparing with limited computation.

tion. This is because GA-DPMOGP is more likely to fall into a local optimum, whereas RS-DPMOGP will not have such a problem.

Therefore, we can conclude that DPMOGP is not suitable for scenarios where computing resources are scarce. When using DPMOGP, GA-DPMOGP should be selected if the amount of computation given to pre-learning is insufficient. If a large amount of computation is allocated to pre-learning, it is recommended to choose RS-DPMOGP.

5 Conclusion and Future Works

In this study, we propose DPMOGP to discuss the impact of the probability of being selected for each operator on genetic programming. We implemented two types of implementations using random search and genetic algorithms.

Through multiple experiments, we demonstrate that different probability mappings of operators significantly affect the performance of GP. Thus, optimizing the algorithm by finding an excellent probability mapping is effective.

However, we identified several limitations of the current method. Firstly, the pre-learning method consumes a significant amount of computation, which may not be a viable approach to finding the probability mapping. Therefore, we propose to integrate the pre-learning of DPMOGP into the GP process to increase its efficiency in the future study. Secondly, RS-DPMOGP and GA-DPMOGP have an important hyperparameter, i.e., the depth of the generated tree, which we currently set based on experience rather than mathematical analysis. If this parameter is too small, it may not fully explore the mapping, while setting it too large will generate numerous invalid nodes. Moreover, the suitable depth varies across different problems. We plan to address this issue in future studies.

References

1. Vapnik, V.: Principles of risk minimization for learning theory. In: Proceedings of Advances in Neural Information Processing Systems, pp. 831–838 (1991)
2. Koza, J.R.: Genetic programming as a means for programming computers by natural selection. Stat. Comput. **4**(2), 87–112 (1994)
3. Ying, B., Xue, B., Zhang, M.: A divide-and-conquer genetic programming algorithm with ensembles for image classification. IEEE Trans. Evol. Comput. **25**(6), 1148–1162 (2021)
4. Gao, S., Yu, Y., Wang, Y., et al.: Chaotic local search-based differential evolution algorithms for optimization. IEEE Trans. Syst. Man Cybern. Syst. **51**(6), 3954–3967 (2021)
5. Pei, Y.: Chaotic evolution: fusion of chaotic ergodicity and evolutionary iteration for optimization. Nat. Comput. **13**(1), 79–96 (2014)
6. Blum, C., Roli, A.: Metaheuristics in combinatorial optimization: Overview and conceptual comparison. ACM Comput. Surv. (CSUR) **35**(3), 268–308 (2003)
7. May, R.M.: Simple mathematical models with very complicated dynamics. In: Hunt, B.R., Li, TY., Kennedy, J.A., Nusse, H.E. (eds.) Proceedings of The Theory of Chaotic Attractors, pp. 85–93. Springer, New York (2004). https://doi.org/10.1007/978-0-387-21830-4_7
8. Xing, H., Salleb-Aouissi, A., Verma, N.: Automated symbolic law discovery: a computer vision approach. In: Proceedings of the AAAI Conference on Artificial Intelligence, vol. 35, no. 1, pp. 508–515 (2021)
9. Lathrop, D.: Nonlinear dynamics and chaos: with applications to physics, biology, chemistry, and engineering. Phys. Today **68**(4), 54 (2015)
10. García, S., Fernández, A., Luengo, J., Herrera, F.: Advanced nonparametric tests for multiple comparisons in the design of experiments in computational intelligence and data mining: experimental analysis of power. Proc. Inform. Sci. **180**(10), 2044–2064 (2010)

Exploring the Potential of Webcam-Based Eye-Tracking for Traditional Eye-Tracking Analysis

Cheng-Hui Chang, Jason C. Hung, and Jia-Wei Chang(✉)

Department of Computer Science and Information Engineering, National Taichung University of Science and Technology, Taichung City, Taiwan
{jhung,jwchang}@nutc.edu.tw

Abstract. Traditional eye-tracking systems can be costly and may pose a barrier to entry for researchers interested in studying gaze behavior. In recent years, there have been significant developments in simulating eye-tracking using webcams. However, little research has explored the use of webcam-based eye-tracking data for traditional eye-tracking analysis. In this paper, we propose a webcam-based eye-tracking system that utilizes an dilated convolutional neural networks to detect point of gaze and calculate a range of analysis indicators, such as duration of first fixation and latency of first fixation. By integrating these indicators, we aim to explore the potential of webcam-based eye-tracking for traditional eye-tracking analysis. This approach could significantly reduce the barrier to entry for researchers in the field of gaze behavior research and open up new avenues for studying gaze behavior.

Keywords: Eye tracking · Analysis indicators · Gaze estimation · Convolutional neural network (CNN)

1 Introduction

The saying "eyes are the windows to the soul" highlights the significant relationship between eye gaze and human cognitive processes. Eye tracking is the entry point to exploring this relationship and the field of eye tracking has made significant progress in recent years. The development of eye-tracking devices or "eye trackers" has played a critical role in improving the accuracy of eye movement tracking. Eye trackers have garnered widespread attention for analyzing and quantifying data and for related applications.

However, the eye-tracking system cannot be widely researched and applied because obtaining high-accuracy data requires powerful eye trackers, which are often expensive. This has made it difficult for novice scholars to enter this field. In the past, some researchers have proposed low-cost solutions to reduce the cost of eye-tracking equipment [3], laying the foundation for the development of eye-tracking devices. On the other hand, the proliferation of mobile devices has made camera lenses ubiquitous, attracting more people to this field, hoping that this technology can become a reality. The advancement of computer vision in deep learning has also made it possible to use these everyday visible devices for eye tracking.

J. C. Hung et al. (Eds.): FC 2023, LNEE 1134, pp. 313–316, 2024.
https://doi.org/10.1007/978-981-99-9342-0_33

2 Related Work

In recent years, with the significant advancements in deep learning techniques, some researchers have proposed using deep learning to track eye movements through webcams and have developed related datasets [2]. These methods have been able to achieve high accuracy while running smoothly on personal computers or mobile devices, greatly increasing the accessibility of eye-tracking applications that use webcams.

Most deep learning-based eye-tracking methods have focused on appearance-based methods. For instance, in [1], researchers utilized a neural network to process facial images, allowing them to capture features even when the head moves slightly, thereby reducing errors. In [2], the authors used a similar concept but input both eye features and facial features, achieving high accuracy even without calibration (Fig. 1).

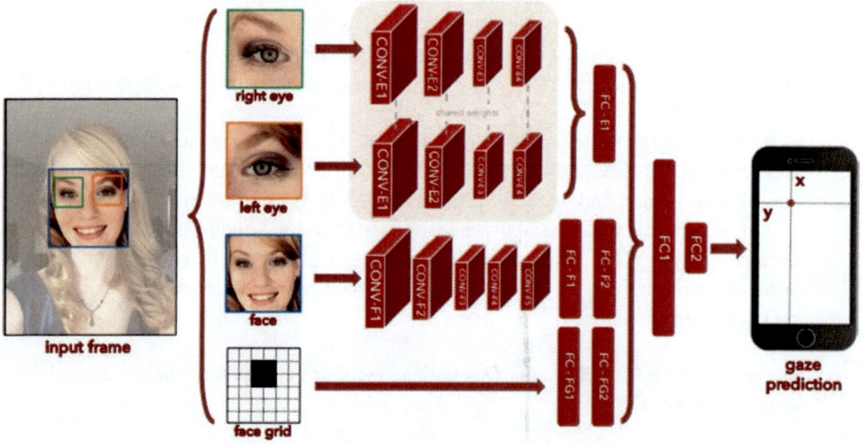

Fig. 1. The model architecture of the dilated convolutional neural network.

In the field of eye-tracking, deep learning methods have mainly been implemented using convolutional neural networks (CNNs), as in [4] and [5]. In [3], the authors proposed using dilated convolutions to reduce the loss of image information when capturing eye features. This method achieved 20.8% higher accuracy than models without dilated convolutions.

3 Design of the System

Although many studies have focused on using webcams to achieve tasks such as gaze estimation, there are few that use this data to perform traditional eye-tracking analysis metrics. This paper aims to investigate whether webcam-based eye-tracking can achieve such metrics.

We will employ the method of extended convolutional neural networks [3] to develop a webcam eye-tracking system that can detect the point of gaze and calculate various analysis metrics, such as Duration of First Fixation (DFF) and Latency of First Fixation (LFF) (Fig. 2).

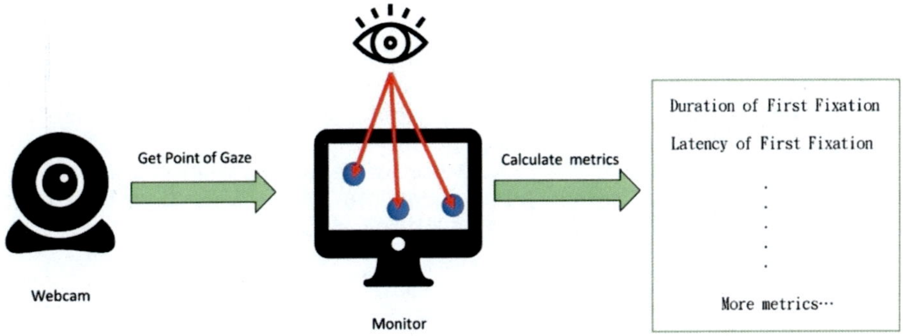

Fig. 2. The architecture and flow of the system design.

4 Conclusion

The high cost of traditional eye-tracking systems has been a barrier for many researchers to enter the field. In recent years, there has been significant development in simulating eye-tracking systems using webcams. This paper aims to explore the future of webcam-based eye-tracking analysis by combining various analysis indicators. By focusing on traditional eye-tracking indicators, we hope to discover new analysis methods and lower the barrier of entry to eye-tracking research using webcams.

5 Future Work

Our next step in research will focus on advancing emotion recognition using webcam eye-tracking technology. Other studies [7] have analyzed various methods for using eye-tracking technology to analyze emotional states, and combining it with other physiological states can lead to even higher accuracy in evaluating gaze patterns and emotions. This will expand the application range of webcam eye-tracking technology.

Acknowledgement. This work was partially supported by the National Science and Technology Council, Taiwan, R.O.C. [grand number MOST 110-2221-E-025-005].

References

1. Zhang, X., et al.: It's written all over your face: full-face appearance-based gaze estimation. In: Proceedings of the IEEE Conference on Computer Vision and Pattern Recognition Workshops (2017)
2. Krafka, K., et al.: Eye Tracking for Everyone (2016)
3. Li, D., Babcock, J., Parkhurst, D.: openEyes: a low-cost head-mounted eye-tracking solution. In: Eye Tracking Research & Application, pp. 95–100 (2006). https://doi.org/10.1145/1117309.1117350
4. Chen, Z., Shi, B.E.: Appearance-based gaze estimation using dilated-convolutions. In: Jawahar, C.V., Li, H., Mori, G., Schindler, K. (eds.) ACCV 2018. LNCS, vol. 11366, pp. 309–324. Springer, Cham (2019). https://doi.org/10.1007/978-3-030-20876-9_20

5. Meng, C., Zhao, X.: Webcam-based eye movement analysis using CNN. IEEE Access **5**, 19581–19587 (2017). https://doi.org/10.1109/ACCESS.2017.2754299
6. Yin, Y., Juan, C., Chakraborty, J., McGuire, M.P.: Classification of eye tracking data using a convolutional neural network. In: 2018 17th IEEE International Conference on Machine Learning and Applications (ICMLA) (2018)
7. Zhan, Z., Zhang, L., Mei ,H., Fong, P.S.: Online learners' reading ability detection based on eye-tracking sensors. Sensors (Basel) **16**(9), 1457 (2016). https://doi.org/10.3390/s16091457. PMID: 27626418; PMCID: PMC5038735

New Group-Key-Based Over the Air (OTA) Update Model Facilitating Security and Efficiency Using MQTT 5

Hung-Yu Chien[1(✉)], Nian -Zu Wang[1], Yuh-Min Tseng[2], and Ruo-Wei Hung[3]

[1] National Chi-Nan University, Nantou County, Taiwan, Republic of China
hychien@mail.ncnu.edu.tw

[2] Department of Mathematics, National Changhua University of Education, Changhua City, Taiwan

[3] Department of Computer Science and Information Engineering, Chaoyang University of Technology, Taichung City, Taiwan

Abstract. The booming development of Internet-of-Things (IoT) has deployed many IoT systems globally, and this trend is continuously accelerating. However, as many IoT devices are widely deployed, the system-update maintenance is a huge challenge. Over The Air (OTA) update is one promising mechanism for securely updating the firmware of the remote IoT devices.

Message Queue Telemetry Transport (MQTT) is one of the most adopted IoT communication protocols globally. It has also been popularly adopted as the communication protocol for delivering the OTA update messages, in addition to delivering normal IoT messages. This paper focuses on MQTT-based OTA models. Even though there exist several MQTT-based OTA models and schemes, we find that no one can simultaneously satisfying user convenience, efficiency and high security. Some sacrifices the privacy against the MQTT broker to achieve user convenience, and some focuses on the privacy while sacrificing the convenience. This paper sorts out the existent models and proposes a new model that distributes the group keys among the manager and the IoT devices, allows the manager deposit the group-key-encrypting firmware on the broker, and then each device can separately access the encrypted OTA images from the broker. We design the scheme using MQTT 5.0 (the new MQTT standard). The analysis and the evaluation show that the new model achieves better privacy protection and gains efficient communication performance.

Keywords: Internet of Things (IoT) · MQTT · Over the Air · privacy · security · Amazon · group key

1 Introduction

IoT systems have been widely deployed globally in many application areas, and the number of deployments is continuously increasing. However, as the number of deployed devices increases very fast, so does the challenge of maintaining these remote devices.

J. C. Hung et al. (Eds.): FC 2023, LNEE 1134, pp. 317–328, 2024.
https://doi.org/10.1007/978-981-99-9342-0_34

In these days, IoT systems need to frequently update the firmware/software to meet the fast new-function release cycle and the requirement of security patches. The mechanisms of the OTA update allow the system manager remotely update the firmware/software of devices via the communications, without recalling the devices back to the companies or personal inspect the devices on-site [1–3]. This approach not only greatly improves the efficiency but also accelerates the product life cycle.

MQTT is a very popular IoT communication protocol [4, 5]. There exist several open-source platforms (like Mosquitto [6], HiveMQ [7], Mosca [8]) and several commercial IoT platforms (for example, Amazon platform [9, 10]). MQTT model adopts the publish-then-forward approach: some devices (called publishers) publish the messages with the specified topic to a broker, and then the broker forwards the messages to those devices (called the subscribers) that have subscribed the same topic. As it is easy-to-use and very efficient, it soon becomes one of the most popular IoT communication protocols. However, the precedent MQTT standards (MQTT 3.1 [4] and its earlier versions) only support account-and-password as their authentication support, and do not provide any encryption by themselves; they assume the deployments would enable SSL/TLS in the underlying layers to protect the privacy of the transmission. The new MQTT standard called MQTT 5.0 [5] adds several new functions (including the User properties and the Enhanced Authentication framework), which greatly improve the flexibility and the security support.

As the MQTT protocol is very efficient, it has also been adopted to deliver the OTA messages, in addition to the normal IoT messages. Several MQTT platforms (like Amazon and Infineon [11]) have included this function in their services. Chien and Wang [12] recently design a new MQTT-based OTA scheme in which a publisher separately builds an End-to-End key with each of its subscribers, and then the publisher securely distributes the encrypted OTA data to these subscribers; in this arrangement, their scheme can protect the privacy against the broker; however, during the OTA update phase, the publisher is required to have a reliable connection with the broker. After surveying the existent MQTT-based OTA solutions, we find that none of the existent models could simultaneously satisfy the requirements of high privacy support, efficiency, and low publisher-burden. Therefore, this paper will propose a new MQTT-based model that our model achieves better privacy protection and gains efficient communication performance. We will leverage the new functions of MQTT 5.0 to achieve this goal.

2 Related Work

Conventionally, MQTT 3.1 [4] and its precedent versions assume that the users would enable SSL/TLS to encrypt the transmission privacy. This approach has several weaknesses. First, the support of SSL/TLS is a burden for some simple IoT devices. Second, even though the transmission between a publisher and its broker and between a subscriber and its broker is encrypted, the broker can still peek at the content of its clients; that is, the SSL/TLS does not protect the clients' privacy against the broker. There exist several publications like [27] elaborating on the performance of MQTT.

In light of the observations on the weak security support, there exist many efforts to improve the security support of MQTT 3.1 and its precedent versions. These proposals could be roughly classified into several categories. One is designing special hardware (for example, Lesjak et al. [13]) to assist IoT devices handle the SSL/TLS overhead. The second category is designing some customized key agreement schemes (like [14–22]) for MQTT systems. The third category, based on the application context, defines the capacities of a IoT device; for example, [23] defines each device's capacity to be publish-only, subscribe-only, and both-publish-subscribe, according to the device's function and location.

The standard organizations also notice the weaknesses and limitations of the precedent MQTT versions, and ratify the new MQTT standard called MQTT 5.0 to enhance both the security support and the flexibility. Regarding the security, the new standard designs the Enhanced Authentication framework in which new Application Programming Interfaces (APIs) and new packet fields are proposed to facilitate users design their own authenticated key agreement schemes and the negotiation of encryption methods. Regarding the flexibility, several new features are introduced; a new packet field called "User Properties" is included to share application data between publisher-broker, between subscriber-broker, and between publisher-subscriber. These application-aware data is very useful for users to embed application-aware message in the MQTT interactions.

Ciou and Chien [24] design a Challenge-Response authentication using the Enhanced Authentication framework of MQTT 5.0; the scheme builds a secure channel between a client with its broker. Chien [12, 25] designs the first End-to-End (E2E) secure channel between a publisher and a subscriber, using the MQTT 5.0 features. SEEMQTT [26] also concerns the end-to-end security where a publisher delegates its encryption authority to a pool of keystores, via secret sharing, so that those designated subscribers can recover the decryption key; this arrangement aims at those scenarios where it is difficult that the publishers can directly verify their subscribers and can protect the privacy against a curious broker. However, in the OTA application, the publishers which release the new firmware are usually resource-abundant devices and can directly verify their subscribers; SEEMQTT model is too complicated and not efficient enough for the OTA applications.

Amazon Web Services (AWS) provides several popular cloud-based services, and the MQTT-based IoT service is one of them. In its MQTT-based IoT service, the broker can deliver both the normal messages and the OTA messages via the MQTT interactions. In the OTA service of the AWS IoT service, an application manager prepares the new firmware, uploads the new firmware to the cloud, and creates an OTA job to take care of the OTA update process; each designated IoT device then interacts with the broker, and the OTA job on the broker is responsible for delivering the OTA messages and the firmware to the device. To ensure the authenticity and the integrity of the firmware, the manager can personally sign the firmware or delegates the authority to the cloud.

Figure 1 shows the AWS MQTT-based OTA model. In this model, the device manager creates Things (that is, the IoT devices in the AWS services) and certificates; he uploads the new firmware; he may sign the firmware on his local computer or delegates the authority to the AWS server; he also creates an AWS IoT job on the broker to handle the OTA process. During the OTA process, the OTA update agent on the device interacts with the broker to process the OTA update. Because the broker handles the OTA process on behalf of the manager and the firmware is already on the server, this model does not require the local computer of the manager be on-line during the OTA update process. However, it has two critical weaknesses: one is the delegation of the signing authority, and the second is the lack of privacy protection against the broker (the broker can peek at the content of the firmware).

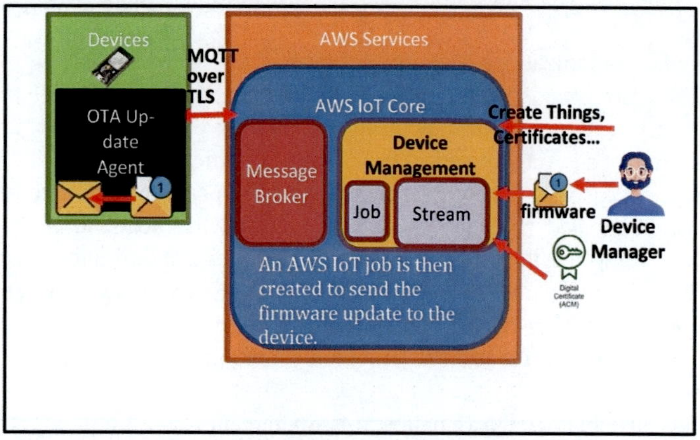

Fig. 1. The AWS MQTT-based OTA model

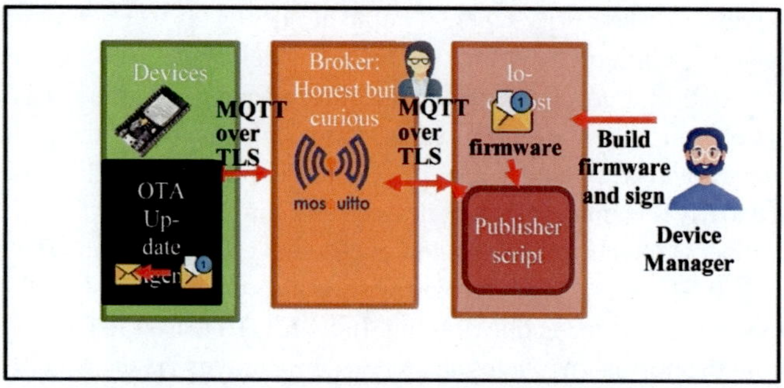

Fig. 2. The Infineon MQTT-based OTA model

Infineon [11] also proposes a MQTT-based OTA model. In the model (depicted in Fig. 2), the manager builds the firmware, signs it, but keeps the firmware on the local host. On the local host, a publisher (which runs the OTA-publisher script) handles the OTA process, and interacts with the IoT devices via the MQTT broker. In the figure, we use the mosquito broker [6] as the broker. Of course, one can implement both the broker and the publisher on the same computer (that is, merging the broker and the local host); but, logically, they are separate machines, as they interact using MQTT and the manager of the broker and the device manager might belong to two different authorities. In this model, the device manager can locally handle the OTA process; however, it requires the localhost be reliably on-line during the process; the broker also can peek at the content of the firmware.

Based on the above survey and observations, we can see that, up to now, there is no one model that simultaneously satisfies all the requirements of high privacy (here it refers to the privacy against the broker), efficiency (less communication overhead during the OTA process), and less publisher-broker connection requirement (it refers to the publisher needs not to be on-line during the OTA process).

3 New MQTT-Group-Key-Based OTA Model

Our new OTA model is based on building a group key among the publisher (here it refers to the device manager's host), using MQTT 5.0. Section 3.1 first introduces the new OTA model. Section 3.2 introduces the new design in details, using MQTT 5.0. Table 1 introduces the notations.

Table 1. The notations.

M.Cert, D1.Cert	M.Cert denotes Publisher M's certificate; D1.Cert denotes D1's certificate M.Cert has the public key as g^a; D1.Cert has its public key as g^b. Here we eliminate the specification of the underlying fields, and any secure fields like Elliptic Curve Cryptographies could be used
DH_{key}	$DH_{key} = g^{ab}$ is the Diffie-Hellman key between the publisher (the device manager) and one of its subscribers
$group_{key}$	The group key chosen by the device manager-in this paper, we use the term the device manager to refer to the device manager (the user) and his host. The group key will be shared among the device manager and its subscribers (the IoT devices)
$Enc_{key}[]$,$Dec_{key}[]$	Encryption/Decryption using the key *key*

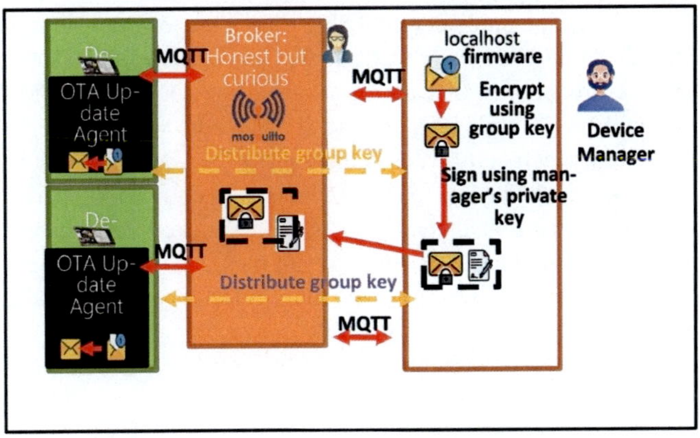

Fig. 3. The group-key-based OTA model

3.1 The Group-Key-Based OTA Model

Figure 3 depicts the new model. The new model consists of four phases: the client-broker authentication phase, the group-key distribution phase, the firmware-encrypt-sign phase, and the OTA update phase. In the client-broker authentication phase, each client and the broker mutually authenticate each other to access the MQTT services. In the group-key distribution phase, the device manager distributes the group key to all of the designated subscribers (the IoT devices). In the firmware-encrypt-sign phase, the device manager first, using the group key, encrypts the firmware, and then sign the encrypted firmware using its private key; he then uploads both the encrypted firmware and the signature to the broker. Finally, in the OTA update phase, the devices interact with the broker to access the encrypted firmware and the signature, verify the signature, and decrypt the encryption.

In this model, we note several improvements. First, the encrypted firmware is deposited on the broker such that the manager's local host is not required to be on-line when the IoT devices perform the OTA update phase. Second, the firmware is encrypted using the group key such that the broker cannot peek at the content of the firmware.

3.2 The Detailed Design Using MQTT 5.0

Figure 4 depicts the message interactions of the four phases.

The Client-Broker Authentication Phase
A client and the broker just perform any secure authentication schemes (for example, SSL/TLS or Chien et al.'s scheme [20]) to mutually authenticate each other.

The Group-Key Distribution Phase
Step 1(a) and 1(b): the publisher (the device manager) and the subscriber (the device) respectively subscribe the topic = OTA/M/devices and the topic = OTA/M to properly receive the expected messages later.

Step 2(a) and 2(b): The publisher publishes the message "Publish(topic = OTA/M, retain = True, Userproperties = {certificate:M.Cert,...}, ResponseTopic = OTA/M/devices)", and the broker forwards the message to the subscribers. "M.Cert" in this message is the publisher's certificate, and "ResponseTopic = OTA/M/devices" notifies the subscribers to reply their certificates in the topic = OTA/M/devices. The field "retain = true" is used to notify the broker to keep the message until the next "retain" message replaces the old one: this mechanism facilitates the designated receivers get the message later even if they were not on-line when the retain message was published.

Step 3: The subscriber D1 subscribes the topic "OTA/M /devices/D1" to receive the group-key message for it.

Step 4(a) and 4(b): the subscriber publishes its certificate D1.Cert to the broker in 4(a), and the broker forwards it to the publisher in 4(b).

When the publisher and the subscriber get the two certificates from each other, they can compute the Diffie-Hellman key 〖DH〗 _key = g^ab.

Step 5(a) and 5(b): the publisher chooses the group key, uses the 〖DH〗 _key to encrypt the group key, and publishes the encryption to the broker; the broker forwards the encryption to the subscriber which decrypts the encryption to get the group key. This completes the group-key distribution phase.

The Firmware-Encrypt-Sign Phase
The publisher prepares the new firmware, encrypts the firmware using the group key, and signs on the encrypted firmware using its private key. He then uploads the encrypted firmware and the signature to the broker.

The OTA Update Phase
All the designated IoT devices (the subscribers) will get the notification from the broker, and directly get the OTA data from the broker.

4 Security Analysis and Performance Evaluation

4.1 Security Analysis

The model consists of four kinds of entities: publishers, subscribers, the broker, and attackers. The broker is assumed to be honest but curious; that is, the broker follows the protocols but is curious about the content. The attackers can actively manipulate any messages on the publisher-broker channel and the subscriber-broker channel, and the goal is to violate the authenticity, the integrity, and the privacy.

The security of the model depends on several modular blocks: the TLS-based client-broker channel, the group-key-based OTA channel, and the encryption-signature firmware protection. Now we analyze the modular design of our scheme as follows.

The TLS-Based Client-Broker Channel. Each client and the broker should mutually authenticate each other before the client can access the MQTT services. The TLS channel protects the authenticity, the privacy, and the integrity of the client-broker channel.

The group-key-based OTA channel. The group-key-based OTA channel is built on top of the publisher-broker-TLS channel, the subscriber-broker-TLS channel, and the publisher-subscriber-DH-key-encryption of the group key. The DH_key is derived from the Diffie-Hellman key using the (public key, private key) pairs of the entities. As long as the TLS channels and the computational Diffie-Hellman problem are secure, this ensures the privacy of the group key against both the attackers and the curious broker.

The Encryption-Signature Firmware Protection. The firmware is encrypted using the group key, and then is signed by the device manager using its private key. This protection protects the privacy and the authenticity of the firmware.

Here, we summarize the security properties. The model provides the authenticity, the integrity, and the privacy of the normal MQTT messages; the attackers cannot access the content of the MQTT messages. The model ensures the privacy of the firmware against a curious broker. Both the broker and the attackers cannot violate the authenticity, the integrity, and the privacy of the firmware.

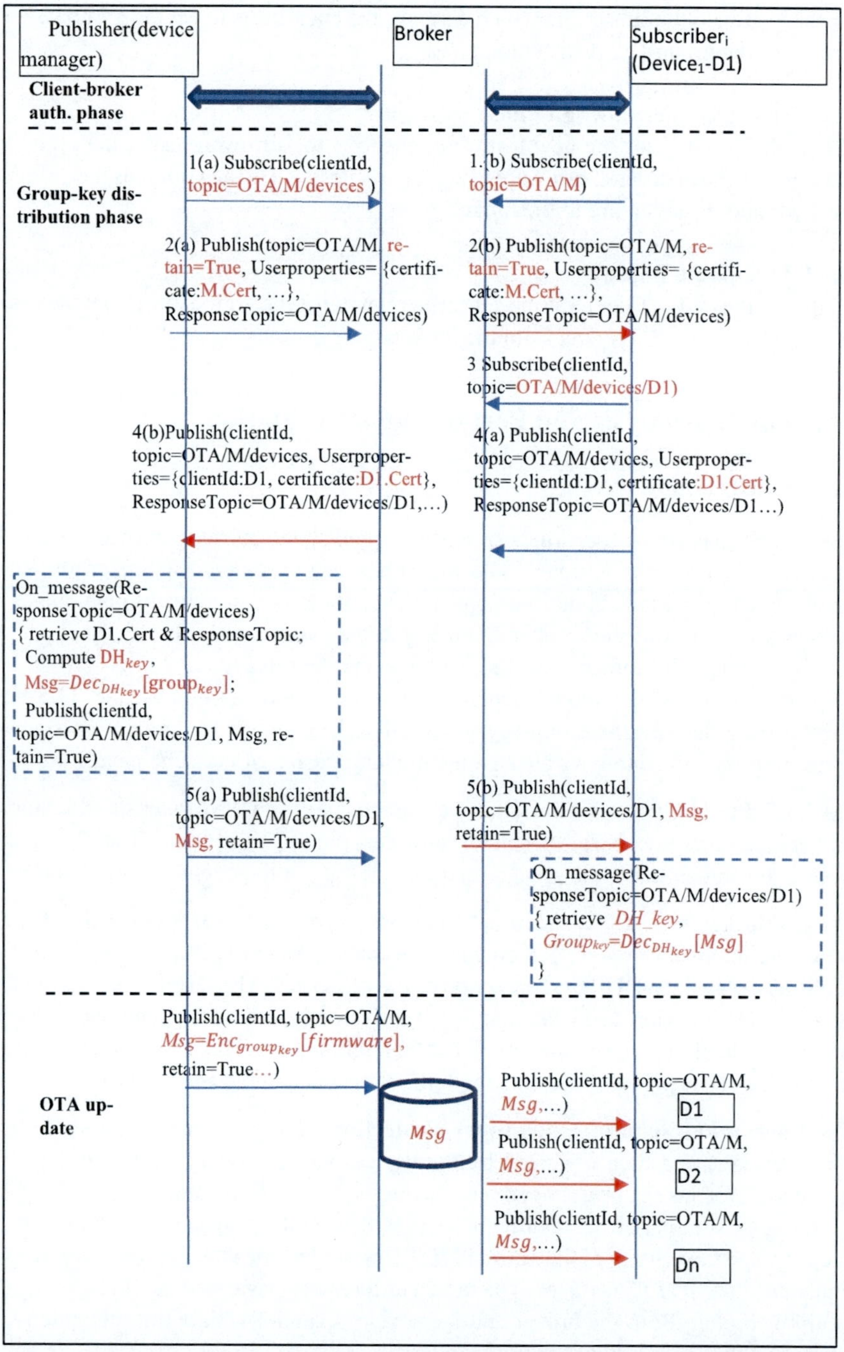

Fig. 4. Publisher distributes E2E key in MQTT 5.0 →: original message; →: forwarded message;

4.2 Performance Evaluation

The new model consists of four phases. The first phase, the MQTT connection phase, could adopt any secure authentication schemes (example, the classic TLS or any other secure authentication) which is not the focus of this paper; therefore, we focus on evaluating the rest three phases. As our new model builds the End-to-End channel, the model can adopt any secure authentication schemes (for example, the identity-password authentication or the Challenge-Response authentication [20]), and can rely on the End-to-End channel to protect the transmissions.

Before examining the schemes, we first define some metric symbols. Let T_{sig} denotes the computation cost for one digital signature, T_{ver} denotes that for one signature verification, T_{enc} denotes that for one symmetric encryption or decryption, and T_{exp} denotes that for one modular exponentiation. Let Lat_{TLS} denotes the authentication time using TLS, Lat_{CR} denotes the authentication time using the Challenge-Response [20], Lat_{E2E} denotes the latency for Chien's End2End channel building [26], Lat_{group} denotes the latency for the group key distribution in our new scheme, Lat_{MQTT} denotes the latency for either publisher-broker MQTT interaction or broker-subscriber interaction, Lat_{OTACTL} denotes the latency for the OTA control message exchanges. The OTA control messages consist of OTA message like notification, response, and confirmation could be exchanged to ensure the smooth OTA interactions; since this part is independent of the model design, we assume this part is the same for all models to simplify the comparison. $Lat_{broker(n)}$ denotes the forwarding delay caused by the broker for n simultaneous subscribers.

Now we examine the second phase. For each pair of (publisher, subscriber), the second phase takes nine MQTT interactions of which four messages involve the publisher, and five messages involve the subscriber. Because these MQTT messages are all simple-and-short MQTT messages, they demand only little overhead.

Now we examine the 3rd phase. The device manager prepares the firmware, encrypts it, signs it, and uploads it to the broker in our model. The AWS model in this phase is similar, but it does not keep privacy against the broker. The Infineon model keeps the firmware on the local host and does not keep privacy against the broker.

The final OTA update phase of our scheme and the AWS model allow the IoT devices directly access the firmware from the broker, while the Infineon model requires the publisher be on-line to handle the firmware delivery. Because the firmware for IoT devices is usually small size, here we assume it takes only one MQTT message to deliver it.

Table 2 summarizes the performance comparison of the related models. We can see that our model out-performs the AWS/Infineon models in terms of firmware privacy protection, at the extra cost of nine MQTT interactions and the extra computations at the second phase. We note that these nine interactions only be executed once, and they are very efficient MQTT interactions. In the last raw, we estimate the rough total latency for n subscribers. Here we can see that our model is expected to have shorter latency; that is because our model's last phase does not require the publisher-broker interactions and the firmware is pre-encrypted once only before the OTA update phase. However, we should note that the inter-impact among the broker's loading, the number of subscribers, and the aggregated latency is quite complicated; here we only capture the possible asymptotic

Table 2. Comparison of the related models

Scheme Properties	AWS	Infineon	Ours
Privacy firmware against the broker	No	No	Yes
Firmware signing authority	Broker/manager	manager	manager
Number of MQTT interactions in the 2nd phase	NA[1]	NA[1]	9
Computations in the 2nd phase on the subscriber side	NA[1]	NA[1]	1 T_{enc}+ 1 T_{ver} ver. + 1 T_{exp}
Computations in the 2nd phase on the publisher side (for n subscribers)	NA[1]	NA[1]	n*(1 T_{enc} + 1 T_{ver} + 1 T_{exp})
Computations in the OTA update phase on the publisher side[2]	NA[1]	n*T_{enc}	NA [4]
Computations in the OTA update phase on the broker side (for n subscribers)[2]	n*T_{enc}	2n*T_{enc}	None
Total comm. Cost for n subscribers[3]	Lat_{TLS}+ Lat_{OTACTL}+$Lat_{broker(n)}$+ n*T_{enc}+ Lat_{MQTT}	Lat_{TLS}+ Lat_{OTACTL}+ $Lat_{broker(n)}$+2n*T_{enc} + 2 Lat_{MQTT}	Lat_{CR}+ Lat_{OTACTL}+ $Lat_{broker(n)}$ + Lat_{MQTT}

1. For the AWS model and the Infineon model, they do not build the End-to-End channel.
2. Here, we assume that the firmware is small-size and can be delivered in one MQTT message.
3. We note that here we can only roughly estimate the total latency, due to it involves three kinds of entities (the broker, the publisher, and n subscribers), and the broker's loading is greatly affected by the number of the subscribers in the AWS model and the Infineon model.
4. In our model, the publisher pre-encrypts the firmware and upload the encrypted firmware before the OTA update phase.

behavior. To have accurate comparison, we plan to implement these models to evaluate their performance in the real scenarios.

5 Conclusions and Future Work

In this paper, we have designed the group-key support for the MQTT5.0 system and the new OTA update model. In the model, the device manager can securely distribute the group key to his IoT devices. The manager encrypts the firmware and uploads it to the broker. During the OTA update phase, the devices directly access the encrypted firmware from the broker. This arrangement greatly enhances the privacy protection of the firmware and keeps the interactions during the TOA update phase simple and efficient.

Through simple analytic comparison, we can see that our model achieves better privacy protection and gains efficient communication performance. In the future work, we plan to implement the four models and evaluate their performance in the real scenarios.

Acknowledgement. This research was funded by the Ministry of Science and Technology, Taiwan, R.O.C. grant number MOST 111-2221-E-260-009-MY3, MOST110-2221-E-018-006-MY2, MOST 110-2221-E-324-007-MY3.

References

1. Karim, H.: Over-the-Air (OTA) updates: what is it and how to do it simply, efficiently with ZDM. https://itskarim.medium.com/over-the-air-ota-updates-what-is-it-and-how-to-do-it-simply-efficiently-with-zdm-db613ea29678. Accessed 30 Aug 2022
2. Mohammad, A.: Implementing over-the-air device firmware update (OTA DFU) – Part 1. https://www.novelbits.io/ota-device-firmware-update-part-1/. Accessed 30 Aug 2022
3. Wikipedia: Over-the-air programming. https://en.wikipedia.org/wiki/Over-the-air_programming. Accessed 30 Aug 2022
4. ISO/IEC 20922:2016, Information technology – message queuing telemetry transport (MQTT) v3.1.1. https://www.iso.org/standard/69466.html. Accessed 25 Mar 2022
5. OASIS, MQTT Version 5.0, 07 March 2019. https://docs.oasis-open.org/mqtt/mqtt/v5.0/mqtt-v5.0.html. Accessed 1 Apr 2022
6. Eclipse mosquitto. https://mosquitto.org/. Accessed 4 Jan 2023
7. HiveMQ homepage, enhanced authentication. https://www.hivemq.com/blog/mqtt5-essentials-part11-enhanced-authentication/. Accessed 2 Apr 2022
8. Mosca. https://github.com/moscajs/mosca. Accessed 4 Jan 2023
9. Amazon: How to perform secondary processor over-the-air updates with FreeRTOS. https://aws.amazon.com/tw/blogs/iot/how-to-perform-secondary-processor-over-the-air-updates-with-freertos/. Accessed 30 Aug 2022
10. Amazon: AWS IoT Over-the-air update. https://aws.github.io/amazon-freertos/202107.00/embedded-csdk/libraries/aws/ota-for-aws-iot-embedded-sdk/docs/doxygen/output/html/ota_design.html. Accessed 30 Aug 2022
11. Implementing MQTT CLIENT USING anycloud libraries. https://community.infineon.com/t5/Blogs/Implementing-MQTT-Client-Using-AnyCloud-Libraries/ba-p/246975. Accessed 4 Jan 2023
12. Hung-Yu, C., Nian-Zu, W.: A novel MQTT 5.0-based over-the-air updating architecture facilitating stronger security, MPDI Electron. **11**(23) (2022). https://www.mdpi.com/2079-9292/11/23/3899
13. Lesjak, C., et al.: Securing smart maintenance services: hardware-security and TLS for MQTT. In: 2015 IEEE 13th International Conference on Industrial Informatics (INDIN), Cambridge, UK, Cambridge, pp. 1243–1250 (2015)
14. Andy, S., Rahardjo, B., Hanindhito, B.: Attack scenarios and security analysis of MQTT communication protocol in IoT System. Proc. EECSI 2017, Yogyakarta, Indonesia, 19–21 September 2017, pp. 19–21 (2017)
15. Firdous, S.N., Baig, Z., Valli, C., Ibrahim, A.: Modelling and evaluation of malicious attacks against the IoT MQTT protocol. In: 2017 IEEE International Conference on Internet of Things (iThings) and IEEE Green Computing and Communications (GreenCom) and IEEE Cyber, Physical and Social Computing (CPSCom) and IEEE Smart Data (SmartData), pp. 748–755 (2017)

16. Rizzardi, A., Sicari, S., Miorandi, D., Coen-Porisini, A.: AUPS: an open source authenticated publish/subscribe system for the internet of things. Inform. Syst. **62**, 29–41 (2016)
17. Neisse, R., Steri, G., Baldini, G.: Enforcement of security policy rules for the internet of things. In 2014 IEEE 10th International Conference on Wireless and Mobile Computing, Networking and Communications (WiMob), Larnaca, pp. 165–172 (2014)
18. Shin, S.H., Kobara, K.: Efficient augmented password-only authentication and key exchange for IKEv2. IETF RFC 6628, Experimental, June (2012). https://tools.ietf.org/rfc/rfc6628.txt. Accessed 5 Feb 2022
19. Chien, H.Y., et al.: A MQTT-API-compatible IoT security-enhanced platform. Int. J. Sens. Netw. **32**(1), 54–68 (2020)
20. Chien, H.-Y., Lin, P.C., Chiang, M.L.: Efficient MQTT platform facilitating secure group communication. J. Internet Technol. **21**(7), 1929–1940 (2020)
21. Chien, H.Y., Qiu, G.H., Hung, R.W., Shih, A.T., Su, C.H.: Hierarchical MQTT with edge computation. In: The 10th International Conference on Awareness Science and Technology (iCAST 2019), Morioka, Japan, pp. 1–5 (2019)
22. Mektoubi, A., Lalaoui, H., Belhadaoui, H., Rifi, M., Zakari, A.: New approach for securing communication over MQTT protocol A comparison between RSA and Elliptic Curve. In: 2016 Third International Conference on Systems of Collaboration (SysCo), Casablanca, Morocco (2016)
23. Prajit Kumar, D., Sandeep, N., Nitin Kumar, S., Anupam, J., Karuna, J., Tim, F.: Context-sensitive policy based security in internet of things. In: 2016 IEEE International Conference on Smart Computing (SMARTCOMP), St. Louis, MO, USA (2016)
24. Ciou, P.-P., Chien, H.-Y.: An implementation of challenge-response authentication for MQTT 5.0 IoT system. In: The 2021 International Conference on Emerging Industry and Health Promotion (EIHP2021), Puli on 3–4 July 2021 (2021)
25. Chien, H.-Y.: Design of end-to-end security for MQTT 5.0. In: The 4th International Conference on Science of Cyber Security - SciSec, Matsue city, Shimane, Japan 10–12 August 2022 (2022)
26. Hamad, M., Finkenzeller, A., Liu, H., Lauinger, J., Prevelakis, V., Steinhorst, S.: SEEMQTT: Secure end-to-end MQTT-based communication for mobile IoT systems using secret sharing and trust delegation. IEEE Internet Things J. **10**(4), 3384–3406 (2023)
27. Seoane, V., Garcia-Rubio, C., Almenares, F., Campo, C.: Performance evaluation of CoAP and MQTT with security support for IoT environments. Comput. Netw. **197**, 108338 (2021)

A Study on the Improvement of Navigation Accuracy with ArUco Markers

Seung-Been Lee[1], Dong-Hyun Jo[1], Min-Ho Kim[1], Hee-Bum Kim[2], and Byeong-Gwon Kang[1(✉)]

[1] Department of Information and Communication Engineering, Soonchunhyang University, Asan, Korea
{soopy6,jiu421,bgkang}@sch.ac.kr

[2] Department of ICT Convergence, Soonchunhyang University, Asan, Korea

Abstract. In this paper, we propose an error decreasing technique using ArUco Marker, in a robot navigation system using SLAM (Simultaneous Localization and Mapping) in which errors occur due to packet loss and time delay. This technique enables more accurate estimation of the position and orientation between the robot and ArUco Marker. Through camera calibration, we convert 3D input values of a real object into undistorted 2D data and establish corresponding relationships between dimensions. Additionally, we use homogeneous transformation matrices to estimate the current direction and degree of rotation of a robot using the marker. Most of robots can reach their destination area through navigation with trial and errors with some time consumption. Therefore, we introduce ArUco Marker to reduce such errors and designed navigation algorithm to enable relatively precise driving with enough fast time. Finally, we compare the navigation accuracy using SLAM of the conventional scheme with the proposed method of twice modifications of the marker information which can reduce the navigation error around actual destination and resulting in accuracy improvement through the position correction process using ArUco Marker recognition.

Keywords: Computer Vision · ArUco Marker · SLAM

1 Introduction

As the increasing interests in the area of artificial intelligence the market with autonomous driving and camera technology in robots has grown in a recent decade. Additionally, robots use SLAM (Simultaneous Localization and Mapping) technology to simultaneously estimate their location and create a map to facilitate navigation. However, errors in location estimation and movement occur due to the robot's operating state, environment, and communication errors. Various attempts have been made to address these issues including GPS technology. However, there are limitations to accurate movement in poor communication environments [1]. Moreover, robots have become commonly used in everyday life, such as robot vacuum cleaners, however, there are some problems of decreased efficiency due to repeated position correction and communication, leading to a significant computational burden in the docking process at charging terminals [2].

J. C. Hung et al. (Eds.): FC 2023, LNEE 1134, pp. 329–333, 2024.
https://doi.org/10.1007/978-981-99-9342-0_35

Thus, we propose an algorithm that simplifies the previous complex processes related to robot docking and charging, enabling the robot to move more quickly and accurately to its destination. Before the algorithm is operated, camera calibration is performed in which the internal parameters, external parameters, and distortion coefficients of the camera are estimated, establishing a correspondence relationship between 2D images and 3D dimensions [3].

The experimental results are presented in the last section. When only conventional navigation technology is used, the error distance to the destination is within 50cm ~ 70cm. However, the error was decreased to less than 1cm using the ArUco marker-based navigation error correction technique proposed in this paper, effectively improving the accuracy of robot movement.

2 Experimental Process

2.1 Movement to Destination Position Using SLAM

To conduct the experiment, the SLAM algorithm was utilized to estimate the current indoor location of the robot and generate a map using data collected from a LIDAR sensor and depth camera [4]. Subsequently, navigation was performed based on the generated map. The configuration of the navigation system is presented in Fig. 1.

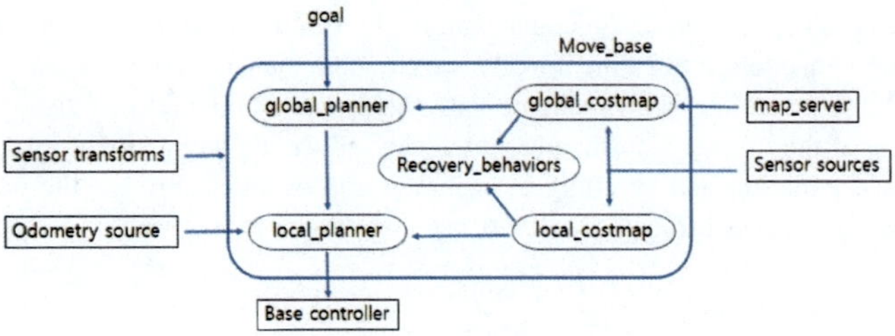

Fig. 1. Navigation system configuration

In order for a robot to move towards a designated destination, it repeats the process of Fig. 1 to estimate its location and identify obstacles to set an optimal path.

2.2 Camera Calibration

To utilize ArUco markers or specific objects for marker recognition, camera calibration is necessary to address measurement errors caused by camera distortion such as position, distance, and direction. In this paper, a camera calibration method is adopted using a 7x10 checkerboard pattern [3]. We capture images of the board at various angles using the camera and detect the corners of the board. The coordinate values of these corners are recorded for subsequent camera calibration procedures. By utilizing the size data of the defined board and the detected coordinate values, distortion coefficients of the camera's intrinsic and extrinsic parameters can be estimated.

2.3 ArUko Marker Tracking and Position Calibration

By utilizing camera calibration to estimate the distortion coefficients of external parameters, the real distance between the camera and marker can be more accurately estimated. This process involves converting the captured image into a binary representation and extracting the marker coordinates using a marker dictionary and its corresponding parameters [5]. Once the marker is detected, information such as the boundary region, data inside the dictionary and the position and orientation vectors between the ArUco marker and the camera can be obtained. In this paper, we propose an algorithm for pathfinding by utilizing the position and orientation vectors. Firstly, a homogeneous transformation matrix is constructed to calculate the distance and rotation direction between the marker and the camera to determine the direction and position (distance) of the robot (camera). The obtained position using the homogeneous transformation matrix [6] can be expressed as follows.

$$\begin{bmatrix} X' \\ Y' \\ Z' \\ 1 \end{bmatrix} = [R|t] \begin{bmatrix} x \\ y \\ z \\ 1 \end{bmatrix} = \begin{bmatrix} R11 & R12 & R13 & t_x \\ R21 & R22 & R23 & t_y \\ R31 & R32 & R33 & t_z \\ 0 & 0 & 0 & 1 \end{bmatrix} \begin{bmatrix} x \\ y \\ z \\ 1 \end{bmatrix} \quad (1)$$

Using the obtained position, it is possible to predict the optimal path. The basic process involves recognizing the marker, stopping, fixing the position of camera to the center of the marker, and then calculating the distance and rotation angle. The path can be approached in various ways thereafter. One method involves moving straight to the destination along the shortest path and then adjusting the direction of the robot based on the pre-calculated rotation angle. Another method uses a Manhattan distance-based path. In this process, using trigonometry, the X value (distance moved forward after exploration) and Y value (distance moved backward after 90-degree rotation from X movement) are calculated based on a triangle with the distance between the camera and marker as the hypotenuse. This allows for more accurate movement in the desired direction. Compared to the conventional method of modifying the position after movement, this method of calculating the path and approaching the destination in reverse direction has the effective advantage of less position modification and errors. Another method involves using rotation vectors. The degree of rotation is monitored in real time to determine the direction in which the robot needs to move. Figure 2 illustrates the process of robot movement using the calculated angle and distance based on straight distance, Manhattan distance, and marker tracking algorithms to set the moving path.

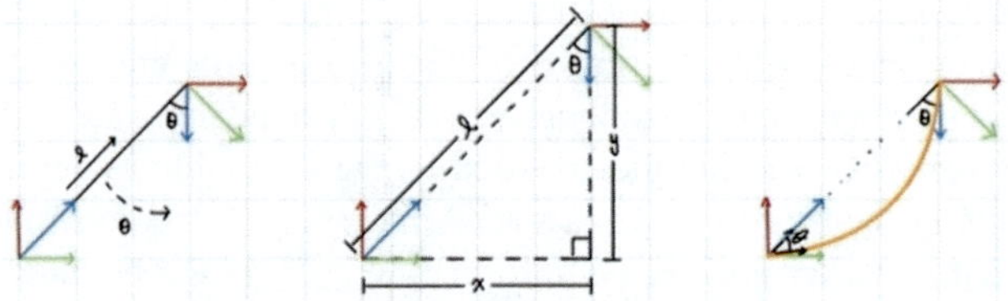

Fig. 2. Marker tracking and moving process.

3 Results and Conclusions

The purpose of this study is to determine the average error distance in destination navigation using only markers and experimentally validate markers with high recognition rates at that distance. Each experiment was performed 10 times, taking into account the environment. Error was measured based on the distance between the center point of the arrival point and the center point of the robot after arrival. Performance evaluation was conducted by comparing with the conventional method using only navigation. Experiment 1 used straight distance, Experiment 2 used trigonometric functions and Manhattan distance, and Experiment 3 used curvature calculation based on the position and direction of the markers. The average results of error distance measurement for each group and the performance improvement rate evaluated by comparing with the control method are shown in Table 1.

Table 1. Performance evaluation results (10 runs).

Test	control method	method 1	method 2	method 3
AVG Error [m]	0.674	0.0625	0.022	0.014
Improvement [%]	0	978.4	2963.636	4714.286

As shown in Table 1, the proposed method demonstrates a performance improvement of over 900% compared to that of conventional method. Particularly, method 2 shows a performance enhancement of 2963%, and method 3 shows a significant improvement of 4714%, highlighting the advantages of the proposed approaches.

Acknowledgements. This work was supported by the National Research Foundation of Korea (NRF) grant funded by the Korea government (MSIT) (No. 2022H1D8A3038040).

References

1. Schleicher, D., Bergasa, L.M., Ocana, M., Barea, R., Lopez, M.E.: Real-time hierarchical outdoor SLAM based on stereovision and GPS fusion. IEEE Trans. Intell. Transp. Syst. **10**(3), 440–452 (2009). https://doi.org/10.1109/TITS.2009.2026317

2. Silverman, M.C.: Staying alive: a docking station for autonomous robot recharging. In: Proceedings 2002 IEEE International Conference on Robotics and Automation (Cat. No.02CH37292), pp. 1050–1055 (2002)
3. Zhang, Z.: A flexible new technique for camera calibration. IEEE Trans. Pattern Anal. Mach. Intell. **22**(11), 1330–1334 (2000)
4. Balasuriya, B.L.E.A.: Outdoor robot navigation using Gmapping based SLAM algorithm. In: 2016 Moratuwa Engineering Research Conference (MERCon), pp. 403–408 (2016)
5. Garrido-Jurado, S.: Automatic generation and detection of highly reliable fiducial markers under occlusion. Pattern Recogn. **47**(6), 2280–2292 (2014)
6. Carlone, L.: A tutorial on SE (3) transformation parameterizations and on-manifold optimization. J. Math. Imaging Vis. **53**(2), 167–190 (2015)

Big Data and Network Analysis in National Innovation Systems: The Roles of Academia, Industry, and Government Research Institutes and Their Interactions

Eun Sun Kim[1], Yunjeong Choi[2], and Jeongeun Byun[2](✉)

[1] Division of Data Analysis, Korea Institute of Science and Technology Information (KISTI), Daejeon, South Korea

[2] Division of Data Analysis, Technology Commercialization Research Center, Korea Institute of Science and Technology Information (KISTI), Daejeon, South Korea

jebyun@kisti.re.kr

Abstract. This study examines the changes that have been made to KNIS since the 2000s, when Korea entered the group of developed countries. Using data on joint research, this study systematically analyzes the interactions among actors representing academia, industry, and government research institutes, and the innovative performance achieved through these interactions. This study argues that the while interactions through joint research generate innovative performance, such interactions have not been occurring as strongly as would be desired, and this has limited the potential for the growth of innovative performance. While reinforcing the capabilities of individual actors is important, this study emphasizes that to build a more effective and systematic NIS, the government should establish policies designed to strengthen the interaction among actors.

Keywords: National innovation systems · Big data · Network analysis · National R&D project · Interactions

1 Introduction

South Korea achieved the quantitative growth of its economy with a national innovation system (NIS) characterized by strong government influence of research and development (R&D) investments, imitative R&D led by government-supported research institutes (GRIs), and the dominance of large firms in utilizing and disseminating the outcomes of R&D. In the 1990s, however, intensified competition through globalization and the emergence of latecomer developing countries such as China and India made it difficult for South Korea to continue to generate innovative performance through imitative R&D. In response to these changes, in the 2000s South Korea redefined its NIS as a system that facilitates the creation, exchange and diffusion of knowledge among academia, industry, and government research institutes and undertook to build a new NIS.

The South Korean national innovation systems (KNIS) had once been considered one of the most successful examples of a NIS in a developing country, but there have been few

J. C. Hung et al. (Eds.): FC 2023, LNEE 1134, pp. 334–337, 2024.
https://doi.org/10.1007/978-981-99-9342-0_36

studies on how the KNIS changed after Korea joined the ranks of the developed countries, and on whether this new KNIS has indeed resulted in innovative performance. This study is a systematic analysis of data on joint research, and examines the interactions among main actors that constitute the KNIS since the 2000s, and the innovative performance that resulted from these interactions.

2 Data and Analytical Method

This study aims to analyze the changes in the KNIS that have taken place since the 2000s by examining the national R&D projects. The dataset of national R&D projects was extracted from the National Science and Technology Information Service (NTIS) which includes information about 540,000 national R&D projects, such as government investment in R&D, joint research projects, and innovative performance.

First, the data of national R&D projects since the 2000s was collected in the NTIS. As the data of joint research projects was only gathered beginning in 2012, the data of 222,812 national R&D projects and 31,762 joint research projects were adopted from 2012 up to the latest available data period, which covered up to June 2017. Next, the data was classified by year and by the main actors that performed the projects. The programming language R, a software environment for statistical computing, was used in the data preprocessing and analysis. Based on the collected data, we examined the joint research network to understand how the main actors in the KNIS create, exchange and diffuse knowledge as they carry out projects. Before analyzing the joint research network, the hypothesis with regard to whether the joint research projects had an effect on innovative performance in the KNIS needed to be validified. It should be noted that we used information on the national R&D performance to analyze the innovative performance achieved by joint research. This is because the NTIS does not separately collect information on the performance of the joint research. Therefore, we hypothesized that the performance of joint research will be correlated to differences in the national R&D performance assessed in terms of the number of papers published, the number of patent applications and grants, technology transfers, royalty income from the technology transfers, the number of commercialized projects, and sales form the commercialized projects. We then performed Levene's homogeneity of variance test and validated the hypothesis with Welch's t-test using R software [1]. Next, we utilized NodeXL, an Excel template provided by Microsoft, to perform social network analysis (SNA) and modularity analysis on the actors that conducted joint research projects.

3 Results

3.1 Innovative Performance in the KNIS

According to the results of the hypothesis tests, the performance of joint research has an effect on innovative performance in the KNIS. In detail, national R&D projects that conducted joint research achieved 0.75 more papers published and 1.16 more cases of patent applications or registrations, on average. The number of technology transfers was analyzed to be statistically significant, and royalty income from the transfers in national

R&D projects that conducted joint research was on average 7.5 million won more. Particularly commercialization, joint research results in more outcomes. The national R&D projects that engaged in joint research were accomplished 0.35 more cases of commercialization, while the sales from the commercialized projects were recorded to be approximately 500 million won higher on average. Since the premise that joint research affects innovative performance in the KNIS has been validated, the joint research network is examined to understand how the main actors in the KNIS interact with each other and create, exchange and diffuse knowledge as they carry out joint research projects.

3.2 Characteristics of the Joint Research Network

To understand the joint research network, SNA was used on a total of 31,762 cases of joint research projects conducted from 2012 to June 2017. Figure 1 is a visual presentation of our analysis the structure of the joint research network. In this diagram, cases with a degree of 2 or less have been omitted, while nodes that have high degree centrality with 10 or more links are shown more prominently. The joint research network consists of 16,760 individual actors and 23,017 relations between actors that proposed the joint research and those that received the proposal.

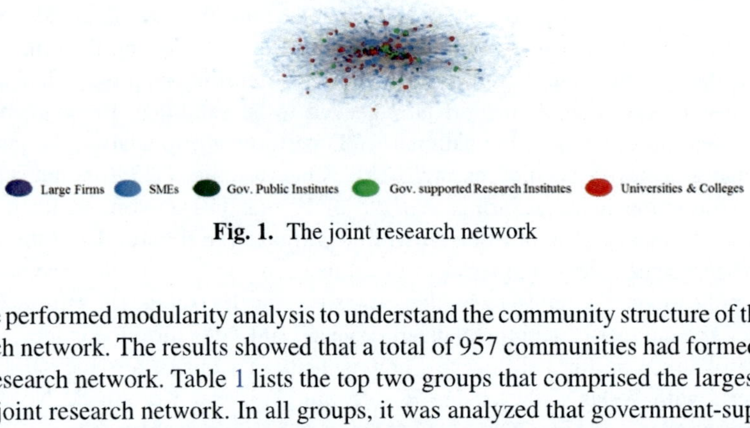

Fig. 1. The joint research network

We performed modularity analysis to understand the community structure of the joint research network. The results showed that a total of 957 communities had formed in the joint research network. Table 1 lists the top two groups that comprised the largest share of the joint research network. In all groups, it was analyzed that government-supported research institutes ranked highest in both in-degree and out-degree centrality but other indices such as betweenness centrality had slightly varying structural characteristics. Therefore, we proceeded to more closely examine government-supported research institutes, which ranked high in centrality, based on the nodes that determine each group's connectivity.

Table 1. The mean centrality of the top two groups

Community		Nodes	Avg. In-degree (A)	Avg. Out-degree (B)	Avg. Betweenness	Avg. Closeness	(A)/node	(B)/node
Group1	Large firms	73	4.70	5.12	0.0001	0.10	0.06	0.07
	SMEs	906	1.33	1.09	0.0000	0.06	0.00	0.00
	Univ.&colleges	82	15.18	11.10	0.0003	0.11	0.19	0.14
	Public institutes	11	2.91	0.09	0.0000	0.11	0.26	0.01
	GRIs	10	21.73	52.27	0.0015	0.11	2.17	5.22
Group2	Large firms	6	1.50	1.17	0.0000	0.10	0.25	0.20
	SMEs	576	0.80	1.14	0.0000	0.05	0.00	0.00
	Univ.&colleges	42	16.86	10.67	0.0004	0.11	0.40	0.25
	Public institutes	11	2.36	0.55	0.0000	0.10	0.21	0.05
	GRIs	6	19.83	23.50	0.0005	0.08	3.31	3.92

4 Conclusion

Upon systematically analyzing the KNIS, we found that actors' interactions do result in improving performance, but these interactions have not been as active as would be necessary to remove the impediments that are hindering stronger performance. Based on the analysis of this study, we offer the following policy recommendations. First, policymakers must encourage more national R&D projects to undertake joint research. Second, in addition to policies encouraging joint research projects, there must be policies designed to facilitate interactions. To promote interactions, the government not only should increase funding for SMEs, but also should induce such firms to engage in exchanges with a wider variety of actors. Finally, to promote balanced growth, rather than reducing the role of GRIs, the government should instead focus on encouraging them to participate in more joint research to expand their exchanges with other actors.

Acknowledgement. This research was supported by Korea Institute of Science and Technology Information (K-23-L03-C03-S01).

Reference

1. Welch, B.L.: The significance of the difference between two means when the population variances are unequal. Biometrika **29**, 350–362 (1937)